AF893900

Leśniewski's Systems

Ontology and Mereology

NIJHOFF INTERNATIONAL PHILOSOPHY SERIES

Volume 13

For a list of other volumes in this series see final page of the volume.

Polish Academy of Sciences
Institute of Philosophy and Sociology

Leśniewski's Systems
Ontology and Mereology

Edited by
Jan T. J. Srzednicki (Melbourne University)
V. F. Rickey (Bowling Green State University)
Assistant Editor: J. Czelakowski (Polish Academy of Sciences)

1984 Martinus Nijhoff Publishers
a member of the Kluwer Academic Publishers Group
The Hague /Boston/ Lancaster

Ossolineum
Publishing House of the Polish Academy of Sciences
Wrocław — Poland

Distributors

for the United States and Canada:
Kluwer Boston, Inc, 190 Old Derby Street,
Hingham, MA 02043, USA

for all other countries:
Kluwer Academic Publishers Group,

Distribution Center,
P. O. Box 322, 3300 AH Dordrecht,
The Netherlands

Library of Congress Cataloging in Publication Data

Main entry under title:
Leśniewski's systems.

(Nijhoff international philosophy series; v. 13)
At head of title: Polish Academy of Sciences.
Institute of Philosophy and Sociology.
Bibliography: v. 13, p.
Include index.
Contents: v. 13. Ontology and mereology.
1. Leśniewski, Stanislaw, 1886–1939 – Addresses,
essyas, lectures. I. Srzednicki, Jan T. II. Rickey,
V. F. III. Czelakowski, J. IV. Instytut Filozofii
i Socjologii (Polska Akademia Nauk)
V. Series. B4691. L44147 1984 199'. 438 84-20711

ISBN 90–247-2331-0 (set)
ISBN 90-247-2879-7 (v. 13)

ISBN 90-247-2879-7 (this volume)
ISBN 90-247-2331-0 (series)
ISBN 83-04-00879-3 Ossolineum

Martinus Nijhoff Publishers,
P. O. Box 566, 2501 CN The Hague,
The Netherlands.

Printed in Poland

Table of Contents

Editorial Note

"Leśniewski's Systems" is an anthology of papers written by various authors and divided in two volumes: A concerned with *Protothetic* and B with *Mereology and Ontology*.

These papers appeared previously in various journals and thanks are due to the editors of these journals for their permission to re-print them here. The papers appeared as follows:

R. Clay "On the Definition of Mereological Class" in *Notre Dame Journal of Formal Logic*, Vol. VII, No. 4 (1966), 359–360;

R. Clay "The Dependence of a Mereological Axiom" in *Notre Dame Journal of Formal Logic*, Vol. XI, No. 4 (1970), 471–472;

R. Clay "Relation of Leśniewski's Mereology to Boolean Algebra" in *The Journal of Symbolic Logic*, Vol. XXXIX, No. 4 (1974), 638–648;

C. Lejewski "Consistency of Leśniewski's Mereology" in *The Journal of Symbolic Logic*, Vol. XXXIV, No. 3 (1969), 321–328;

J. Canty "Ontology: Leśniewski's Logical Language" in *International Journal of Language and Philosophy*, Vol. V (1969), 455–469;

B. Iwanuś "On Leśniewski's Elementary Ontology" in *Studia Logica*, Vol. XXXI (1973), 73–119;

J. Słupecki "S. Leśniewski's Calculus of Names" in *Studia Logica*, Vol. III (1955), 7–72;

Z. Kruszewski "Ontologia bez aksjomatów" in *Przegląd Filozoficzny*, Vol. XXVIII (1925), 136;

C. Lejewski "Logic and Existence" in *The British Journal for the Philosophy of Science*, Vol. V (1954), 104–119;

C. Lejewski "On Leśniewski's Ontology" in *Ratio*, Vol. I, No. 2 (1958), 150–176;

B. Sobociński "Studies in Leśniewski's Mereology" in *Yearbook for 1954 — 55 of the Polish Society of Arts and Sciences Abroad*, Vol. V (1954), 34–48;

B. Sobociński "L'analyse de l'antinomie Russellienne par Leśniewski" in *Methodos*, Vol. I (1949): 1, 94–107; 2, 220–228; 3, 308–316; Vol. II, (1950): 6–7, 237–257.

The editors are grateful to Professor Ryszard Wójcicki, the editor of "Studia Logica", who initiated the present project and helped generously throughout its preparation. Further thanks are due to R. Clay who kindly translated Sobociński's "L'analyse de l'antinomie Russellienne par Leśniewski" from the French and also to Dr. Lynne M. Broughton and Mr. R Ładniak for the work that they contributed to the present volume.

Z. Kruszewski

Ontology without Axioms*

"Ontology" is here treated as a part of the general theory of logica functions. Approaching ontology as a separate system with its own set of axioms and agreed principles, we avoid difficulties consequent on operating functions having different and varying domains. Such difficulties ought to be solved once and for all within the theory of apparent variables so as to make free operations with functions of different categories possible within the framework of one and the same system of logic. Not expecting essential troubles in the problem of constructing such a theory of apparent variables, and for the time being appealing to intuition as to what can be substituted with respect to which quantifier (A criterion of the meaningfulness of expressions obtained through substitution), the speaker begins to analyse the meaning of a propositional function, i.e., a function which as the result of the substitution of a constant for the variable becomes a sentence. — Consider some expressions of type $F(x)$, e.g.: x is divisible by 5, x is running, it is not true that x, etc. All these expressions attribute some property to x, and when the expression is written in symbols, the property is represented by the symbol placed before the left bracket, that is by the exponent of the function. On the other hand, if we consider the set of all x's satisfying the expression $F(x)$, then the exponent F may be treated as a symbol of some property common to all these objects, and consequently it may have in symbolic logic a function similar to that of general names in colloquial language. Moreover, due to one-one correspondence between the exponent of a function and the set of all x's satisfying the function $F(x)$, we can use the exponent F as the symbol of the set of all such x's. This possibility of the dual interpretation of the exponent of a function of the one variable allows us to build both ontology and set theory in the framework of a general logical theory of functions. Moreover, neither ontology nor set theory so constructed require any new special axioms. A set theory built in this way is free of antinomies and not a single of the numerous false theorems usually proved by *reductio*

---

* *Editorial Note*: This is an abstract of Kruszewski's lecture delivered at the meeting of the Warsaw Institute of Philosophy on December 20, 1924. The report was published by B. Gawecki in "Przegląd Filozoficzny", Vol. XXVIII (1925) in Polish — see Kruszewski [1925]. Translated from the Polish by Ewa Jansen.

ad absurdum where the absurd is not an actual contradiction but an antinomy ingeniously introduced into the proof, can be proved in it. – The speaker defines all fundamental concepts of ontology and proves as theorems the axiom and all equivalences formulated as definitions in Leśniewski's ontology. – With respect to ontological definitions, i.e., definitions formulated by the use of the word "is" (e.g.: x is even $= x$ is a natural number and x is divisible by 2), it is possible to give a general method by means of which equivalences of that sort are obtained straightforwardly given an appropriate logical definition.

Bolesław Sobociński

Leśniewski's Analysis of Russell's Paradox*

The purpose of this article is to give, with a minimum of symbolism, a simple, accessible, unpolemical exposition of Leśniewski's analysis of Russell's Paradox.[1]

This analysis merits attention for several reasons. Mainly, because it was the point of departure for the construction of Leśniewski's system for the foundation of mathematics. His method of overcoming the paradox in question is very different from those employed by others. From the start it forced Leśniewski to take a path on which he had to overcome great difficulties related to the problem of the paradoxes; it determined the character of the theories which constitute his system. This system, which differs in many ways from contemporary systems, is non-contradictory (which is easy to prove), and is an adequate base for the construction of contemporary mathematics. However, it is not very easy to get the feel of the system, nor is it easy to penetrate the psychology from which it arose – what precisely were Leśniewski's thoughts about Russell's Paradox.[2]

I. Introduction

Before giving the basic presupposition, it is necessary to give some clarifications and terminological explanations; in addition we must sketch the general character of the logic we use, of the methods utilized in the proofs and of the symbolism that will be adopted.

I. As is known, this paradox of Mr. B. Russell was formulated in 1903, in the framework of contemporary set theory.[3] His discovery is a turning

---

* *Editorial Note*: Translated from the French by Robert E. Clay.

[1] Stanisław Leśniewski (born March 18, 1886, died May 13, 1939) was professor of philosophy of mathematics at the University of Warsaw from 1919 until his death.

[2] Leśniewski presented the essentials of his views on Russell's paradox – Leśniewski [1927–1931, Chs. II–III] (*Editorial Note*: English translation – Leśniewski [1983].) However, the formal reasonings given here have never been published.

[3] Russell's paradox was not the first one discovered within the framework of Cantor's set theory: in 1895 G. Cantor himself had already constructed the paradox known as "the set of all sets" ("Menge aller Mengen") and in 1897 C. Burali-Forti published a new paradox in the theory of ordinal numbers of that time. However, general attention was only drawn to the problem thanks to Mr. Russell's result.

point in the evolution of modern logic and the theory of the foundations of mathematics because it necessitated a profound study and discussion of the presuppositions of the deductive sciences. This study has led to the discovery of many other paradoxes and to evidence that the problem of the paradoxes had already appeared in Antiquity and in the Middle Ages, (e.g. the liar paradox); it was also realized that if these problems were not as startling then as now, it was only due to the fact that they were not associated with formalized deductive systems. On the other hand, the discovery provoked a whole series of attempts to reconstruct the foundations of mathematics in order to avoid the paradoxes. These attempts, numerous and of varying degrees of success, have not led to the formulation of a universally recognized system for the foundations of mathematics — this despite the fact that among them are theories of great interest such as that of "Principia Mathematica" in the mathematical-logical domain and E. Zermelo's set theory in the purely mathematical domain.

It seems to us that the weakness of these attempts lies in the fact that they are generally limited to modifications of the foundations which outlaw the formulation of the known paradoxes within the system. But such a procedure is no protection against unanticipated paradoxes which might appear in the system. Because after all, what is a paradox? In our opinion it is more than a simple contradiction. We employ the word "paradox" here in the sense used by L. Nelson and which can be described as follows: a paradox is a contradiction which we deduce starting from presuppositions whose truth we believe and using methods which we recognize us valid. Consequently, not every contradiction that results in a deductive theory is necessarily a paradox. Indeed it is trivial to give a theory which is correctly constructed and formalized according to all the rules of the art, but contradictory; but there will be no paradox in this case, because among the axioms or admissible values of deduction there will be false propositions.

According to L. Nelson, a paradox also has a psychological component, notably the belief in the truth of the presuppositions and in the correctness of the rules of reasoning. Therefore, the modification of a deductive system so that the paradoxes no longer appear in the modified system is not a resolution of the paradox. Such modification, usually done in an artificial and "painful" manner with the sole aim of eliminating the paradoxes does not alter the fact that there are paradoxes; it only enables us to construct, in a rather mechanical manner, a system free of contradictions. The paradox will only be resolved when we become convinced that we have used either incorrect rules of reasoning or false presuppositions in the construction of the paradox, for the its "tragic" aspect, caused by our above-mentioned belief, will be eliminated: we will cease to believe in the truth of one of the presuppositions of the paradox.

In this case there will be nothing strange or incomprehensible that this supposition, included in a system of postulates, causes the appearance of a contradiction. There will be no necessity to make "sacrifices" or to give up anything whatever, because the simple statement that we used to believe in the truth of a false thesis will suffice — and this statement will represent a deepening of our knowledge of reality. What then remains to be done is either for sake the theory whose presuppositions contain a false element or introduce new presuppositions sufficient for the foundation of the theory and of whose truth we are convinced.

Such an interpretation of paradox also enables us to make a perfectly clear distinction between real paradoxes (e.g. the liar paradox and Russell's paradox) and the simple contradictions which, in a more or less camouflaged form, arise from insufficiently formalized deductive reasoning.

2. As we have already said, Russell's paradox appeared in the framework of the set theory formulated by G. Cantor. For the rigorous construction of this paradox one needs a certain logical base in addition to the terms and presuppositions of this theory. This logical base will be extremely simple. The symbolism that we will use will be understandable to anyone acquainted with "Peano-Russell" symbolism. We also believe that the fact that all variables of our theses will be bound (i.e. every variable will appear in a quantifier) should cause no problems of understanding for anyone. All definitions will be in equivalential form; there will be no "definition by equations". The manner of presenting proofs will be similar to that of "Principia Mathematica". The proofs will be based on the theorems of the theory of deduction and on the laws concerning the use of quantifiers. In addition to these theorems we are going to use in our proofs some very simple and intuitive theses belonging to the theory of relations between names. These last contain individual (proper) names and determine certain relations between what are called "individuals"; they belong to the "ontology" of S. Leśniewski.[4]

Certain of our readers may find that some of these "ontological" theses are not valid in some other contemporary system, and may think that it is these theses and not those of Cantor's set theory itself, that cause the appearance of paradoxes.

---

[4] One finds a general description of ontology in Leśniewski [1930]. (*Editorial Note*: English translation in Leśniewski [1984].) The only work in which one finds a certain number of ontological theses deduced from its axiom, is my article Sobociński [1934].

Leśniewski's ontology is the most general theory of the relations between names. Its presuppositions, which can be formulated in a single short axiom, permit the deduction of theorems corresponding to all the theses of syllogistic, the algebra of names, the theory of relations and that of sets — with the sole exception of those which require certain existential presuppositions.

However, we believe that:

a) the theses in question can be proved in any sufficiently developed logical system, and this without making use of specifically "ontological" terms;

b) there exists a demonstration of the consistency of ontology, of the theory of deduction and of the laws concerning the use of quantifiers.

Therefore, the contradictions cannot be caused by the above-mentioned theses.

In order to facilitate the understanding of the proofs, it is also necessary to clarify the meaning "individual proposition" which we use here. It is important to note that we use the expression "A is b" (in symbols, "$A \varepsilon b$") in exactly the same way that one uses propositions such as "Socrates is mortal". This proposition says that a concrete, individual, really existing object is something or other. Thus it is necessary to distinguish individual propositions in this strict sense from the apparently individual propositions often employed in ordinary language, but which are really abbreviations of proposition of another type, e.g. universally affirmative propositions. This is the case when one says, for example, "Man is mortal" instead of saying "Every man is mortal." Consequently, for a proposition of the type, "$A \varepsilon b$" to be correctly employed, one can only substitute names of individual for the subject, i.e. "A"; that is only names like Socrates, Napoleon I, "this table", etc. can be substituted for "A", while for "b", that is to say, for the complement of an individual proposition, we also permit — and this conforms to ordinary language — the substitution of names of individuals; for example "St. Peter, the Apostle, is the first pope" is a correctly constructed proposition. In other words, we suppose that the subject and the complement of an individual proposition belong to the same logical type (according to recent terminology) or to the same semantical category (according to Leśniewski's terminology). Thus we employ the word "is" ("ε") as it was used before the reform of G. Peano. We believe we are thus conforming not only with the intuitions associated with ordinary language, but also (and this does not seem to have been noted by historians of logic) with the Aristotelian and Scholastic tradition[5].

The reason for using these theses in the discussion which follows is not only that they permit the construction of extreme simple schemes of proof, but above all, our desire to present the development of Leśniewski's thought and reasoning as accurately as possible. As we have already said, Leśniewski overcame the difficulties caused by the appearance of

---

[5] We point out that the "ε" of ontology differs from the "ε" utilized in Russell and Whitehead [1913] and from the "$\in$" of set theory.

Russell's paradox (and overcame them in a definitive manner, at least for him and for those he has convinced) long before the construct of his system for the foundations of mathematics. However, during this period he was using intuitive rules of reasoning, which when later formalized, corresponded to ontological theses. If one attempts to transfer this reasoning into the framework of some other logical system, the understanding and intuitive clarity of this line of reasoning become considerably obscured. What is more, as we shall see at the end of this artictle, the mere formulation of Russell's paradox already imposes certain conceptualizations which will be formulated in the foundations of ontology.

3. At this moment we are not interested in whether set theory, in whose framework the above-mentioned paradox appeared, is a logical theory or a mathematical theory. We shall limit our discussion to those terms of the theory which will be utilized in our analysis. There will be two. They will be used here in a manner which seems to be in agreement with Cantor's set theory.

The first of these terms is "the set of objects a"; which will also be called "class of objects a" (in symbols, "$\mathrm{Kl}(a)$"). Therefore, "$A \varepsilon \mathrm{Kl}(a)$" we mean "$A$ is the set of all the objects a" or "A is the set formed by all the objects a".

The second term is "element of a set A" (in symbols "$\mathrm{el}(A)$") — By "$B \varepsilon \mathrm{el}(A)$" we mean therefore that there exist such object(s) a that: $A \varepsilon \mathrm{Kl}(a)$ and $B \varepsilon a$. It seems that this description should not cause any misunderstanding. One can correctly define the term "element" by means of the term "class" in the following manner:

$$[AB]\colon B \varepsilon \mathrm{el}(A). \equiv . [\exists a]. A \varepsilon \mathrm{Kl}(a). B \varepsilon a$$ [6]

II. Formulation of the Paradox

1. We now begin the formulation of the paradox and of the problems which it entails. This formulation is the very simple result of two presuppositions to which no one raised objections during the time preceding the appearance of this paradox. These two presuppositions — which contain only one set-theoretical term, namely, "class" — are the following:

A1 $[a]. [\exists A]. A \varepsilon \mathrm{Kl}(a)$

A2 $[ABab]\colon A \varepsilon \mathrm{Kl}(a). A \varepsilon \mathrm{Kl}(b). B \varepsilon b. \supset . B \varepsilon a$

[6] We employ capital letters as individual variables and lower case letters as universal variables. It is only a written convention which seems to render the theorems more intuitive; it is not an obligatory rule. We employ the symbol "Kl" instead of "cl" to avoid confusion with the same symbol of Russell and Whitehead [1913]. The quantifiers and the symbols from the propositional calculus are the same as those of Peano-Russell symbolism.

The presupposition A1 says that for each a (be it a name, an object, etc.) there exists a set (class) formed of these objects. The presupposition A2 postulates that for each B and a, if B is an element of set composed of the objects a, then B is a. For the moment we will not consider whether these presuppositions are true or false, we are only going to prove that they lead to a contradiction. To this end we introduce the following definition:

D1 $[A] \therefore A \varepsilon * . \equiv : A \varepsilon A : [a] : A \varepsilon \mathrm{Kl}(a) . \supset . \sim (A \varepsilon a)$[7]

[*Translator's note.* Recalling the formulation of element in terms of class in the preceding section:

$[AB] : B \varepsilon \mathrm{el}(A) . \equiv . [\exists a] . A \varepsilon \mathrm{Kl}(a) . B \varepsilon a$

we see that

$[A] : A \varepsilon \mathrm{el}(A) . \equiv . [\exists a] . A \varepsilon \mathrm{Kl}(a) . A \varepsilon a$

and thus

$[A] \therefore \sim (A \varepsilon \mathrm{el}(A)) . \equiv : [a] : A \varepsilon \mathrm{Kl}(a) . \supset . \sim (A \varepsilon a)$]

According to this definition we can say that "$A \varepsilon *$" means that "A is a set (class) which is not an element of itself" or in other words "A is a class which is not subordinated to itself".

[*Translator's note.*

$[A] \therefore A \varepsilon * . \supset : A \varepsilon \mathrm{Kl}(*) . \supset . \sim (A \varepsilon *)$ [D1, $a/*$]
$[A] \therefore \sim (A \varepsilon *) . \supset : A \varepsilon \mathrm{Kl}(*) . \supset . \sim (A \varepsilon *)$ [Propositional Calculus]

A3 $[A] : A \varepsilon \mathrm{Kl}(*) . \supset . \sim (A \varepsilon *)$ [D1, $a/*$[8]]

This theorem says that if A is the class of all classes which are elements of themselves, then it is not true that A is not an element of itself.

A4 $[Aa] : A \varepsilon \mathrm{Kl}(*) . A \varepsilon \mathrm{Kl}(a) . \supset . \sim (A \varepsilon a)$ [A2, $a/*$, B/A, b/a; A3]
A5 $[A] : A \varepsilon \mathrm{Kl}(*) . \supset . A \varepsilon *$

[*Translator's note.* In the proof of this thesis and in those which follow, "Hyp(n)" will mean that the hypothesis of the thesis is the conjunction of n terms, and these terms are numbered $1, 2, \ldots, n$ in the order of their appearance in the statement of the thesis].

---

[7] The factor "$A \varepsilon A$" is introduced in definition D1 to guarantee that "A" is an individual name, as is explicitly required by the definiendum. The omission of this factor in definitions of this type can lead to contradictions known by the name of "paradox of the contradictory object".

[8] By the notation "a/x" we mean that the expression "x" is substituted for the variable "a". The notation "$a//b$" means that we substitute the variable "b" for the variable "a" in an interior quantifier of the thesis in question.

PF $[A]\therefore \mathrm{Hyp}(1). \supset :$

2) $A \varepsilon A:$ [1[9]]

3) $[a]: A \varepsilon \mathrm{Kl}(a). \supset . \sim (A \varepsilon a)$ [A4; 1]

$A \varepsilon *$ [D1; 2; 3]

This theorem says that if A is the class of all classes which are not elements of themselves, then A is not an element of itself.

A6 $[A]. \sim (A \varepsilon \mathrm{Kl}(*))$ [A3; A5]

This theorem says that there is no class of the classes which are not elements of themselves.

A7 $[\exists A]. A \varepsilon \mathrm{Kl}(*)$ [A1, $a/*$]

A6 and A7 contradict each other; thus we see that A1 and A2 lead to a contradiction.

2. The formulation of Russell's paradox which we have just presented is the simplest among all the known formulations. Usually, one formulates it using "element" instead of class. So before beginning the discussion of the causes of the paradox, we will give another formulation in which the term "element" will be utilized.

[*Translator's note.* The final conclusion of this section is correct; but the proof is in error. What follows is the translator's repair of the proof.]*

Consider the following two theorems:

B1 $[AaB]: A \varepsilon \mathrm{Kl}(a). B \varepsilon a. \supset . B \varepsilon \mathrm{el}(A)$

B2 $[AaB]: A \varepsilon \mathrm{Kl}(a). B \varepsilon \mathrm{el}(A). \supset . B \varepsilon a$

Theorem B1 postulates that if A is a class of the objects a, then all of the a's are elements of the object A. Theorem B2 postulates that if A is the class of the objects a, then every element of A is a.

A2 is easily derived from B1 and B2 as follows:

A2 $[ABab]: A \varepsilon \mathrm{Kl}(a). A \varepsilon \mathrm{Kl}(b). B \varepsilon b. \supset . B \varepsilon a$

PF $[ABab]: \mathrm{Hyp}(3). \supset .$

4) $B \varepsilon \mathrm{el}(A).$ [B1, a/b; 2; 3]

$B \varepsilon a$ [B2; 1; 4]

---

[9] By virtue of ontological thesis: $[Aa]: A \varepsilon a. \supset . A \varepsilon A$, cf. Sobociński [1934, Th. IV].

* *Editorial Note*: The following definition occurring in the original proof will be employed in subsequent sections:

D2 $[AB]: A \varepsilon \mathrm{el}(B). \equiv [\exists a]. B \varepsilon \mathrm{Kl}(a). A \varepsilon a$

Therefore, A1 and A2 are derivable in the system {A1, B1, B2}. Since A1 and A2 lead to a contradiction, the system {A1, B1, B2} is also contradictory.

III. Analysis of the Presupposition A1

We are now going to analyze the presuppositions A1 and A2 (respectively A1, B1, B2). One is immediately struck by the doubtful nature of A1. Indeed this presupposition states that for every a there is a class formed of these objects; as a result one can form a really existing class of contradictory objects. It is true that this supposition is admitted, in one form or another by all set-theoreticians, to wit it may be compared with Axiom VII of the axiomatic system of Zermelo. However, it might not be accepted by everybody, because it gives rise to paradoxical consequences. For example, without believing in any Egyptian god, one accepts the existence of the class of Egyptian gods. Without discussing the truth or falsity of A1 we will attempt to show that A1 is not the cause of the appearance of the paradox. This procedure is especially in order since the truth of A1 is not universally accepted.

So in the systems {A1, A2} and {A1, B1, B2} let us replace the presupposition A1 by the weaker statement:

C1 $[Ba]:B \varepsilon a. \supset .[\exists A].A \varepsilon \mathrm{Kl}(a)$

which is acceptable to everyone.

This presupposition states that if there exist some a's, then there exists a class (set) formed by these a's. It raise no doubts, since everyone concedes that if there exists at least one object of a given kind, then the class of this kind also exists —even if it contains only one element. It is also obvious that A1 inferentially implies C1 but not vice-versa. — Now since we have no objections to C1, let us examine the systems {C1, A2} and {C1, B1, B2}. This examination will show that these systems also entail a contradiction, at least if one also accepts another undeniable proposition, namely: that there exist at least two distinct objects in the world. We formulate this supplementary presupposition as follows:

C2 $[\exists AB].A \varepsilon A.B \varepsilon B. \sim(A = B)$

This means that there exist A and B such that A is an object, B is an object and A is not identical to B.

We begin with an analysis of the system {C1, A2, C2}. As before we introduce the definition D1. A3, A4, A5 and A6 now follow as before. We call them C3, C4, C5, C6 respectively. Then:

C7 $[B]. \sim(B \varepsilon *)$ [C1, $a/*$; C6]

C8 $[Aa]:A \varepsilon \mathrm{Kl}(a). \supset .A \varepsilon a$

PF $[Aa]:\mathrm{Hyp}(1).\supset$

2) $A \varepsilon A.$ [1]

$[\exists b].$

3) $A \varepsilon \mathrm{Kl}(b).$
4) $A \varepsilon b.$ [D1, a//b; C7, B/A; 2]

$A \varepsilon a$ [A2, B/A; 1; 3; 4]

C9 $[ABC]:A \varepsilon A.C \varepsilon \mathrm{Kl}(A \cup B).B \varepsilon C.\supset.A = B$[10]

PF $[ABC]:\mathrm{Hyp}(3).\supset.$

4) $B \varepsilon B.$ [3]

5) $A \varepsilon A \cup B.$ [1[11]]

$[\exists D].$

6) $D \varepsilon \mathrm{Kl}(B).$ [C1, a/B, A//D; 4]

7) $D \varepsilon B.$ [C8, A/D, a/B; 6]

8) $D \varepsilon C.$ [7; 3]

9) $D \varepsilon \mathrm{Kl}(A \cup B).$ [8; 2]

10) $A \varepsilon B.$ [A2, A/D, B/A, a/B, b/A ∪ B; 6; 9; 5]

$A = B$ [10; 4[12]]

C10 $[AB]:A \varepsilon A.B \varepsilon B.\supset.A = B$

PF $[AB]::\mathrm{Hyp}(2).\supset\therefore$

3) $A \varepsilon A \cup B\therefore$ [1]

$[\exists C]\therefore$

4) $C \varepsilon \mathrm{Kl}(A \cup B)$ [C1, B/A, a/A ∪ B, A//C; 3]

5) $C \varepsilon A \cup B:$ [C8, A/C, a/A ∪ B; 4]

6) $C \varepsilon A.\vee.C \varepsilon B:$ [5]

---

10 Here the symbol "∪" means the "or" between names. In ontology it is defined by the universally accepted definition in the algebra of names:

$$[Aab]\therefore A \varepsilon a \cup b. = :A \varepsilon a.\vee.A \varepsilon b$$

The symbol "=" denotes identity (between individuals). It is defined in ontology as follows:

$$[AB]:A = B. \equiv .A \varepsilon B.B \varepsilon A$$

Cf. Sobociński [1934, Df. I].

11 We deduce this by the universally recognized law which is valid in ontology

$$[Aab]:A \varepsilon a.\supset.A \varepsilon a \cup b$$

12 By the ontological theorem:

$$[AB]:A \varepsilon B.B \varepsilon B.\supset.A =$$

7) $A \varepsilon C . \vee . B \varepsilon C :$ [6; 1; 2[13]]
8) $C \varepsilon \mathrm{Kl}(A \cup B) :.$ [4[14]]
$A = B$ [7; C9, A/B, B/A; 2; 8; C9; 1; 4]

Theorem C10 is inferentially implied by C1 and A2 only. It contradicts C2 because it says that whenever A is an object and B is an object, A is identical to B; thus only one object can exist in the world. Therefore {C1, A2, C2} (and *a fortiori* {C1, B1, B2, C2}) yields a paradox — while both {C1, A2} and {C1, B1, B2} yield the rather paradoxical thesis C10.

IV. Frege's Correction

The discussion in Chapter III shows that Russell's paradox is caused neither by A1 nor the weaker C1. Indeed, if one thought that the contradiction was caused by the admission of A1 one would be obliged to conclude that the replacement of A1 by C1 — which is certainly true — would eliminate the paradox. But this is not the case: the cause then must be in the acceptance of A2.

G. Frege, who was aware of Russell's paradox which destroyed his system for the foundations of mathematics, proposed (in the supplement to the second volume of Frege [1903][15] the replacement of A2 by a weaker presupposition which can be formulated as follows:

E1 $[ABab] : A \varepsilon \mathrm{Kl}(a) . A \varepsilon \mathrm{Kl}(b) . B \varepsilon b . \sim (B \varepsilon \mathrm{Kl}(b)) . \supset B \varepsilon a$

Since the second and fourth parts of the antecedent imply that "$\sim (A = B)$", the meaning of E1 is: for all B and a, if B is a proper element of the set formed by the objects a, then $B \varepsilon a$.[16]

This proposition of Frege was hardly noticed and caused no interest. Not until 1938 did Leśniewski show that E1 also leads to a contradiction, whether one assumes A1 or C1. We will begin with the more interesting case of the system {C1, E1} since C1 is not open to question. Later we will consider the system {A1, E1}.

---

[13] By the ontological thesis:

$$[ABa] : A \varepsilon B . B \varepsilon a . \supset . B \varepsilon A$$

cf. Sobociński [1934, Th. III] (*Editorial Note*: English translation in McCall [1967].)

[14] By the ontological thesis:

$$[ab] \therefore [A] : A \varepsilon a . \equiv . A \varepsilon b : \equiv : [f] : f(a) . \equiv . f(b)$$

(cf. Sobociński [1934, Th. Ω]. This thesis establishes extensionality for name variables.

[15] Cf. pp. 262–265 of the work cited.

[16] Compare thesis E1 with the later thesis E5, which is inferentially equivalent to it on the basis of the axioms assumed here. By "A is a proper element of B" we mean "$A \varepsilon \mathrm{el}(B)$ and $\sim (A = B)$".

1. The system C1 and E1 leads to a contradiction if one accepts two other propositions which are undoubtedly true. We state them in the following manner:

E2 $[ABa]{:}A \varepsilon \mathrm{Kl}(a).B \varepsilon \mathrm{Kl}(a). \supset .A = B$

E3 $[\exists ABC].A \varepsilon A.B \varepsilon B.C \varepsilon C.\sim(A = B).\sim(A = C).\sim(B = C)$

E2 postulates that the set of all objects *a* is unique; E3 states that there are at least three distinct objects in the world.

Now let us suppose C1, E1, E2, E3 and proceed with the proof of

E4 $[Ba]{:}B \varepsilon a. \supset .[\exists A].A \varepsilon \mathrm{Kl}(\mathrm{Kl}(a))$

PF $[Ba]{:}\mathrm{Hyp}(1).\ .$

$[\exists A].$

2) $A \varepsilon \mathrm{Kl}(a).$ [C1; 1]

$[\exists A].A \varepsilon \mathrm{Kl}(\mathrm{Kl}(a))$ [C1, B/A, $a/\mathrm{Kl}(a)$; 2]

E5 $[ABab]{\therefore}B \varepsilon b.A \varepsilon \mathrm{Kl}(a).A \varepsilon \mathrm{Kl}(b). \supset {:}A = B. \lor .B \varepsilon a$

PF $[ABab]{\therefore}\mathrm{Hyp}(3). \supset {:}$

4) $B \varepsilon \mathrm{Kl}(b). \lor .B \varepsilon a{:}$ [E1; 2; 3; 1]

$A = B. \lor .B \varepsilon a$ [4, E2, a/b; 3]

E6 $[BCab]{::}B \varepsilon a.C \varepsilon \mathrm{Kl}(\mathrm{Kl}(a)).C \varepsilon \mathrm{Kl}(\mathrm{Kl}(b)). \supset {\therefore}[\exists A]{\therefore}$
$A \varepsilon \mathrm{Kl}(a){:}C = A. \lor .A \varepsilon \mathrm{Kl}(b)$

PF $[BCab]{::}\mathrm{Hyp}(3). \supset {\therefore}$

$[\exists A]{\therefore}$

4) $A \varepsilon \mathrm{Kl}(a){:}$ [C1; 1]

5) $C = A. \lor .A \varepsilon \mathrm{Kl}(b){\therefore}$ [E5, A/C, B/A, $a/\mathrm{Kl}(b)$, $b/\mathrm{Kl}(a)$; 4; 3; 2]

$[\exists A]{\therefore}A \varepsilon \mathrm{Kl}(a){:}C = A. \lor .A \varepsilon \mathrm{Kl}(b)$ [4; 5]

E7 $[ABCb]{:}B \varepsilon b.A \varepsilon \mathrm{Kl}(\mathrm{Kl}(b)).C \varepsilon \mathrm{Kl}(b).A = C. \supset .A \varepsilon b$

PF $[ABCb]{:}\mathrm{Hyp}(4).\ .$

5) $A \varepsilon \mathrm{Kl}(b).$ [3; 4]

6) $B \varepsilon \mathrm{Kl}(b).$ [E1, $a/\mathrm{Kl}(b)$; 2; 5; 1]

7) $A = B.$ [E2, a/b; 5; 6]

$A \varepsilon b$ [1; 7]

E8 $[BCab]{:}B \varepsilon a.C \varepsilon \mathrm{Kl}(\mathrm{Kl}(a)).C \varepsilon \mathrm{Kl}(\mathrm{Kl}(b)).C \varepsilon b. \supset .C \varepsilon a$

PF $[BCab]{::}\mathrm{Hyp}(4). \supset {\therefore}$

$[\exists A]{\therefore}$

5) $A \varepsilon \mathrm{Kl}(a){:}$
6) $C = A. \lor .A \varepsilon \mathrm{Kl}(b){:}$ } [E6; 1; 2; 3;]

7) $C = A. \lor .C \varepsilon a{:}$ [6; E5, B/C; 4; 5]

$C \varepsilon a$ [7; E7, A/C, C/A, b/a; 1; 2; 5]

We now introduce D3 which is analogous to D1,

D3 $[A] \therefore A \,\varepsilon\, \divideontimes . \equiv : A \,\varepsilon\, A : [a] : A \,\varepsilon\, \mathrm{Kl}(\mathrm{Kl}(a)) . \supset . \sim (A \,\varepsilon\, a)$

and proceed with the deduction:

E9 $[AB] : A \,\varepsilon\, \mathrm{Kl}(\mathrm{Kl}(\divideontimes)) . \supset . \sim (B\, (B \,\varepsilon\, \divideontimes)$

PF $[AB] : \mathrm{Hyp}(1) . \supset .$

2) $A \,\varepsilon\, A .$ [1]

3) $\sim (A \,\varepsilon\, \divideontimes) .$ [D3, $a/\divideontimes$; 1]

$[\exists a] .$

4) $A \,\varepsilon\, \mathrm{Kl}(\mathrm{Kl}(a)) .$ [D3; 2; 3]

5) $A \,\varepsilon\, a .$

$\sim (B \,\varepsilon\, \divideontimes)$ [E8, C/A, $a/\divideontimes$, b/a; 1; 4; 5; 3]

E10 $[B] . \sim (B \,\varepsilon\, \divideontimes)$ [E4, $a/\divideontimes$; E9]

E11 $[Ba] : B \,\varepsilon\, a . \supset . [\exists A] . A \,\varepsilon\, \mathrm{Kl}(\mathrm{Kl}(a)) . A \,\varepsilon\, a$

PF $[Ba] :: \mathrm{Hyp}(1) . \supset \therefore$

$[\exists A] \therefore$

2) $A \,\varepsilon\, \mathrm{Kl}(\mathrm{Kl}(a) .$ [E4; 1]

3) $A \,\varepsilon\, A .$ [2]

$[\exists b] :$

4) $A \,\varepsilon\, \mathrm{Kl}(\mathrm{Kl}(b)) .$ [D3, a//b; E10, B/A; 3]

5) $A \,\varepsilon\, b :$

6) $A \,\varepsilon\, a \therefore$ [E8, C/A; 1; 2; 4; 5]

$[\exists A] . A \,\varepsilon\, \mathrm{Kl}(\mathrm{Kl}(a)) . A \,\varepsilon\, a$ [2; 6]

E12 $[BDE] : B \,\varepsilon\, B . \sim (B = D) . E \,\varepsilon\, \mathrm{Kl}(\mathrm{Kl}(B \cup D)) . D \,\varepsilon\, E .$
$\supset . B \,\varepsilon\, \mathrm{Kl}(D)$

PF $[BDE] :: \mathrm{Hyp}(4) . \supset ::$

5) $B \,\varepsilon\, B \cup D .$ [1]

6) $\sim (B \,\varepsilon\, D) .$ [2; 4]

7) $D \,\varepsilon\, D ::$ [4]

$[\exists C] ::$

8) $C \,\varepsilon\, \mathrm{Kl}(\mathrm{Kl}(D)) .$ [E11, B/D, a/D, $A//C$; 7]

9) $C \,\varepsilon\, D .$

10) $C \,\varepsilon\, E .$ [9; 4[17]]

11) $C \,\varepsilon\, \mathrm{Kl}(\mathrm{Kl}(B \cup D)) ::$ [10; 3]

$[\exists A] ::$

12) $A \,\varepsilon\, \mathrm{Kl}(D) :$ [E6, B/C, a/D, $b/B \cup D$; 9; 8; 11]

13) $C = A . \vee . A \,\varepsilon\, \mathrm{Kl}(B \cup D) \therefore$

$[\exists F) \therefore$

[17] By the ontological thesis:

$$[ABa] : A \,\varepsilon\, B . B \,\varepsilon\, a . \supset . A \,\varepsilon\, a$$

(Cf. Sobociński [1934, Th. D]).

14) $F \varepsilon \mathrm{Kl}(B \cup D):$
15) $C = F . \vee . F \varepsilon \mathrm{Kl}(D):$ } [E6, $a/B \cup D$, b/D, $A//F$; 5; 11; 8]

16) $A = F.$ [13; 15; E2, B/F, $a/B \cup D$; 14; E2, B/F, a/D; 12]

17) $F \varepsilon \mathrm{Kl}(D)$ [12; 16]

18) $F = B{:}{\cdot}{:}$ [E5, A/F, $b/B \cup D$, a/D; 5; 17; 14; 6]

$B \varepsilon \mathrm{Kl}(D)$ [17; 18]

E13 $[BDE]{:} B \varepsilon B . \sim(B = D) . E \varepsilon \mathrm{Kl}(\mathrm{Kl}(B \cup D)) . D \varepsilon E . \supset$
$\supset . D \varepsilon \mathrm{Kl}(B)$

PF $[BDE] \therefore \mathrm{Hyp}(4) . \supset :$

5) $B \varepsilon \mathrm{Kl}(D):$ [E12; 1; 2; 3; 4]

$[\exists A]:$

6) $A \varepsilon \mathrm{Kl}(\mathrm{Kl}(B)).$
7) $A \varepsilon B.$ } [E11; a/B; 1]

8) $A \varepsilon \mathrm{Kl}(D).$ [7; 5]

9) $B = A.$ [E2, A/B, B/A, a/D; 5; 8]

10) $\sim(A = D):$ [9; 2]

$D \varepsilon \mathrm{Kl}(B)$ [E5, B/D, $a/\mathrm{Kl}(B)$, b/D; 4; 6; 8; 10]

E14 $[BD]{:} B \varepsilon B . D \varepsilon D . \sim(B = D) . \supset . B \varepsilon \mathrm{Kl}(D)$

PF $[BD]{::}\mathrm{Hyp}(3) . \supset \therefore$

4) $B \varepsilon B \cup D \therefore$ [1]

$[\exists E] \therefore$

5) $E \varepsilon \mathrm{Kl}(\mathrm{Kl}(B \cup D)).$
6) $E \varepsilon B \cup D.$ } [E11, $a/B \cup D$, $A//E$; 4]

7) $E \varepsilon \mathrm{Kl}(\mathrm{Kl}(D \cup B)):$ [5]

8) $E \varepsilon B . \vee . E \varepsilon D:$ [6]

9) $B \varepsilon E . \vee . D \varepsilon E \therefore$ [8; 1; 2]

$B \varepsilon \mathrm{Kl}(D)$ [9; E13, B/D, D/B; 2; 3; 7; E12; 1; 3; 5]

E15 $[ABC]{:} A \varepsilon A . B \varepsilon B . C \varepsilon C . \sim(A = B) . \sim(B = C).$
$\supset . A = C$

PF $[ABC]{:}\mathrm{Hyp}(5) . \supset$

6) $A \varepsilon \mathrm{Kl}(B).$ [E14, B/A, D/B; 1; 2; 4]

7) $C \varepsilon \mathrm{Kl}(B).$ [E14, B/C, D/B; 3; 2; 5]

$A = C$ [E2, B/C, a/B; 6; 7]

E15 is the negation of E3. Therefore the system {C1, E1, E2, E3} leads to a contradiction.

2. Since A1 inferentially implies C1 and E15 was proved starting from just C1, E1 and E2, it is evident that the system {A1, E1, E2, E3} also yields a contradiction. It is worth noting that if one uses A1 instead of C1, one obtains a contradiction by assuming C2 instead of E3, that is to say, it suffices in this case to postulate a world with at least two different objects. The proof is very simple.

We suppose A1, E1, E2, C2, and proceed as follows: C1 [A1], E4 [C1], E5 [E1; E2], E6 [C1; E5], E7 [E1; E2], E8 [E5; E6; E7]. Then we introduce D3 and proceed with the deduction: E9 [E8, D3], E10 [E4; E9, E11 [E4; E8; E10; D3], E12 [E2; E5; E6; E11], E13 [E2; E5; E11; E12], E14 [E11; E12; E13], E15 [E2; E14].

Next we obtain:

E16 $[ABC]:A \varepsilon A.B \varepsilon B.C \varepsilon \mathrm{Kl}(\ast).C = A.\supset A = B$

PF $[ABC] \therefore \mathrm{Hyp}(4).\supset:$

5) $A = B.\vee.A \varepsilon \mathrm{Kl}(B):$ $[\mathrm{E14}, B/A, D/B; 1; 2]$

6) $A \varepsilon \mathrm{Kl}(\ast):$ $[3; 4]$

$A = B$ $[5; \mathrm{E5}, a/\ast, b/B; 2; 6; \mathrm{E10}]$

E17 $[AB]:A \varepsilon A.B \varepsilon B.\supset.A = B$

PF $[AB]::\mathrm{Hyp}(2).\supset\therefore$

$[\exists C]\therefore$

3) $C \varepsilon \mathrm{Kl}(\ast).$ $[\mathrm{A1}, a/\ast, A//C]$

4) $C \varepsilon C:$ $[3]$

5) $C = A.\vee.C = B.\vee.A = B:$ $[\mathrm{E15}; 1; 2; 4]$

$A = B$ $[5; \mathrm{E16}; 1; 2; 3; \mathrm{E16}, A/B, B/A; 1; 2; 3]$

Theorem E17, which was proved from A1, E1 and E2 contradicts C2. Therefore the system {A1, E1, E2, C2} leads to a contradiction.

Thus we see that the systems {C1, E1, E2, E3} and {A1, E1, E2, C2} are both contradictory. Therefore the correction of G. Frege is insufficient.

V. Conditions for the Consistency of the Systems Studied Thus Far

Before considering anew the causes of Russell's paradox, it will be worthwhile to summarize certain results which can be deduced about the systems we have studied thus far; this will help to give a more complete picture of the problem. In Ch. III and IV we saw that these systems lead to a contradiction if one adds the presuppositions C2 and E3, which establish the existence of a universe with at least two objects and one with at least three objects respectively. By means of the well-known method of interpretation, we will now prove that the systems are consistent if we do not introduce C2 and C3.

It is easy to verify that, preserving the whole system of logic, as well the rules of reasoning on which we have based it, just as they are:

a) the system {C1, E1, E2, C2} does not lead to a contradiction, if the contradiction is not implied by the following system of presuppositions:

F1 $\mathfrak{A} \varepsilon \mathfrak{A}$
F2 $\mathfrak{B} \varepsilon \mathfrak{B}$
F3 $\sim(\mathfrak{A} \varepsilon \mathfrak{B})$
E15;

b) the system {A1, E1, E2} does not lead to a contradiction, if F1 and C10 do not lead to a contradiction.

c) the system {C1, A2} (as well as {C1, B1, B2}) does not lead to a contradiction if C10 does not lead to a contradiction.

1. The system {C1, E1, E2, C2} can be interpreted in the system {F1, F2, F3, E15}[18] in the following fashion:

We introduce the term "class" by means of the definition:

DI $[Aa] \therefore A \varepsilon \mathrm{Kl}_1(a) . = : A \varepsilon A : A = \mathfrak{A} . \mathfrak{B} \varepsilon a . \vee . A \varepsilon a . \sim (A \varepsilon a);$

we then deduce:

F4 $[Ba] \therefore B \varepsilon a . \mathfrak{A} = B . \supset : \mathfrak{A} \varepsilon \mathrm{Kl}_1(a) . \vee . \mathfrak{B} \varepsilon \mathrm{Kl}_1(a)$

PF $[Ba] \therefore \mathrm{Hyp}(2) . \supset :$
3) $\mathfrak{A} \varepsilon a :$ [1; 2]
$\mathfrak{A} \varepsilon \mathrm{Kl}_1(a) . \vee . \mathfrak{B} \varepsilon \mathrm{Kl}_1(a)$ [DI, $A/\mathfrak{A}$; F1; DI, $A/\mathfrak{B}$; 3]

---

[18] It is easy to establish that the system of presuppositions {F1, F2, F3, E15} is inferentially equivalent to the presuppositions: F1, F2, F3 and

E15* $[A] : A \varepsilon A . \sim(\mathfrak{A} = A) . \supset . \mathfrak{B} = A$

a) We assume F1, F2, F3, E15 and deduce:

E15* [E15, $A/\mathfrak{A}$, $B/\mathfrak{B}$, C/A; F1; F2; F3]

b) Now we assume F1, F2, F3, E15* and deduce:

E15 $[ABC] : A \varepsilon A . B \varepsilon B . C \varepsilon C . \sim(A = B) . \sim(A = C) . \supset . B = C$

PF $[ABC] \therefore \mathrm{Hyp}(5) . \supset :$
6) $A = \mathfrak{A} . \vee . A = \mathfrak{B} :$ [E15*; 1]
7) $B = \mathfrak{A} . \vee . B = \mathfrak{B} :$ [E15*, A/B; 2]
8) $A = \mathfrak{B} . B = \mathfrak{A} . \vee . A = \mathfrak{A} . B = \mathfrak{B} :$ [6, 7, 4]
9) $C = \mathfrak{A} . \vee . C = \mathfrak{B} :$ [E15*, A/C; 3]
$B = C$ [8; 9: 5]

F5 $[Ba]:B \varepsilon a.\mathfrak{B} = B. \supset .\mathfrak{A} \varepsilon \mathrm{Kl}_1(a)$

PF $[Ba]:\mathrm{Hyp}(2). \supset .$

3) $\mathfrak{B} \varepsilon a.$ [1; 2]

$\mathfrak{A} \varepsilon \mathrm{Kl}_1(a)$ [DI, $A/\mathfrak{A}$; F1; 3]

F6 $[Ba]:B \varepsilon a. \supset .[\exists A].A \varepsilon \mathrm{Kl}_1(a)$

PF $[Ba]\therefore \mathrm{Hyp}(1). \supset :$

2) $B \varepsilon B:$ [1]

3) $\mathfrak{A} = B.\vee .\mathfrak{B} = B:$ [E15, $A/\mathfrak{B}$, $B/\mathfrak{A}$, C/B; F2; F1; 2; F3]

4) $\mathfrak{A} \varepsilon \mathrm{Kl}_1(a).\vee .\mathfrak{B} \varepsilon \mathrm{Kl}_1(a):$ [3; F4; F5]

$[\exists A].A \varepsilon \mathrm{Kl}_1(a)$ [4]

The thesis F6 is the interpretation of C1.

F7 $[ABa]:B \varepsilon B.A = \mathfrak{A}.\mathfrak{B} \varepsilon a. \sim(A = B). \supset .B \varepsilon a$

PF $[ABa]:\mathrm{Hyp}(4). \supset .$

5) $\sim(A = B).$ [2; 4]

6) $\mathfrak{B} = B.$ [E15, $A/\mathfrak{B}$, $B/\mathfrak{A}$, C/B; F2; F1; 1; F3; 5]

$B \varepsilon a$ [3; 6]

F8 $[ABa]:A \varepsilon A.B \varepsilon B.\mathfrak{A} \varepsilon a. \sim(A \varepsilon a). \sim(A = B). \supset. B \varepsilon a$

PF $[ABa]:\mathrm{Hyp}(5). \supset .$

6) $\sim(\mathfrak{A} = A)$ [3; 4]

7) $\mathfrak{A} = B.$ [E15, $A/\mathfrak{A}$, B/A, C/B; F1; 1; 2; 6; 5]

$B \varepsilon a$ [3; 7]

F9 $[ABab]:A \varepsilon \mathrm{Kl}_1(a).A \varepsilon \mathrm{Kl}_1(b).B \varepsilon b. \sim(B \varepsilon \mathrm{Kl}_1(b)). \supset .B \varepsilon a$

PF $[ABab]\therefore \mathrm{Hyp}(4). \supset .$

5) $A = \mathfrak{A}.\mathfrak{B} \varepsilon a.\vee .\mathfrak{A} \varepsilon a. \sim(A \varepsilon a):$ [DI; 1]

6) $\sim(A = B).$ [2; 4]

7) $B \varepsilon B.$ [3]

8) $A \varepsilon A.$ [1]

$B \varepsilon a$ [5, F7; 7; 6; F8; 8; 7; 6]

The thesis F9 is the interpretation of E1.

F10 $[ABa]:A \varepsilon \mathrm{Kl}_1(a).B \varepsilon \mathrm{Kl}_1(a). \supset .A = B$

PF $[ABa]\therefore \mathrm{Hyp}(2). \supset :$

3) $A = \mathfrak{A}.\mathfrak{B} \varepsilon a.\vee.\mathfrak{A} \varepsilon a. \sim(A \varepsilon a):$ [DI; 1]

4) $B = \mathfrak{A}.\mathfrak{B} \varepsilon a.\vee.\mathfrak{A} \varepsilon a. \sim(B \varepsilon a):$ [DI, A/B; 2]

5) $A \varepsilon A.$ [1]

6) $B \varepsilon B.$ [2]

$A = B$ [3; 4; F7, A/B, B/A; 5; F7; 6; F8; 5; 6]

F10 is the interpretation of E2.

F11 $[\exists AB].A \varepsilon A.B \varepsilon B. \sim(A = B)$ [F1; F2; F3]

F11 is the interpretation of C2.

The above interpretation of {C1, E1, E2, C2} in the system {F1, F2, F3, E15} demonstrates that {C1, E1, E2, C2} is not contradictory in a universe in which there are exactly two different objects.

2. The system {A1, E1, E2} can be interpreted in the system {F1, C10}[19] as follows:

We introduce the term "class" by means of the definition:

DII $[Aa]:A \varepsilon \mathrm{Kl}_2(a). \equiv .A \varepsilon A.a \subset a$[20]

and we then deduce:

G1 $[a].[\exists A].A \varepsilon \mathrm{Kl}_2(a)$ [DII, $A/\mathfrak{A}$; F1]

The thesis G1 is the interpretation of A1.

G2 $[ABb]:A \varepsilon \mathrm{Kl}_2(b).B \varepsilon b. \supset .B \varepsilon \mathrm{Kl}_2(b)$

PF $[ABb]$:Hyp(2). ⊃ .
3) $A \varepsilon A.$ [1]
4) $B \varepsilon B.$ [2]
5) $A = B.$ [C10; 3; 4]
$B \varepsilon \mathrm{Kl}_2(b)$ [1; 5]

G3 $[ABab]:A \varepsilon \mathrm{Kl}_2(a).A \varepsilon \mathrm{Kl}_2(b).B \varepsilon b. \sim(B \varepsilon \mathrm{Kl}_2(b)). \supset .B \varepsilon a$ [G2]

---

[19] It is easy to prove that the system {F1, C10} is inferentially equivalent to: F1 and

C10* $[A]:A \varepsilon A. \supset .A = \mathfrak{A}$

a) We assume F1 and C10 and deduce:

C10* [C10, $B/\mathfrak{A}$; F1]

b) Now we assume F1 and C10* and deduce:

C10 $[AB]:A \varepsilon A.B \varepsilon B. \supset .A = B$

PF $[AB]$:Hyp(2). ⊃ .
3) $A = \mathfrak{A}.$ [C10*; 1]
4) $B = \mathfrak{A}.$ [C10*, A/B; 2]
$A = B$ [3; 4]

[20] The symbol "⊂" denotes inclusion for names; it is introduced in ontology by the generally accepted definition from the algebra of names:

$$[ab]\therefore a \subset b. = :[A]:A \varepsilon a. \supset .A \varepsilon b$$

We have used the expression "$a \subset a$" in the definiens of the definition DII in order to satisfy the requirements of the rules of reasoning obligatory in ontology. It is one of the conditions for a correctly constructed definition that every variable which occurs in the general quantifier preceding the whole definition must appear in both the definiens and the definiendum.

We also note that the expression: $[a].a \subset a$, which is a well-known theorem (cf. Russell and Whitehead [1913, Vol. I, *22.42]) is also a thesis of ontology.

The thesis G3 is the interpretation of E1.

G4 $[ABa]:A \varepsilon \mathrm{Kl}_2(a).B \varepsilon \mathrm{Kl}_2(a). \supset .A = B$

PF $[ABa]:\mathrm{Hyp}(2). \supset .$

3) $A \varepsilon A.$ [1]
4) $B \varepsilon B.$ [2]
$A = B$ [C10; 3; 4]

G4 is the interpretation of E2.

This interpretation of {A1, E1, E2} in the system {F1, C10} demonstrates that {A1, E1, E2} is not contradictory in a universe in which there is exactly one object.

3. The system {C1, A2} (and also {C1, B1, B2}) can be interpreted in the system consisting of C10 in the following manner.

We define the term "class" as follows:

DIII $[Aa]:A \varepsilon \mathrm{Kl}_3(a). \equiv :A \varepsilon a$

and we then deduce:

H1 $[Ba]:B \varepsilon a. \supset .[\exists A].A \varepsilon \mathrm{Kl}_3(a)$ [DIII, A/B]

The thesis H1 is the interpretation of C1.

H2 $[ABab]:A \varepsilon \mathrm{Kl}_3(a).A \varepsilon \mathrm{Kl}_3(b).B \varepsilon b. \supset .B \varepsilon a$

PF $[ABab]:\mathrm{Hyp}(3). \supset .$

4) $A \varepsilon a.$ [DIII; 1]
5) $A \varepsilon A.$ [4]
6) $B \varepsilon B.$ [3]
7) $A = B.$ [C10; 5; 6]
$B \varepsilon a$ [4; 7]

The thesis H2 is the interpretation of $A2$. Thus {C1, A2} is not contradictory if one assumes that there is at most one object in the world.

[*Translator's note.* For the system {C1, B1, B2} make the additional definition:

D $[AB]:A \varepsilon \mathrm{el}_3(B). \equiv .A \varepsilon B$

The proofs of the interpretations of B1, B2 are easy.]

All this reasoning sheds some interesting light on the systems that we have just analyzed, to wit:

If one supposes that there exist exactly two different objects in the universe, then neither {C1, E1, E2} nor {C1, E1, E2, C2} leads to a contradiction.

If one supposes that there exists exactly one object in the universe, then {A1, E1, E2} does not lead to a contradiction.

If one supposes that there exists at most one object in the world, then {C1, A2} does not lead to a contradiction.

On the contrary, we have not given a similar interpretation of {A1, A2} because this system must be considered absolutely contradictory.

VI. Analysis of the Presupposition A2

1. We showed in Chapter IV that the Frege modification of the presuppositions (which consisted of replacing A2 by the weaker thesis E1) did not eliminate the paradox. We must therefore return to the analysis of the systems of presuppositions {A1, A2} and {C1, A2, C2} which was begun in Chapters II and III. As we indicated in Chapter III, A1 was unquestionable false in Leśniewski's view, since he did not believe in a class consisting of empty or contradictory elements. Indeed, to one who does not believe in the existence of the class of mythical Greek centaurs because he does not believe in centaurs, it is clear that the admission of {A1, A2} might lead to a contradiction. But here there is no paradox, because one does not believe in the truth of one of the presuppositions. Consequently, from the viewpoint of Russell's paradox, this system no longer interests us.

But this does not resolve the longstanding question of the system {C1, A2, C2}, which as we know, is also contradictory. The theses C1 and C2 contain nothing which provokes any objection from our viewpoint. In admitting C2 one admits that there exist at least two objects in the universe; in admitting C1, one recognizes that if there is one object in the universe which is an *a* (e.g. an elephant), then there also exists a class of these objects *a* (e.g. the class of elephants). So if one admits the existence of centaurs, he would admit the existence of a class of centaurs. Thus C1 does not permit one to recognize the existence of a class of centaurs without first believing in the existence of centaurs. But the theses C1 and C2, which are completely believable, when combined with A2 yield a contradiction in the non-contradictory system of logic which we presuppose. It is also easy to prove that this logical system together with C1 and C2 is not contradictory. This is done with the aid of the method of interpretation.

At the same time, we know (we are not going to prove it here) that the system {C1, C2} does not lead to a contradiction if we add the presupposition that there exist *n* (*n* a natural number) distinct objects in the universe and at the same time interpret "Kl" as "Kl_3" of DIII (Chapter V), i.e.

$$[Aa]:A \varepsilon \mathrm{Kl}_3(a) . \equiv . A \varepsilon a$$

It is easy to prove that C1 and C2 are valid but A2 is not. Indeed, in this interpretation, A2 becomes

J1 $[ABab]:A \varepsilon a . A \varepsilon b . B \varepsilon b . \supset . B \varepsilon a$

But J1 is obviously false (e.g. assuming that Socrates is a Greek, Socrates is a man and Louis XIV is a man, we can conclude Louis XIV is a Greek). Thus it is A2 that is the cause of the appearance of the contradiction.

Now consider the system {B1, B2}. The terms "class" and "element" both occur in B1 and B2. The relation between these two terms can be established by D2. As we know, the thesis B1 follows directly from this definition. We also know that a correctly constructed definition can never lead to a contradiction. Consequently, the appearance of a contradiction within the framework of the system {C1, B1, B2, C2} does not depend on the acceptance of B1.

The statement of the fact that these systems are contradictory as well as the proofs establishing which of the presupposition provoke this contradiction, does not as yet resolve Russell's paradox. Up to now, we have given nothing intuitive that would permit us, as we have done with the presupposition A1, to reject either A2 or B2 or to modify them adequately. We know from Chapter IV, that the considerable weakening of A2, proposed by Frege, did not make the contradiction dissappear.

The fact of the existence of the paradox and the fact that the replacement of A1, recognized by him to be false, by C2, intuitively clear and true, did not make the paradox disappear, forced Leśniewski to submit the terms which appear in the presuppositions to a fundamental analysis. Since he was sure of the laws of logic on which our reasoning rests as well as the manner of understanding the logical constants contained in these laws, he analyzed the terms "class" and "element" which appear in the presuppositions. The examination of this problem led him to the conclusion that a superficial determination of these terms such as the one sketched in Chapter I. 3, is insufficient and mistaken and, in fact, provokes the appearance of the contradiction.

According to Leśniewski, there are two different concepts encompassed by the term "class":

a) distributive class
b) collective class.

This distinction made by Leśniewski, enabled him to realize that the illusory intuitiveness of A2 (respectively B2) is the consequence of the confusion caused by the use of the same noun for the two different concepts. The precise determination of these terms suggests, if one accepts this distinction, the necessity of rejecting A2 (resp. B2) as invalid and contrary to previously accepted intuition. Thanks to this the paradox will disappear, since the belief in the validity of the presuppositions which led to the contradictions will disappear. Only the proof (accepted without disappointment) that the systems {A1, A2}, {C1, A2, C2}, {C1, B1, B2, C2}, {C1, E1, E2, E3}, {A1, E1, E2, C2} are contradictory will remain. Moreover, as we shall see farther on, this distinction became Leśniewski's point of departure for the construction of the deductive theories which constitute his system for the foundations of mathematics. Thus it is important to understand the distinction between these two meanings of the term "class".

2. The expression "class(a)" in the distributive sense is nothing more than a fictitious name which replaces the well-known term of classical logic, "the extension of the objects a". If one takes the term in this sense, the formula "$A \varepsilon \mathrm{Kl}(a)$" means the same thing as "A is an element of the extension of the objects a" or, more briefly, "A is a". In which case, the formula "Socrates is Kl(white)" means the same thing as "Socrates is an element of the extension of white objects"; in other words: "Socrates is white". Thus, the understanding of "class" in the distributive sense would reduce the formula "$A \varepsilon \mathrm{Kl}(a)$" to the purely logical formula "$A \varepsilon a$", where "ε" is understood as the connective of the individual proposition. Setting as for the moment the question if it is possible, under this assumption, to construct Russell's paradox using only this connective, we wish to emphasize that in this present article we are using "ε" exclusively in the sense of Leśniewski's ontology.

We see now that among the presuppositions of the systems under discussion, some are true and some are false, if one understands the term "class" in the distributive sense. It is clear that A2 will take the form J1 and thus is false. A1 will take the form:

J2 $[a]:[\exists A].A \varepsilon a$

which is false. C1 will take the form

J3 $[Ba]:B \varepsilon a. \supset .[\exists A].A \varepsilon a$

which is true. The definition D2 will become:

J4 $[AB]:A \varepsilon \mathrm{el}(B). \equiv [\exists a].B \varepsilon a.A \varepsilon a$

which is a correctly formed definition. B1 will take the form:

J5 $[ABa]:A \varepsilon a.B \varepsilon a. \supset .B \varepsilon \mathrm{el}(A)$

It one accepts J4, J5 is inferentially equivalent to:

J6 $[ABa]:A \varepsilon a.B \varepsilon a. \supset .[\exists a].A \varepsilon a.B \varepsilon a$

which is true. B2 will take the form:

J7 $[ABa]:A \varepsilon a.B \varepsilon \mathrm{el}(A). \supset .B \varepsilon a$

which is false since it is inferentially equivalent to J1 (if one accepts J4). E1 becomes:

J8 $[ABa]:A \varepsilon a.A \varepsilon b.B \varepsilon b. \sim(B \varepsilon b). \supset B \varepsilon a$

which is true since it follows directly from the propositional calculus. E2 will take the form:

J9 $[ABa]{:}A \varepsilon a . B \varepsilon a . \supset . A = B$

which is false.

In summary we can state that if one understands the term "class" in the distributive sense we have the following false presuppositions in the systems under discussion:

{A1, A2} – both,
{A1, B1, B2} – the first and the third,
{C1, A2, C2} – the second,
{C1, B1, B2, C2} – the third,
{C1, E1, E2, E3} – the third,
{A1, E1, E2, C2} – the first and the third.

Therefore, if the term "class" has this meaning, Russell's paradox disappears, because we do not believe in the truth of all of the presuppositions in any of the above systems.

[*Translator's note.* Since D2 is not a definition in any system containing B1 and B2 as presuppositions (indeed it cannot be a definition in such a system since "el" already appears as a primitive term) one's acceptance or rejection of J5, J7 (the interpretations of B1, B2) should not be based on J4 (the interpretation of D2). My intuition says that at least J5 is false. Thus each of the systems above still has at least one false presupposition.]

3. On the other hand, the term "collective class" is a real name; one cannot eliminate it by reducing it to some logical concept. In the collective sense, the expression "class(a)" is a really existing object, which is composed of all of the objects of the domain of objects a. In other words, if one is given any domain of objects a, with the aid of the term "class" taken in the collective sense, one can obtain an object composed exactly of the objects appearing in the given domain, that is, of the objects a. But the indispensable condition for the existence of an object "Kl(a)" is that the domain of objects a be non-empty, i.e. there exists an object a. Obviously, the objects from which the collective class is constructed may be disjoint, for example, in space or time. Thus if there exist some objects whatever, for example, some bricks, the "class of bricks", in the collective sense, is an absolutely real existing object composed of all bricks no matter where they are found. An example: the books that are on my desk at this moment constitute an object which is the "collective class" of all the books that are on my desk at this moment, whether they are in physical contact with each other or not. "Class" defined in this way clear has

the following properties: (1) it is unique, (2) if it is the class of one object only, it is identical with that object, (3) the class of a class of objects is identical with that class. E.g., for all A, B and a, if A is Kl(a) and B is Kl(a), then $A = B$; Kl(Julius Caesar) = Julius Caesar; Kl(Kl(bricks)) = = Kl(bricks).

If the term "class" is taken in the collective sense, definition D2 takes on an entirely new meaning. If an object B is the collective class constructed from the objects a (in the limiting case a would be B) and the object A is an element of B, then A need not necessarily be one of the elementary objects a which form the given class, i.e. the object B. To clarify this let us return to the example of the books above. We will consider books as consisting only of printed pages and we will suppose that there are no printed pages on my desk except those in the books. In this case, B is the collective class of all the books found on my desk at this instant, and the same B is also the collective class of all the printed pages found on my desk at this instant. Thus the collective class has the property that for all A, a and b, if A is Kl(a) and A is Kl(b), then Kl(a) = Kl(b). But it does not follow that the objects a are the same as the objects b. From which one concludes that if an object A is an element of an object B an B is the collective class of the objects a, then A is not necessarily an a. So if A is the class of books found on my desk at this instant and B is an element of A, then B is not necessarily a book. B could be the fifth page of one of these books or the object which is the collective class of all the illustrated pages contained in the books on my desk (provided of course that there is at least one illustrated page). In a word, if B is an element of the object A, then B may be any piece whatever of A.

The most characteristic and elementary properties of the terms "class" and "element" taken in this sense are given in the following theses.

K1 $[a]:[\exists A].A \varepsilon a. \equiv .[\exists A].A \varepsilon \mathrm{Kl}(a)$

The thesis K1 states that if there are some objects a, then Kl(a) exists and conversely if Kl(a) exists, then there are some objects a.

K2 $[ABa]:A \varepsilon \mathrm{Kl}(a).B \varepsilon \mathrm{Kl}(a). \supset .A = B$

K2 says that "Kl(a)" is unique.

K3 $[A]:A \varepsilon A. \supset .A = \mathrm{Kl}(A)$
K4 $[AB]:A \varepsilon \mathrm{Kl}(B).B \varepsilon B. \supset .A = B$
K5 $[Aa]:A \varepsilon \mathrm{Kl}(a). \equiv .A = \mathrm{Kl}(a)$
K6 $[Aa]:A \varepsilon \mathrm{Kl}(a). \equiv .A \varepsilon \mathrm{Kl}(\mathrm{Kl}(a))$
K7 $[Aab]:A \varepsilon \mathrm{Kl}(a).A \varepsilon \mathrm{Kl}(b). \supset .\mathrm{Kl}(a) = \mathrm{Kl}(b)$
K8 $[AB]:A \varepsilon \mathrm{el}(B). \equiv .[\exists a].B \varepsilon \mathrm{Kl}(a).A \varepsilon a$

The thesis K8 is the equivalent of D2.

K9 $[Aa]:A \varepsilon \mathrm{Kl}(\mathrm{el}(\mathrm{Kl}(a))). \equiv .A \varepsilon \mathrm{Kl}(a)$

K10 $[A]:A \varepsilon A. \supset .A \varepsilon \mathrm{el}(A)$

The thesis K10 states that each ob ject is an element of itself.

K11 $[ABC]:A \varepsilon \mathrm{el}(B).B \varepsilon \mathrm{el}(C). \supset .A \varepsilon \mathrm{el}(C)$

K11 gives the transitivity of "el".

K12 $[AB]:A \varepsilon \mathrm{el}(B). \supset .B \varepsilon B$

K13 $[AB]:A \varepsilon \mathrm{el}(B).B \varepsilon \mathrm{el}(A). \supset .A = B$

K14 $[Aa]\therefore A \varepsilon \mathrm{Kl}(a). \equiv :A \varepsilon A:[\exists B].B \varepsilon a:[B]:B \varepsilon a. \supset .B \varepsilon \mathrm{el}(A):$
$[B]:B \varepsilon \mathrm{el}(A). \supset .[\exists CD].C \varepsilon a.D \varepsilon \mathrm{el}(C).D \varepsilon \mathrm{el}(B)$

The thesis K14 shows that one can give a well-constructed definition of "Kl" in terms of "el". We cannot present an intuitive justification of this formula here because it would cause usto digress too far from the subject of this article.

These theses characterize with sufficient clarity the elementary properties of the collective class. One immediately sees that the term "class" thus conceived is not a logical term and that the theses which characterize it do not follow from logical presuppositions[21].

Let us now consider the systems under discussion and see which of the presuppositions are true and which are false, according to the collective meaning of "class". A1 is clearly false, since together with K1, it establishes the existence of contradictory objects. K8 shows that both D2 and B1 are true. The truth of C1 and E2 follows from K1 and K2. On the contrary, A2, B2 and E1 are false because, according to the properties of the collective class, if *A* is the class of all propositions which compose this article and *B* is an element of *A* and this same *B* is not identical to *A*, it does not necessarily follow that *B* is one of the propositions cited. It could be, for example, the first of the letters "r" appearing in this article.

In summary, if the term "class" is taken in the collective sense, the following propositions are false in the given systems:

{A1, A2} – both,
{A1, B1, B2} – the first and the third,
{C1, A2, C2} – the second,
{C1, B1, B2, C2} – the third,
{C1, E1, E2, E3} – the second,
{A1, E1, E2, C2} – the first and the second.

---

[21] We emphasize, to avoid misunderstanding, that the term "class" in the collective sense says absolutely nothing about whether or not certain objects exist in the universe. One can employ it to relate different objects, physical or not, on condition that one believes in the existence of these objects. Thus each person can use the term "class" in the collective sense, no matter what metaphysical opinions he holds.

Since we no longer believe in at least one of the presuppositions in each of the systems examined above, the contradictions are no longer paradoxical and we have eliminated the problem of Russell's paradox. We have just shown how the distinction of the two meanings of the term "class" (distributive and collective) eliminates the paradox. The result of our deductions is that the belief in the truth of certain presuppositions which are the cause of the appearance of the paradox, is misplaced. This belief, as we have seen, arises from the confusion of our intuitive notions of distributive and collective class; the presupposition B1 seems to be particularly noteworthy in this regard. On the other hand, the clear understanding of the distinction between the two notions given by Leśniewski dissolves this confusion and along with it, the belief in the presuppositions that we have discussed[22]. It frees us from the phantom of the paradox.

VII. The Usual Formulation of Russell's Paradox and Leśniewski's Analysis

The formulation of Russell's paradox, presented in Chapters II-V and the analysis concerning it, differ considerably from what one generally finds in the literature. The above-mentioned considerations will clarify why the usual formulation of the paradox is not, according to Leśniewski, paradoxical in character. It is nothing more than a formal construction provoking contradiction or non-sense by violating the principles of correct construction for deductive systems. An analysis of it will permit us to clarify certain principal points of Leśniewski's logical systems.

1. One usually formulates Russell's paradox in the following manner: Without carefully specifying the meaning of the terms "set" ("class") and "element" one supposes that one can divide all sets in two categories. To the first belong all and only those sets which are not elements of themselves. Sets of this category are called "normal sets". To the second belong all and only those sets which are elements of themselves; these are called "abnormal sets". Thus if we do not take into consideration the restrictions imposed by some theory of logical types and as a consequence we act in the manner permitted before their introduction into logic, we can, with the use of the symbol "ε" (understood as the informal expression "is an element of") assert that:

a) A is a normal set if and only if: $\sim(A \varepsilon A)$;
b) A is an abnormal set if and only if: $A \varepsilon A$.

[22] Outside of the work cited we know of only one other author who mentions this distinction as made by Leśniewski: Kotarbiński [1929]. (*Editorial Note*: English translation – Kotarbiński [1966].)

We can now define the expression "set of all normal sets" as follows if we continue to ignore the theory of types:

L1 $[A]: A \varepsilon \wedge . \equiv . \sim (A \varepsilon A)$

The formula immediately yields:

L2 $\wedge \varepsilon \wedge . \equiv \sim (\wedge \varepsilon \wedge)$

L2 is contradictory according to the propositional calculus.

Leśniewski thought that the problem thus formulated was not paradoxical in character because, not believing in L1, he felt that this construction did not capture the total essence of Russell's paradox. The contradiction we obtain by L1 results from the fact that L1 is an incorrectly constructed definition. This incorrectness remains no matter how one understands the term " ε " appearing in L1, be it as the term "is an element of" of the majority of the present systems of logic or set theory or as the term "is" of ontology. As one knows, the first understanding of the term, rather imprecise and confused, contains two equally admissible forms for expressing the idea of "A is an element of B", namely: "$B\{A\}$" and "$A \varepsilon B$". If we examine each of these separately and make the abstraction that each of them individually designates the same thing in the systems mentioned, whether taken from the point of view of ontology or of a totally different logical construction, then for the form "$B\{A\}$", the equivalent of L1 will be:

L3 $[A]: \wedge\{A\} . \equiv . \sim (A\{A\})$

But L3 is, for Leśniewski, an incorrectly constructed expression for the same reasons as it is for present-day logicians. It is a non-sense expression because there is a confusion of semantical categories (logical types) in the form "$A\{A\}$" where the variable "A" is employed simultaneously in two different categories. Thus there is no paradox because one does not have the right to admit the expression in any system, not because a paradox would appear, but because this expression is non-sense. In the construction of deductive systems one has to take semantical categories (logical types) into account in the same way that in ordinary meaningful language one must take the grammatical parts of speech into consideration.

If, with regard to L3, we can assert that there is complete agreement between Leśniewski's point of view and that of the great majority of present-day logicians, it is completely otherwise for the expression L1. In order to avoid the paradox, these latter have asserted that the subject and the complement of an individual proposition belong to two different logical types. Thus for them, the expression "$A \varepsilon A$" is, like "$A\{A\}$", devoid of meaning, thus L1 is non-sense. From this viewpoint, the individual proposition whose complement is an object cannot be expressed without the aid of the concept "the set formed of exactly one element", i.e. "$A \varepsilon \iota' A$". Obviously, from this viewpoint, instead of the meaningless L1, one can accept:

L4 $[A]:A \varepsilon \Lambda . \equiv . \sim (A \varepsilon \iota' A)$

which does not give a contradiction.

But, as we know from Chapter I, in Leśniewski's ontology the subject and the complement of an individual proposition belong to the same semantical category. We also know that in this theory the expression "$A \varepsilon A$" is correctly constructed and meaningful. This was the position taken by Leśniewski long before he constructed ontology. According to him, the set formed of only one element (in the distributive or collective sense of the word) is identical with that element. From this viewpoint, L4 does not differ from L1 and this latter is not in any way meaningless. It is just an invalid thesis because L1, being a definition whose definiendum has the form of an individual proposition, does not assert in its definiens that the subject of this proposition is an object. Failure to observe this essential condition in the construction of definitions of this type (Leśniewski calls them "definitions of ontological type") often leads to a contradiction which one might call, for lack of a better term, "paradox of the contradictory object". In the construction of a definition, if the projected definiens does not guarantee in an obvious manner that the subject of the individual proposition (i.e. the definiendum) is an object, we must add a conjunct to the definiens which guarantees this without asserting anything more. This conjunct can have either of the following two forms: "$A \varepsilon A$" or "$[\exists a].A \varepsilon a$" because in ontology the term "object" can be defined by:

D4 $[A]:A \varepsilon \Lambda . \equiv .A \varepsilon A$

which is inferentially equivalent to the thesis

M1 $[A]:A \varepsilon \bar{\Lambda} . \equiv .[\exists a].A \varepsilon a$ [D4; Th IV, see note 9]

This fully explains why Leśniewski considers L1 to be a false but not meaningless thesis. In L1 there is no confusion of semantical categories, but there is a violation of the rules of correct construction for ontological type definitions. It is not at all surprising that a contradiction should appear, but it is certainly not paradoxical. Therefore, a correctly constructed definition of the term "$\bar{\Lambda}$" is:

D5 $[A]:A \varepsilon \Lambda . \equiv .A \varepsilon A . \sim (A \varepsilon A)$

which will not yield a contradiction. On the contrary, it is a definition of the term "contradictory object".[23]

2. The preceding considerations also explain the role played by definitions D1 and D3 in the systems of presuppositions analyzed in Chapters II-V. They are correctly constructed definitions and as such, cannot cause contradictions. In these systems, the contradiction does not result from these definitions but from the acceptance of false presuppositions. If on the other hand, instead of D1 and D3 we would accept:

L5 $[A] \therefore A \varepsilon *. \equiv :[a]:A \varepsilon \mathrm{Kl}(a). \supset . \sim(A \varepsilon a)$
L6 $[A] \therefore A \varepsilon \overset{*}{*}. \equiv :[a]:A \varepsilon \mathrm{Kl}(\mathrm{Kl}(a)). \supset . \sim(A \varepsilon a)$

then, analogous to the case of L1, we would immediately obtain a contradiction without the add of supplementary presuppositions.

From L5 we can deduce:

L7 $[A]: \sim(A \varepsilon *). \supset .A \varepsilon A$

PF $[A:\mathrm{Hyp}(1). \supset$
$[\exists a].$
2) $A \varepsilon a.$ [L5; 1]
$A \varepsilon A$ [2]

L8 $[A]:A \varepsilon A$ [L7; M1; D4]
L1 $[A]:A \varepsilon \Lambda. \equiv . \sim(A \varepsilon A)$ [D5; L8]

Consequently, we would have a contradiction. We can proceed in a completely analogous manner for L6.

We remark further, that in the distributive and collective sense of the term "class", the two defined terms "$*$" and "$\overset{*}{*}$" mean the same as "Λ". In the first place we can state by DIII (in the distributive sense) and by K6 (in the collective sense) that D3 can be modified to D1. Assuming the distributive sense of the term "class", D1 will give:

---

[23] Clearly one can construct at will formulas analogous to L1, L5, L6, as far as faulty construction and leading immediately to a contradiction are concerned. One sometimes cites the thesis

$$[A]::A \varepsilon \Lambda . \equiv \therefore [B] \therefore [a]:A \varepsilon a. \equiv .B \varepsilon a: \supset . \sim(B \varepsilon A)$$

(cf. Mostowski [1948, p. 219]) which immediately leads to a contradiction, and claims that it gives a paradox analogous to that of Russell. But this thesis produces a contradiction because it is an incorrectly constructed definition. The corresponding thesis, correctly constructed, is:

$$[A]::A \varepsilon \Lambda . = \therefore A \varepsilon A \therefore [B] \therefore [a]:A \varepsilon a. \equiv .B \varepsilon a : \supset . \sim(B \varepsilon A)$$

This does not lead to a contradiction. It is easy to prove that it is one of the possible forms of the definition of the term "contradictory object".

M2 $[A]:A \varepsilon *. \supset . \sim(A \varepsilon A)$

PF $[A]$:Hyp(1).$\supset$

2) $A \varepsilon A.$ [1]

3) $A \varepsilon \mathrm{Kl}_3(A)$ [DIII, a/A; 2]

$\sim(A \varepsilon A)$ [D1, a/A; 1; 3]

M3 $[A]:A \varepsilon *. \equiv .A \varepsilon \Lambda$ [M2; D5]

One uses K3 instead of DIII in an analogous proof of M3 for the collective sense of the term "class".

As we know, D1 and D3 are used in the proofs of the contradictions in the systems under consideration. However, one may not assume that they are the cause of the contradictions in these systems. The rules of reasoning of the logical system we are using always permits the addition of a correctly constructed definition to any deductive system based on it. Thus if the use of a definition within the framework of accepted presuppositions leads to a contradiction, the fault lies with the presuppositions and not with the definition, because each correctly constructed definition, when added to a deductive system, is always a valid thesis. It cannot be a contributing factor in the contradictoriness of a system of presuppositions within the framework of a complete system of logic.[24]

3. Consequently according to Leśniewski, the usual formulation of Russell's paradox only shows that for one thing, the failure to observe the principles of semantical categories leads to meaningless formulas (e.g. L3), and for another, that the use of incorrectly constructed definitions leads to a contradiction. But the essence of Russell's paradox lies elsewhere, namely in the determination of the relation between a set (class) and its elements. The lack of a precise understanding of this relation leads to B2 which together with B1, permits us to describe this relation in the following manner:

B3 $[Aa] \therefore A \varepsilon \mathrm{Kl}(a). \supset :[B]:B \varepsilon \mathrm{el}(A). \equiv .B \varepsilon a$ [B1; B2]

i.e. each element of the set of objects a is one of the objects a, and each object a is an element of the set of objects a. As we know, the acceptance of this relation leads to a contradiction within the framework of a complete system of logic. It is precisely the acceptance of the distinction between the ditributive class and the collective class that permitted Leśniewski to overcome the paradox, because this distinction eliminates the relation given in B3.

---

24 As Mr. J. Łukasiewicz observed (letter to the author, Dublin, April 8, 1949), the system composed of A1, A2, D1 and the ontological thesis Th. IV (cited in footnote 9) yield a contradiction within the framework of propositional calculus, enriched with the rules of quantification. At the same time, Mr. J. Łukasiewicz proved that in the framework of this system, any combination of three of these four presuppositions does not yield a contradiction.

We cannot show here how one can explain and eliminate the other known paradoxes within the framework of Leśniewski's system. This would necessitate a very complete presentation of the principles on which his system for the foundations of mathematics is based, a detailed description of the respective theories (which constitute it) and in addition, an exact analysis of certain metalogical properties characteristic of this system. We will give just one example, namely, how to overcome the executioner (respectively barber) paradox. It is also of interest to show that not all the known paradoxes result from a confussion of semantical categories, a violation of the rules for the construction of definitions, or the unconscious acceptance of false presuppositions.

As we are going to see, the paradox in question results from none of these. It can be presented in the following manner:

One can say that each man is (a man who commits suicide) or he is not (a man who commits suicide). Suppose there is a man who kills and must kill all men and only those men who do not commit suicide. At the moment when he has killed all the non-suicides he will be faced with a dilemma: what must he do with himself? If he kills himself he will be a suicide and thus he does not have the right to kill himself. If he does not kill himself, he will not be a suicide, thus he will have to kill himself. He will be confounded by the appearance of a contradiction with respect to his own person.

A succinct demonstration will convince us that there is no paradox here, since it will be established that in general, objects such as our "executioner" or "barber", etc. cannot exist because they are contradictory. To this end we begin by attempting to construct a definition of the objects. In the case of the "executioner" (we will call him "X") one can determine the relations utilized in the reasoning with the aid of the terms "man" and "kill". Thus we have:

a) for each S, S is a suicide. $\equiv$ .S is a man and S kills S

b) for each S, S is a non-suicide. $\equiv$ .S is a man and it is not true that S kills S.

Then one obtains:

c) for each A, A is X. $\equiv$ .A is a man and for each S, if S is a suicide it is not true that A kills S and for each S, if S is a non-suicide, then A kills S.

With the help of this example, the general definition of objects such as "executioner", "barber", etc. can be constructed as follows:

D6 $[A]::A \varepsilon \overset{*}{**}. \equiv .\therefore [\exists \varphi a] .\therefore A \varepsilon a : [B] : B \varepsilon a . \varphi\{BB\} . \supset . \sim (\varphi\{AB\}) : [B] : B \varepsilon a . \sim (\varphi\{BB\}) . \supset . \varphi\{AB\}$

But there exists the logical thesis:

M4 $[A\ \varphi\ a] \therefore A\ \varepsilon\ a : [B] : B\ \varepsilon\ a . \varphi\{BB\} . \supset . \sim(\varphi\{AB\}) :$
$\supset . [\exists B] . B\ \varepsilon\ a . \sim(\varphi\{BB\}) . \sim(\varphi\{AB\})$

PF $[A\ \varphi\ a] : \mathrm{Hyp}(2) . \supset .$
3) $\sim(\varphi\{AA\}) .$ [2, B/A; 1]
$[\exists B] . B\ \varepsilon\ a . \sim(\varphi\{BB\}) . \sim(\varphi\{AB\})$ [1; 3]

Thus:

M5 $[A] : A\ \varepsilon\ \overset{*}{**} . \supset . A\ \varepsilon\ \wedge$

PF $[A] :: \mathrm{Hyp}(1) . \supset \therefore$
$[\exists \varphi a] \therefore$
2) $A\ \varepsilon\ a :$
3) $[B] : B\ \varepsilon\ a . \varphi\{BB\} . \supset . \sim(\varphi\{AB\}) :$ [D6; 1]
4) $[B] : B\ \varepsilon\ a . \sim(\varphi\{BB\}) . \supset . \varphi\{AB\} :$
$[\exists B] .$
5) $B\ \varepsilon\ a .$
6) $\sim(\varphi\{BB\}) .$ [M4; 2; 3]
7) $\sim(\varphi\{AB\}) .$
8) $\varphi\{AB\} \therefore$ [4; 5; 6]
$A\ \varepsilon\ \wedge$ [7; 8; D5 $(p . \sim p . \supset . q)$]

M6 $[A] : A\ \varepsilon\ \overset{*}{**} . = . A\ \varepsilon\ \wedge$ [D5; M5]

Thus objects such as our "executioner" do not exist in the universe because by their very definition they are "contradictory objects". Thus the paradox of the executioner (respectively, barber) is only an apparent paradox. It results from our ignorance of the fact that D6 is one of the possible definitions of the "contradictory object". The establishment of this fact totally eliminates the problem.

VIII. Leśniewski's Analysis and his System for the Foundations of Mathematics

The formal considerations given in this article are based on certain logical system, namely on the system constructed by Leśniewski. During the period when he was occupied with the analysis of Russell's paradox he had not yet constructed his own logical system. In analyzing the problem posed by Russell, the deductions he made were based only on his logical intuition. Therefore, as has been emphasized in many opportunities, this intuition agreed completely with the conception of logic which much later found precise expression in his logical system. As a consequence, our formulation of proofs does not differ essentially from the deductions he made at the time he distinguished the two notions of "distributive class" and "collective class".

The determination of the properties of the collective class, characterized in this work with the aid of K1–K14, led Leśniewski, during the course of the years 1914-1917, to construct a deductive theory which he later called "mereology". Without presupposing what or how many objects exist in the world, he made precise the most general relations between possible existing objects with respect to the properties of the term "collective class" as well as those of other terms definable by means of the term "collective class" within the framework of mereology. Mereology is not a logical theory because its terms and axioms cannot be deduced from the principles of logic. On this theory, together with logic, Leśniewski based the construction of certain geometrical systems. Because of its very general nature, one cannot consider this theory as part of geometry. According to our personal conviction, a great many of its theorems bear a certain resemblance to the different conceptions found in that part of Scholastic metaphysics known as "ontology".

The construction of mereology presented Leśniewski with the problem of giving form to the logical reasoning which until then he had used intuitively in his deductions, and also with the problem of formally determining the meaning of the individual proposition, meaning established by him and to which he reduced the notion of "distributive class". The study of this problem led him to construct, during the years 1919–1921, a general theory of names which he called "ontology". It encompasses the syllogistic, the algebra of logic and the theory of relations. It constitutes, thanks to the richness and diversity of its forms, a system of logic clearly superior to the other systems presently known. It is possible to give only a very general description of this extremely rich theory here. We will indicate only those characteristic points to which we have referred in our considerations and which differentiate it from other theories:

a) Expressions of ordinary language such as, for example, the propositions "John loves" and "John is American", symbolically rendered by "$\varphi\{A\}$" and "$A \varepsilon a$" are two different logical structures in ontology.

b) The subject and the complement of the individual proposition "$A \varepsilon a$" always belong to the same semantical category. The copula "ε" is the sole primitive term of ontology.

c) For the individual proposition "$A \varepsilon a$" to be true, it is necessary (and sufficient) to assert: 1) for any B, if $B \varepsilon A$ then $B \varepsilon a$; 2) A is not an "empty object" ("contradictory object"); 3) A is unique.

d) If the complement is an object, the subject of the individual proposition is identical with its complement (cf. note 12).

e) The notion "set" is not different from the grammatical notion "common noun".

f) The number of semantical categories is unlimited (as in all of Leśniewski's deductive theories). New semantical categories, necessary for the construction of theorems one wishes to prove are gradually introduced into the theory in a manner strictly determined by the rules of the system.

The construction of protothetic in the year 1923 marked the next step in the formation of this system for the foundations of mathematics. The creation of this theory, the most general one for interpropositional relationships, resulted from the necessity of basing the reasoning of ontology and mereology on the laws of the propositional calculus. It also resulted from the decision of Leśniewski to replace the rules for operating with quantifiers which, according to him, were not formulated in an exact enough fashion. He reached this goal by enlarging the theory of deduction, enriching it with quantifiers and introducing variable functors, i.e. variables for which one can substitute the constant terms of the propositional calculus. In the framework of protothetic, thanks to its richness of forms — if one introduces in it or any other deductive theory based on it, a new semantical category, one can then prove a series of theses which replace all the rules concerning the usage of quantifiers which bind the variables of this semantical category. In short, protothetic, of which the propositional calculus is only a fragment, allows one to omit all the rules of quantifiers from the system for the foundations of mathematics based on it.[25]

In this fashion, Leśniewski constructed a system for the foundations of mathematics composed of three deductive theories hierarchically superimposed: protothetic, ontology and mereology. Protothetic is not based on any other theory; ontology is based on protothetic; mereology and a whole series of different deductive theories (e.g. arithmetic) are based on protothetic and ontology. There are geometrical systems and other theories which are based on all three.

The later research of Leśniewski and those of his school consisted first of all in formulating the rules of procedure in a precise fashion, as well as proving that all three theories are not contradictory. Consequently, one cannot construct paradoxes in this system. Also, the simplest forms of the axioms were sought. We cannot give a detailed description of this research here, we will just close this article with the statements of the

---

[25] A careful exposition of protothetic is given by Leśniewski in his works, to wit: Leśniewski [1929a] and Leśniewski [1938]. (*Editorial Note*: Also in Leśniewski [1984].)

As for ontology, cf. the works cited in footnote 4; for mereology, cf. footnote 2. Compare also the works Leśniewski [1931], Leśniewski [1929], Leśniewski [1929b].

simplest axioms of these three theories. They are the result of long years of research by Leśniewski and his followers. Each of these theories can be based on a single axiom which satisfies the conditions set forth by Leśniewski for well-constructed axioms.

If we take as single primitives, in protothetic: equivalence ($\equiv$); in ontology: is (ε); in mereology: element (el), the following theses will be single axioms for these three theories:

I. $[p\,q]::p \equiv q. \equiv .\therefore[f]\therefore f(p\,f(p\,[u].u)). \equiv :[r]:f(q\,r). \equiv .q \equiv p$[26]

II. $[Aa]:A \;\varepsilon\; a. \equiv .[\exists B].A \;\varepsilon\; B\,.\,B \;\varepsilon\; a$[27]

III. $[AB]::A \;\varepsilon\; \mathrm{el}(B). \equiv ::B \;\varepsilon\; B::[fa]::[C]\therefore C \;\varepsilon\; f(a). \equiv :[D]:D \;\varepsilon\; a. \supset .D \;\varepsilon\; \mathrm{el}(C):[D]:D \;\varepsilon\; \mathrm{el}(C). \supset .[\exists EF].E \;\varepsilon\; a.F \;\varepsilon\; \mathrm{el}(D).F \;\varepsilon\; \mathrm{el}(E) \therefore B \;\varepsilon\; \mathrm{el}(B).B \;\varepsilon\; a.\therefore \supset .A \;\varepsilon\; \mathrm{el}(f(a))$[28, 29].

---

[26] I obtained this axiom in 1945, it is based on previous research by S. Leśniewski and my own studies. This result has not been published yet. Cf. Sobociński [1949].

[27] This axiom was obtained by S. Leśniewski in 1929; it is based on my research on the axiomatics of ontology undertaken during the course of the same year. Cf. Sobociński [1934, Th. H].

[28] I obtained this axiom in 1949. It is a simplification of a result obtained by Mr. A. Grzegorczyk in 1946.

[29] The axiom of protothetic has only the primitive term "$\equiv$". The axiom of ontology contains the propositional terms "$\equiv$" and "." as well as its primitive term "ε". The axiom for mereology contains the propositional terms "$\equiv$", ".", "$\supset$" and the ontological term "ε" as well as its primitive term "el". This is understandable since protothetic is not based on any other system, whereas ontology is based on protothetic, and mereology is based on both protothetic and ontology. Concerning the manner in which one defines "." and "$\supset$" in terms of "$\equiv$" in protothetic, cf. Tarski [1922].

Czesław Lejewski

Logic and Existence[1]

I have given my essay this title because it roughly indicates the boundaries of the topic to be discussed and at the same time hints at the method that will be adopted in my analysis. The problem of existence will interest me only to the extent to which it enters the province of logical enquiry and I shall try to disentangle it a little by departing from the generally accepted interpretation of the quantifiers and by bringing in other concepts related to that of existence.

When we have to commit ourselves to asserting or rejecting propositions like

(1) electrons exist
(2) minds exist
(3) Pegasus exists,

our hesitation can be traced to twofold causes. In the first place we may not be willing to give our judgment because we are not quite certain what we mean by 'electrons' or 'minds', or we may not understand the word 'Pegasus'. In the second place we may be confused as regards the meaning of the term 'exist(s)'. It is the latter cause of our embarrassment that calls for closer attention. Let the physicist, the psychologist, and the mythologist deal with the meaning of the words 'electrons', 'minds', and 'Pegasus' respectively. The logician's task, as I understand it, will be to establish the meaning of the constant term 'exist(s)' as it occurs in the function 'x exist(s)' where 'x' is a variable for which any noun-expression can be substituted.

I hope that it will be a permissible simplification to say that in recent times the discussion over the logical side of the problem of existence centres around what Professor Quine has written on the subject.[2] In presenting

---

[1] The first draft of this paper was presented to a post-graduate seminar at the London School of Economics on 12th November 1953, and was also read and criticised by Professors J. Łukasiewicz, K. R. Popper, W. V. Quine and J. H. Woodger, from whose generous comments I have benefited much.

[2] Quine's most important contributions in this connection are the following: Quine [1940a]. This paper was read at the Fifth International Congress for the

the views of this author I shall have to use a little more quotation than is customary as the whole matter is of exceptional subtlety.

On page 150 of Quine's *Mathematical Logic* (= Quine [1947] —*Editorial Note*) we read:

To say that *something* does not exist, or that there *is* something which *is not*, is clearly a contradiction in terms; hence '$(x)(x$ exists$)$' must be true.

Let us translate this argument into a symbolic language so that its logical structure may become more perspicuous. I think that the following translation stands a fair chance of being acceptable to Quine:

(4) $(\exists x)(x \text{ does not exist}) \equiv (\exists x)(x \text{ exists}. \sim(x \text{ exists}))$.

If instead of 'x does not exist' in the antecedent of (4) we write '$\sim(x$ exists$)$' then from (4) and from the circumstance that the consequent of (4), being a contradiction, is false, we get immediately

(5) $\sim(\exists x) \sim (x \text{ exists})$,

which in accordance with the law relating the existential quantifier to its universal counterpart is equivalent to

(6) $(x)(x \text{ exists})$.

This result seems to confirm our interpretation of the passage, which is in complete harmony with the opening paragraph of Quine [1948] as this paragraph runs as follows:

A curious thing about the ontological problem is its simplicity. It can be put into three Anglo-Saxon monosyllables: 'What is there?' It can be answered moreover in a word – 'Everything' – and everyone will accept this answer as true.[3]

But let us revert to page 150 of *Mathematical Logic*. The passage which we began to analyse just now continues thus:

Moreover we should certainly expect leave to put any primitive name of our language for the 'x' of any matrix '$\ldots x \ldots$', and to infer the resulting singular statement from '$(x)(\ldots x \ldots)$'; it is difficult to contemplate any alternative logical rule for reasoning with names.

Unity of Science, Cambridge (Mass.), 1939; Quine [1939], also Quine [1949]; Quine [1943]; Quine [1948]; also Quine [1951a].

This list would have to be supplemented with the titles of several, more technical papers, published by Quine in *The Journal of Symbolic Logic*, and also with some passages from Quine [1947] and Quine [1952].

[3] See Quine [1948, p.21].

This logical rule, which in Quine [1952] is referred to by Quine as the rule of universal instantiation, owes its validity to a certain logical law which with the aid of symbols can be expressed in the following way:

(7) $(y)((x)(Fx) \supset Fy)$.

Now difficulties begin to appear and Quine sets them out as follows:

But this rule of inference leads from the truth '(x)(x exists)' not only to the true conclusion 'Europe exists' but also to the controversial conclusion 'God exists' and the false conclusion 'Pegasus exists' if we accept 'Europe', 'God', and 'Pegasus' as primitive names in our language.[4]

From the whole passage quoted from page 150 of *Mathematical Logic* we seem to be entitled to conclude that for Quine (6), i.e. '(x)(x exists)', is a truth while (3), i.e. 'Pegasus exists', is a falsehood. Regarding the rule which allows us to infer 'Fy' from '$(x)(Fx)$', Quine cautiously remarks that "it is difficult to contemplate any alternative logical rule for reasoning with names". He is more outspoken in Quine [1943], to which we now turn.

In that paper Quine discusses inferences which would be exemplified by the one whereby from

(8) Pegasus does not exist

we infer

(9) $(\exists x)$(x does not exist)

in virtue of the rule which allows us to infer '$(\exists x)(Fx)$' from 'Fy'[5]. In Quine [1943] and in Quine [1952] this rule is described as the rule of existential generalisation, and we may add at once that it derives its validity from the following logical law:

(10) $(y)(Fy \supset (\exists x)(Fx))$.

According to Quine (8) would be true but (9) would be false. Regarding the rule he observes the following:

The idea behind such inference is that whatever is true of the object designated by a given substantive is true of something; and clearly the inference loses its justification when the substantive in question does not happen to designate.[6]

I think that we have come to a point where a brief summary of Quine's argument may not appear to be superfluous. We have two inferences:

---

[4] See Quine [1947, p. 150].

[5] See Quine [1943]. The inference has been rephrased so that it may conform with example taken from Quine [1947]. The original propositions are: 'There is no such thing as Pegasus' and '$\exists x$ (there is no such thing as x)', respectively. See Quine [1947, p. 116].

[6] See Quine [1947, p. 116].

Inference I

From

$(x)(x \text{ exists})$

we infer

Pegasus exists

by universal instantiation.

Inference II

From

Pegasus does not exist

we infer

$(\exists x)(x \text{ does not exist})$

by existential generalisation.

According to Quine both these inferences are objectionable to our intuitions. In his opinion they lead from truths, '$(x)(x$ exists)' and 'Pegasus does not exist', to falsehoods, 'Pegasus exists' and '$(\exists x)(x$ does not exist)'. In other words the logical laws (7) and (10), which we thought to be behind the rule of universal instantiation and the rule of existential generalisation, do not hold for every interpretation of 'F' and every substitution for the variable. If we interpret 'F' as 'exists' and substitute 'Pegasus' for 'y', (7) turns 'y' out to be false. Similarly, (10) turns out to be false for the same substitution if we interpret 'F' as 'does not exist'. But no difficulty arises if the noun-expressions substituted for 'y' in (7) and (10) are non-empty:

(11) $(x)(x \text{ exists}) \supset \text{Socrates exists}$

(12) $\text{Socrates does not exist} \supset (\exists x)(x \text{ does not exist})$

are true propositions.

A remedy that might suggest itself to an unscrupulous mind would be to ban the use of empty noun-expressions and consider them as meaningless. Quine is right in not following this course. One may disagree as to the truth-value of the proposition 'Pegasus exists' but one would have to have attained an exceptionally high degree of sophistication to contend that the expression was meaningless. Quine does not think that empty noun-expressions are meaningless just because they do not designate anything. He allows for the use of such words as 'Pegasus', 'Cerberus', 'centaur', etc. under certain restrictions and tries to distinguish between logical laws which prove to be true for any noun-expressions, empty or non-empty, and those which hold for non-empty noun-expressions only. It follows from his remarks that before we can safely use certain laws established by logic, we have to find out whether the noun-expressions we may like to employ, are empty or not. This, however, seems to be a purely empirical question. Furthermore, all the restrictions which according to Quine must be observed whenever we reason with empty noun-expressions, will have to be observed also in the case of noun-expressions of which we do not know whether they are empty or not.

This state of affairs does not seem to be very satisfactory. The idea that some of our rules of inference should depend on empirical information, which may or may not be forthcoming, is so foreign to the character of logical enquiry that a thorough re-examination of the two inferences may prove to be worth our while. Let us then try to find out on what grounds (6) is asserted as true while (9) is rejected as false, and let us find out also on what grounds the rules of universal instantiation and existential generalisation are regarded as inapplicable to reasoning with empty names. In seeking answers to the above questions we shall have to turn to the interpretation of the quantifiers.

Regarding Quine's interpretation of the quantifiers, which is the one accepted by the majority of modern logicians, we have a very useful passage in Quine [1952]. It reads as follows:

> If we think of the universe as limited to a finite set of objects $a, b, \ldots, h$, we can expand existential quantifications into alternations and universal quantifications into conjunctions; '$(\exists x)Fx$' and '$(x)Fx$' become respectively:
>
> $$Fa \vee Fb \vee \ldots \vee Fh, \qquad Fa \cdot Fb \cdot \ldots \cdot Fh.$$ [7]

To have a still simpler though fictitious example let us think of the universe as limited to two objects **a** and **b**. Then the corresponding expansions would be:

$$F\mathbf{a} \vee F\mathbf{b} \quad \text{and} \quad F\mathbf{a} \cdot F\mathbf{b}.$$

Our language, which for reasons of simplicity needs not synonyms, may leave room for noun-expressions other than the singular names '**a**' and '**b**'. We may wish to have a noun-expression '**c**' which would designate neither of the two objects, in other words which would be empty, and also a noun-expression '**d**' which would designate either. Introducing noun-expressions is a linguistic matter. It does not affect our assumed universe, which continues to consist of **a** and **b** only. The new noun-expressions can now be put to use. For instance we can form the following true proposition; '**c** does not exist', and on turning '**d**' into a predicate-expression '**D**' we can further assert that '$(x)(\mathbf{D}x)$' is true. To say that something exists means the same as to say that it belongs to the universe. Thus '**a** exists' and '**b** exists' are true propositions. From the expansion of the existential quantification we see that either of these two propositions implies '$(\exists x)(x$ exists$)$'. But we have no ground to contend that

[7] See Quine [1952, p. 88].

'**c** does not exist' implies '$(\exists x)(x$ does not exist$)$'. Again since '**a** does not exist or **b** does not exist' is false, we conclude from the expansion that '$(\exists x)(x$ does not exist$)$' is also false. Hence '$(x)(x$ exists$)$' is true. This is confirmed by the expansion of the universal quantification and in view of the circumstance that '**a** exists and **b** exists' is true. We also learn from this expansion that '$(x)(x$ exists$)$' does not imply '**c** exists'. Within our fictitious universe the rule of universal instantiation and the rule of existential generalisation fail if applied to '**c**' or '**d**'. They are valid rules of inference if their application is restricted to reasoning with '**a**' and '**b**'.

But this is not the only possible interpretation of the quantifiers. With the aid of the same fictitious example I shall now present an interpretation which as far as I can judge, is in harmony with the one adopted for instance by Leśniewski of the Warsaw School.

Our universe consisting of **a** and **b** remains the same but the quantifiers read differently. Under the present interpretation '$(\exists x)(Fx)$' becomes:

$$F\mathbf{a} \vee F\mathbf{b} \vee F\mathbf{c} \vee F\mathbf{d}.$$

Correspondingly, '$(x)(Fx)$' is to mean:

$$F\mathbf{a} \cdot F\mathbf{b} \cdot F\mathbf{c} \cdot F\mathbf{d}.$$

Thus one can argue that '$(\exists x)(x$ does not exist$)$' is true because '**c** does not exist', which under the new interpretation is one of the components of the existential expansion, is true. Hence '$(x)(x$ exists$)$' is false. This is confirmed by the corresponding universal expansion, which contains a false component, namely '**c** exists'. Under this interpretation the rule of universal instantiation and the rule of existential generalisation are valid without any restrictions. They can be safely applied in reasoning with any noun-expressions: singular non-empty like '**a**' and '**b**', empty like '**c**', or general non-empty like '**d**'. The noun-expression '**d**' need not be changed into the corresponding predicate-expression '**D**'.

These two different interpretations of the quantifiers, which in what follows, will be referred to as the *restricted* interpretation and the *unrestricted* interpretation respectively, can now be generalised to apply to a universe with any number of objects In the case of a finite universe we have finite expansions which are equivalent to their respective quantifications. If, however, we think of the universe as consisting of infinite or unknown number of objects then we cannot have equivalences for the simple reason that we can never form complete expansions of our quantifications. Consequently we abandon equivalences in favour of implications. We say that an existential quantification is implied by any component of its infinite or unknown expansion and that a universal quantification implies any component of its infinite or unknown expansion. Now the expansions will vary depending on how we choose to interpret the quantifiers. Under the restricted interpretation every component of an expansion contains a noun-expression which designates only one of the objects belonging to the universe. Under the unrestricted interpretation every component of an expansion contains an expression of which we can only say that it is a meaningful noun-expression. It may designate only one of the objects belonging to the universe, it may designate more than one, or it may designate nothing at all.

The two interpretations of the quantifiers give rise to two different theories of quantification and we may well be expected to say a few words on the relation in which one theory stands to the other. In this respect the most important point is that whatever is said with the aid of the theory of restricted quantification, can be easily expressed in terms of the unrestricted quantification provided we are allowed to use the notion of existence. A few examples set out below in the form of two lists will suffice to illustrate the procedure:

I. Expressions to be understood in the light of the restricted interpretation:

(7) $$(x)(Fx) \supset Fy$$

(10) $$Fy \supset (\exists x)(Fx)$$

(13) $$(x)(Fx) \equiv \sim(\exists x)(\sim Fx).$$

II. Corresponding translations to be understood in the light of the unrestricted interpretation:

(7a) $$(x)(x \text{ exists} \supset Fx) \supset Fy$$

(10a) $$Fy \supset (\exists x)(x \text{ exists} \cdot Fx)$$

(13a) $$(x)(x \text{ exists} \supset Fx) \equiv \sim(\exists x)(x \text{ exists} \cdot \sim Fx).$$

It is not difficult to check that under their respective interpretations the corresponding expressions in the two lists yield the same truth value for the same substitutions performed on the free variables regardless of how we choose to interpret the predicate letters. The general rule for translating expressions is simple: expressions of type '$(x)(Fx)$' and '$(\exists x)(Fx)$' become expressions of type '$(x)(x$ exists $\supset Fx)$' and '$(\exists x)(x$ exists$\cdot$ $\cdot Fx)$' respectively; other expressions remain unchanged.

Thus, for instance, (6) and (9) translated for the purpose of the unrestricted interpretation become

(14) $(x)(x$ exists $\supset x$ exists$)$ and

(15) $(\exists x)(x$ exists$\cdot x$ does not exist$)$

respectively. These translations fully account for the assertion of (6) and the rejection of (9) under the restricted interpretation of the quantifiers. Similarly (7a) and (10a) make it clear why (7) and (10) do not turn out true for all interpretations of 'F' and for all substitutions for the free variable. For if we interpret 'F' as 'exists' and substitute say 'Pegasus' for 'y', the antecedent of (7a) becomes a tautology but at the same time the consequent turns out to be false. And again, if we interpret 'F' as 'does not exist' and substitute 'Pegasus' for 'y' then the antecedent of (10a) comes out true but the consequent must be rejected as a contradiction. Thus for certain interpretations of 'F' (7) and (10) turn out to be false if we substitute an empty noun-expression for the free variable. Consequently the rules of universal instantiation and existential generalisation, which derive their validity from (7) and (10), can no longer be applied without restrictions.

The position is different if we choose to understand the quantifiers in the light of the unrestricted interpretation. (3), which is 'Pegasus exists', and (8), which is 'Pegasus does not exist', are meaningful propositions either of which contains a noun-expression, viz. 'Pegasus'. Thus (3) and (8) may be regarded as components of quantificational expansions. Now, (3) being false, the corresponding universal quantification, i.e. '(x)(x exists)', which ought to imply any component of its expression, must also be false. On the other hand, (8) being a true proposition the corresponding existential quantification, i.e. '$(\exists x)$(x does not exist)', which is implied by any component of its expansion, must also be true. Thus under the unrestricted interpretation of the quantifiers the two inferences on p. 48 above cannot be used as counterexamples to disprove the validity of the rules of universal instantiation and existential generalisation in application to reasoning with empty noun-expressions. In Inference I both the premise and the conclusion are false, while in Inference II both the premise and the conclusion are true. There is nothing wrong with the inferences provided we adopt the unrestricted interpretation of the quantifiers. Furthermore, under the unrestricted interpretation the logical laws (7) and (10) turn out to be universally true. For every proposition of type 'Fa' where 'a' stands for any noun-expression, empty or non-empty, is now regarded as a component of the quantificational expansions and consequently is implied by the corresponding proposition of type '$(x)(Fx)$' and implies, in turn, the corresponding proposition of type '$(\exists x)(Fx)$'. (7) and (10) being universally true, the rule of universal instantiation and the rule of existential generalisation are universally valid as the principles that are behind them are no longer principles by courtesy.[8]

The unrestricted interpretation of the quantifiers seems to remove yet another difficulty from quantification theory. It has been argued by several authors that

(16) $(\exists x)(Fx \vee \sim Fx)$ and

(17) $(x)(Fx) \supset (\exists x)(Fx)$,

---

[8] See Quine [1943, p. 118].

which are valid if the universe is not empty, fail for the empty universe as their truth depends on there being something.[9] When discussing these laws Quine tries to dismiss the case of the empty universe as relatively pointless and reminds us that in arguments worthy of quantification theory the universe is known or confidently believed to be non-empty.[10] This contention, however, does not quite remove our uneasiness particularly as (16) and (17), not unlike (7) and (10), are demonstrable in quantification theory.

On considering (16) and (17) we readily admit that these two formulae fail for the empty universe if we understand the quantifiers in accordance with the restricted interpretation. This becomes evident from

(16a) $(\exists x)(x \text{ exists} \cdot (Fx \vee \sim Fx))$ and

(17a) $(x)(x \text{ exists} \supset Fx) \supset (\exists x)(x \text{ exists} \cdot Fx)$,

which are the corresponding translations of (16) and (17) to be understood in the light of the unrestricted interpretation. If there exists nothing then (16a) and the consequent of (17a) are obviously false while the antecedent of (17a) is obviously true. Under the unrestricted interpretation, however, (16) and (17) come out to be true irrespective of whether the universe is empty or non-empty. For (16) is implied by any component of type '$Fa \vee \sim Fa$' where 'a' stands for a noun-expression. In particular it is implied by a component '$Fa \vee \sim Fa$' in which 'a' stands for an empty noun-expression. Such a component is true for all choices of universe and so is (16). In the case of (17) we argue as follows: if we assume that the antecedent of (17) is true then a proposition of type 'Fa' where 'a' stands for an empty noun-expression must also be true in harmony with the unrestricted interpretation of the universal quantifier. Now any such proposition implies the proposition of type '$(\exists x)(Fx)$', which again must be true. Thus in the establishing of the truth value of (16) and (17) the problem of whether the universe is empty or non-empty is altogether irrelevant on condition, of course, that we adopt the unrestricted interpretation of the quantifiers.

It ought to be evident from what has already been said that under the unrestricted interpretation existential quantifications have no existential import. In fact it would be misleading to read '$(\exists x)(Fx)$' as 'there exists an x such that Fx'. The non-committal 'for some x, Fx' seems to be more appropriate. Similarly the terms 'existential quantification' and 'existential quantifier' no longer apply and could be conveniently replaced by such expressions as 'particular quantification' and 'particular quantifier'. The rule of existential generalisation could perhaps be referred to as the rule of 'particular generalisation'.

---

[9] For details see Quine [1953, 160 sq].

[10] See Quine [1952, p. 96].

Finally, the unrestricted interpretation in comparison with the restricted one appears to me to be a nearer approximation to ordinary usage. Somehow we do not believe that *everything exists* and we do not see a contradiction in saying that *something does not exist*. It is only from logicians who favour the restricted interpretation that we learn that things are the other way round. We may further add that the unrestricted interpretation of the quantifiers is in complete harmony with the formal quantification theory. I do not know of any formulae which are demonstrable in the formal quantification theory and which, under the unrestricted interpretation, are not applicable to reasoning with empty noun-expressions or do not hold for universes of some specific size.

When we consider (7), (10), (13), (16), and (17) as understood in the light of the restricted interpretation and compare them with their corresponding translations for the purpose of the unrestricted interpretation, i.e. with (7a), (10a), (13a), (16a), and (17a), we cannot fail to notice that the idea of the restricted quantification is not a simple one. The translations reveal that it can be analysed into two separate constituents: the idea of the unrestricted quantification on the one hand and the notion of existence on the other. In my opinion the most serious disadvantage of the theory of the restricted quantification is that by merging the idea of quantification with the notion of existence it has put logicians and philosophers on a wrong track in their endeavours to elucidate the problem of existence in logic. In what follows we shall adhere to the theory of unrestricted quantification and we shall attack the problem of existence in logic by determining the meaning of the constant term 'exist(s)' as used in the function 'x exist(s)'.

From the logical point of view there are two satisfactory methods of determining the meaning of a constant term. The one consists in setting forth a set of axioms for the term in question. The other is adopted whenever we give a definition of the term in question with the aid of other terms whose meaning has already been determined axiomatically.[11] We may add at once that we shall employ the latter method.

The meaning of 'exist(s)' can best be determined on the basis of the logic of noun-expressions constructed as a deductive system by Leśniewski in Warsaw in 1920 and called by him 'Ontology'.[12] The original system of Leśniewski's Ontology is based on singular inclusion (a is b or in symbols $a \varepsilon b$) as the only primitive function. For various reasons, however, I prefer to continue my analysis of 'exist(s)' with reference to a system of Ontology based on ordinary inclusion, with I shall write in the following manner:

(18) $a \subset b$.

[11] See Łukasiewicz [1953].

[12] See Leśniewski [1930]. For a brief account of Ontology see Ajdukiewicz [1951, 8 sq.] or Łukasiewicz [1953, 77 sq.].

I shall read it 'all a is b' or 'all a's are b's'. I prefer doing this because ordinary inclusion seems to be more intuitive to an English speaking reader than Leśniewski's singular inclusion. Thus for instance ordinary inclusion has recently been used by Woodger in his 'Science without Properties"[13] for the purpose of constructing a language whose general tendency approximates the tendencies embodied in Ontology.

The functor of ordinary inclusion is a proposition forming functor for two arguments either of which is a noun-expression. If in (18) we substitute constant noun-expressions for the variables 'a' and 'b' then the result of the substitution will be true if and only if every thing named, or designated, by the noun-expression substituted for 'a' is also named by the noun--expression substituted for 'b'. It may be of some historical interest to mention that the above semantic characterisation of inclusion can be traced back to Hobbes who used it in order to determine the meaning of the copula 'est' in propositions such as 'homo est animal'.[14]

From the semantic characterisation of inclusion[15] it is evident that the following propositions are true:

$$\begin{array}{l} \text{man} \subset \text{animal} \\ \text{man} \subset \text{man} \\ \text{Socrates} \subset \text{man} \\ \text{Socrates} \subset \text{Socrates} \\ \text{Pegasus} \subset \text{animal} \\ \text{Pegasus} \subset \text{Socrates} \\ \text{Pegasus} \subset \text{Pegasus}. \end{array}$$

The last three propositions are true because nothing is designated by 'Pegasus'. Thus whatever is designated by 'Pegasus', is designated by anything you like. The corresponding proposition can be formulated in symbols as follows:

(19) $(a)(\text{Pegasus} \subset a)$

---

[13] See Woodger [1952].

[14] Hobbes wrote: 'Ut qui dicit *homo est animal* intellegi ita vult ac si dixisset "si, quem recte *hominem* dicimus eundem etiam *animal* recte dicimus".' See Hobbes [1839–1845b, p. 497]; see also Hobbes [1839–1845a, p. 27].

[15] Strictly speaking the meaning of inclusion ought to have been determined axiomatically. I understand from Dr. B. Sobociński that Leśniewski had an axiom for inclusion. It has never been published and I never saw it when I was studying with Leśniewski in Warsaw before the war. But I found some time ago that a system of Ontology can be built up on the basis of the following single axiom:

$$(a, b)((a \subset b) \equiv (c, d)(\sim(c \subset d)\cdot(c \subset a) \supset (\exists e, f)(\sim(e \subset f)\cdot(e \subset c)\cdot(e \subset b)\cdot (g, h, i)(\sim(h \subset i)\cdot(g \subset e)\cdot(h \subset e) \supset (g \subset h))))).$$

If instead of 'Pegasus' in (19) we write the constant noun-expression 'Λ', which designates nothing, then we shall get the following ontological thesis:

(20) $(a)(\Lambda \subset a)$

'Λ' can be defined in terms of inclusion but for the sake of simplicity I prefer to introduce it as undefined term with the aid of (20).

In order to determine the meaning of 'exist(s)' we shall need three definitions, which I write below in the form of equivalences:

(21) $(a)(\mathrm{ex}(a) \equiv (\exists b)(\sim(a \subset b)))$

(22) $(a)(\mathrm{sol}(a) \equiv (b, c, d)(\sim(c \subset d) \cdot (b \subset a) \cdot (c \subset a) \supset (b \subset c)))$

(23) $(a)(\mathrm{ob}(a) \equiv \mathrm{ex}(a) \cdot \mathrm{sol}(a))$.[16]

For the present we need not trouble ourselves with the question how to read the newly, defined functors. We can proceed straight on to the consequences which can be deduced from (20) and the three definitions.

Thus from (21) we immediately get

(24) $\mathrm{ex}(\Lambda) \equiv (\exists a)(\sim(\Lambda \subset a))$.

Since from (20) we know that

(25) $\sim(\exists a)(\sim(\Lambda \subset a))$

we use (25) and (24) to show that

(26) $\sim(\mathrm{ex}(\Lambda))$.

From (26) we obtain, by particular generalization,

(27) $(\exists a)(\sim(\mathrm{ex}(a)))$

which is equivalent to

(28) $\sim(a)(\mathrm{ex}(a))$.

From (23) and (26) we conclude that

(29) $\sim(\mathrm{ob}(\Lambda))$.

---

[16] In Leśniewski's original system of Ontology the three functors are defined as follows:

$(a)(\mathrm{ex}(a) \equiv (\exists b)(b \varepsilon a))$
$(a)(\mathrm{sol}(a) \equiv (b, c)((b \varepsilon a) \cdot (c \varepsilon a) \supset (b \varepsilon c)))$
$(a)(\mathrm{ob}(a) \equiv (\exists b)(a \varepsilon b))$.

See Kotarbiński [1929, 235 sq.] (*Editorial Note*: English translation – Kotarbiński [1966, 197 sq.]), see also Ajdukiewicz [1951, p. 8] and Łukasiewicz [1953, 79 sq.].

From (29) we get

(30) $(\exists a)(\sim(\mathrm{ob}(a)))$

which is equivalent to

(31) $\sim(a)(\mathrm{ob}(a))$.

We can now draw Pegasus into our deductions. From (19) we immediately obtain

(32) $\sim(\exists a)(\sim(\text{Pegasus} \subset a))$

and use it together with (21) to show that

(33) $\sim(\mathrm{ex}(\text{Pegasus}))$.

From (23) and (33) we derive

(34) $\sim(\mathrm{ob}(\text{Pegasus}))$.

Now (27), (28), (30), (31), (33), and (34) show that the functors 'ex' and 'ob' are very close approximations of 'exist(s)'. We remember that under the unrestricted interpretation of the quantifiers '$(x)(x$ exists)' is false and so is '$(a)(\mathrm{ex}(a))$' and '$(a)(\mathrm{ob}(a))$' as is evident from (28) and (31). Under the same interpretation '$(\exists x)(x$ does not exist)' comes out true and so does '$(\exists a)(\sim(\mathrm{ex}(a)))$' and '$(\exists a)(\sim(\mathrm{ob}(a)))$' as is evident from (27) and (30).

Further evidence is supplied by (33) and (34) from which it follows that 'ex(Pegasus)' and 'ob(Pegasus)' are false just as 'Pegasus exists' is admittedly false. It remains then to be explained what is the difference in the meaning of 'ex' and 'ob'. We can find the required explanation if we consider the meaning of (22), which is a definition of 'sol'. The right hand side of this definition turns out to be true in two cases: if there is no such thing as *a* or if there is only one *a*. Thus 'sol(a)' can be read as 'there is at most one *a*'. Now if we agree to read 'ex(a)' as '*a* exists' or '*a*'s exist' then in accordance with (23) 'ob(a)' will have to be read as 'there exists exactly one *a*', which is equivalent to '*a* is an object' or to '*a* is an individual'. Thus the distinction between 'ex' and 'ob' roughly corresponds to the one made by Quine in his "Designation and Existence", where he talks about *general* existence statements and *singular* existence statements.[17]

We have already remarked that under the restricted interpretation every component of quantificational expansions contains a noun-expression which designats only one of the objects as belonging to the universe. It therefore follows that the function 'x exists' as used by us when we discussed the two interpretations of the quantifiers, means in fact the same as 'there exists exactly one x' which is the rendering of the symbolic

---

[17] See Quine [1949, 44 sq.].

'ob(x)'. The functor 'ex', on the other hand, appears to be a very close approximation of the 'exist(s)' as used in ordinary language as it forms true propositions with noun-expressions which may designate more objects than one.

I wish to conclude with a brief summary of the results. The aim of the paper was to analyse rather than criticise. I started by examining two inferences which appeared to disprove the validity of the rules of universal instantiation and existential generalisation in application to reasoning with empty noun-expressions. Then I distinguished two different interpretations of the quantifiers and argued that under what I called the unrestricted interpretation the two inferences were correct. Further arguments in favour of the unrestricted interpretation of the quantifiers were brought in, and in particular it was found that by adopting the unrestricted interpretation it was possible to separate the notion of existence from the idea of quantification. With the aid of the functor of inclusion two functors were defined of which one expressed the notion of existence as underlying the theory of restricted quantification while the other approximated the term 'exist(s)' as used in ordinary language.

It may be useful to supplement this summary by indicating some aspects of the problem of existence which have not been included in the discussion. I analysed the theory of quantification so far as it was applied in connection with variables for which noun-expressions could be substituted and my enquiry into the meaning of 'exist(s)' was limited to cases where this functor was used with noun-expressions designating concrete objects or with noun-expressions that were empty. It remains to explore, among other things, in what sense the quantifiers can be used to bind predicate variables and what we mean when we say that colours exist or that numbers exist. These are far more difficult problems, which may call for a separate paper or rather for a number of separate papers.

Jerzy Słupecki

S. Leśniewski's Calculus of Names

Introduction

The only primitive term in Leśniewski's system of the Calculus of Names is the verb 'is' for which the participle 'being' corresponds to the Greek '*ον*'[1] (gen. '*όντος*'). This was by no means the only reason for Leśniewski's use for his system a name indicating one of the main branches of philosophy. Thus in Leśniewski's article "On the Foundations of Mathematics"[2] we read:

... I used the term 'ontology' for the theory I developed, as this was not opposed to my 'linguistic intuition', just in view of the fact that I formulated in that theory a sort of 'general principles of being'.

In the title of this paper I nevertheless thought it better to use "Calculus of Names", as 'ontology' might cause some misunderstanding, but I shall be using throughout the text 'ontology' for Leśniewski's system.

This paper, like my article on Leśniewski's protothetics[3], is based chiefly on notes from Leśniewski's lectures held at the Warsaw University. These notes were made by Leśniewski's students and were collected after the last war by Professor T. Kotarbiński. Leśniewski's original manuscripts were destroyed in Warsaw Uprising. The notes from a course of lectures under the title "An outline of elementary ontology"[4] held by Professor Leśniewski in 1929/30, form the basis of this paper.

---

[1] I have taken this observation from Kotarbiński [1929]. (*Editorial Note*: English translation – Kotarbiński [1966]).

[2] Leśniewski [1927–1931, Ch. XI, p. 163]. (*Editorial Note*: English translation – Leśniewski [1983].)

[3] Słupecki [1953].

[4] The first loose observations on ontology were made in exercises conducted by Leśniewski in the course of 1919/20 under the heading "Exercises in Cantor's Theory of Sets". Further, systematic and detailed observations on ontology were presented in a course of lectures on "The Foundations of Arithmetics", held during the years 1920–23. Leśniewski devoted the following lectures entirely, or to a considerable extent, to ontology:

"Foundations of Ontology", in the years 1925–27,

Leśniewski dedicated entirely to ontology only his work [1930]. In the twenty-two pages of that paper there is but a fragmentary description of the system. Some extremely interesting remarks on the intuitive foundations of ontology are included in chapter XI of Leśniewski [1931]. This treatise was, however, not completed and the remarks on ontology contained therein are clearly introductory. An exact though necessarily brief exposition of Leśniewski's system is given in Kotarbiński [1929, pp. 227-247][5]; this was the only source of information for the Polish reader not versed in Logic. It contains the following valuation of ontology[6]:

> We preferred to base the calculus of names on Leśniewski's system, known to us in manuscript form and offered to a wider circle of students in a course of lectures; in our opinion this system of the calculus of names is the most mature, the most natural and in its applications the most practical of all the systems we know. At the same time it is closely connected with the traditional Aristotelian formal logic of which it is an improvement and an extension, although on the other hand it is the final point of the attempts to construct a calculus of names within mathematical logic.

Relevant information on ontology is contained also in Sobociński [1934].

I have subdivided this paper into four sections. Section I discusses the intuitive and formal foundations of ontology. Section II deals with theorems of that part of Leśniewski's system to which I shall refer as elementary ontology[7] and which contains the simplest theorems of the whole system and those nearest to intuition. It is in this section too that I discuss the relation between ontology and traditional logic as well as the algebra of sets. Section III deals with the remaining part of the system, i.e. the non-elementary ontology. However, I shall not give a systematic presentation of non-elementary ontology, as this could only be done in a separate and large treatise. I shall only adduce the theorems and definitions I consider most characteristic of non-elementary ontology. Further, I shall present all those theorems and definitions contained in the notes from Leśniewski's lectures which, though exceeding the limits of elementary ontology, are nevertheless closely related to its theorems and notions. Section IV contains brief methodological considerations on the system.

"An Outline of Elementary Ontology", in 1929–32,

"Traditional Formal Logic and Traditional Theory of Sets from the Point of View of Ontology", in 1937/38,

"An Outline of Ontology", in 1938/39.

[5] *Editorial Note*: Also Kotarbiński [1966, pp. 190–204].

[6] Kotarbiński [1929, p. 253]. (*Editorial Note*: Also Kotarbiński [1966, p. 210]).

[7] The definition of elementary ontology is given in II, § 1.

I

§ 1. The Language of Ontology

Leśniewski subdivides all correctly constructed expressions of his logical systems, with the exception of quantifiers and parentheses, into classes named semantic categories.[8] The fundamental semantic categories are those of propositions and of names. The former category comprises not only propositions but also propositional functions, i.e. expressions containing free variables and becoming propositions if appropriate constants are substituted for all variables. The category of names possesses an analogous property. Leśniewski includes all names whether empty, singular or general into one semantic category. I draw attention to the fact that when Leśniewski distinguished empty, singular and general names he took into account solely their extension.

I shall call expressions included in the semantic categories of propositions and names propositional and nominal expressions, respectively.

All logical expressions included in semantic categories other than those of propositions and names are seen as functors. This notion of functor is broader than that usually given for it covers also variables representing functors, other than functors in narrower sense, i.e. certain constant symbols. Along with Leśniewski I shall call such variables variable functors. Every functor in Leśniewski's systems of logic constitutes together with its arguments an expression belonging either to the category of propositions or to that of names. In the former case we denote the functor as proposition-forming, in the latter as name-forming.[9] We obtain a further subdivision of functors by putting into different semantic categories functors differing as to the number or the semantic category of their arguments.

In theorems of ontology, as in those of all more highly developed deductive systems, appear the constants of the calculus of propositions.[10] In order to determine the semantic categories of functors appearing in theorems of ontology and not appearing in the calculus of propositions I shall apply the notion of the order of a functor; I shall define this notion by induction.

---

[8] Observations on semantic categories in accordance to Leśniewski's theory are contained in Ajdukiewicz [1928, p. 9], Kotarbiński [1929, p. 67] (*Editorial Note*: Also Kotarbiński [1966, p. 56]) and Słupecki [1953, p. 45].

[9] The language of ontology contains therefore no functor-formative functors.

[10] In this paper, in accordance with the usual terminology, I denote as calculus of propositions any system, the only constants of which are proposition-formative functors of propositional arguments.

A functor of order 1 is any functor all the arguments of which are expressions belonging to the category of names. A functor of this kind may be proposition-forming or name-forming; it may be constant or variable. The number of its arguments is arbitrary.

I am assuming now that any order of a functor of a lower order than k is determined, k being an arbitrary natural number > 1. A functor of order k is any functor all the arguments of which are either expressions belonging to the category of names or else functors of order not higher than $k-1$, one of these arguments at least being a functor of order $k-1$.

A statement analogous to that referring to functors of order 1 may be made about functors of order k.

Two functors of order 1 belong to the same semantic category if both are proposition-forming or both are name-forming, and if the numbers of their arguments are equal.

Let us assume now that we have determined in which case two functors of orders lower than k (k being any natural number > 1) belong to the same semantic category. Two functors of order k belong to the same semantic category if both are proposition-forming or if both are name-forming, the numbers of their arguments are equal, and if their respective arguments belong to the same semantic category.

In theorems of ontology appear functors of arbitrary semantic categories having an arbitrary finite number of arguments.

If a constant term which is neither a constant term of the calculus of propositions nor a quantifier appears in theorems of ontology, then these theorems contain also variables belonging to the same semantic category as that constant term. Any variable may be bound by either the general or the existential quantifier.[11] Let us also note that propositional variables do not appear in theorems of ontology.

It is a fundamental property of the subdivision of expressions into semantic categories that every proposition and every propositional function always remains a proposition or a propositional function if an expression forming part of it is replaced by an expression belonging to the same semantic category. Should the replaced and the replacing expressions belong to different semantic categories, we obtain a meaningless expression.

Applying the notion of the order of a functor I shall define the language of that part of Leśniewski's system which I call elementary ontology. In the theorems of elementary ontology apart from functors of the calculus

---

[11] For reasons of principle, unknown to me, Leśniewski did not introduce the existential quantifier into his systems, but replaced it by a combination of negation and the general quantifier. For didactical reasons, however, in some of his publications and in all his lectures Leśniewski employed this quantifier in the notation of theorems of his systems. This is the notation I shall be employing in this paper for theorems of ontology.

of propositions there appear only constant functors of order 1. Thus, the only variables appearing in these theorems are nominal variables. These variables may be bound by quantifiers.[12] The theorems of elementary ontology also contain nominal constants.

§ 2. Symbols Employed

I shall now discuss the symbolism of elementary ontology; a discussion of the symbolism of the entire system is given in III, § 1.

The following functors of the calculus of propositions appear in the theorems of ontology considered in this paper:

the sign of implication $\supset$,
" " " conjunction $\wedge$,
" " " alternation $\vee$,
" " " equivalence $\equiv$,
" " " negation $\sim$.

The sign of negation will be written before the argument and the latter will be enclosed in round parentheses; functors of two arguments will be placed between the arguments, and the latter may be enclosed only in round parentheses.

The general quantifier is symbolized by square brackets enclosing the bound variables. Thus I write $[x, y]$ for the expression: 'for every x and every y'. The existential quantifier is symbolized by square brackets and the sign '$\exists$' enclosed therein and preceding the bound variable. Thus the symbol $[\exists x, y]$ should be read: 'for a certain x and for every y'. I shall always enclose the expression covered by a quantifier in brackets.

I put proposition-formative and name-forming functors of one nominal-argument before the arguments. However, the sign of complementation forms an exception: I write the name 'not-S' as 'S''. I place the argument of a proposition-forming functor between oblique lines. Thus I write the expression 'There exists at least one S' as follows: '*ex* $/S/$'. I enclose the argument of a name-forming functor in vertical lines. Here again the functor of complementation forms an exception: its argument, when a non-composite symbol, is not enclosed in parenthesets.

Proposition-formative and name-forming functors of two nominal arguments are placed between the latter, just as the functors of two arguments in the calculus of propositions. Functors of more than two arguments do not appear in theorems put forth in this paper, nor do they appear in any of the theorems contained in the notes from Leśniewski's lectures on

---

[12] Leśniewski included in elementary ontology certain theorems containing variable functors of the first order or constant functors of the second order. These theorems are given in III, § 2.

which this paper is based. It should be mentioned, however, that the rule of adding definitions to the system which holds in the system makes it possible to define functors of any finite number of arguments. When arguments of name-forming functors are complex expressions, I enclose them in vertical lines.

Leśniewski coordinates the symbol 'ε' with 'is', the primitive term of ontology.

The only variables of elementary ontology are nominal variables. These are the letters 'x', 'y', 'z', ..., 't', 'n', 'M', 'P', 'S', 'X' 'Y', 'Z', ... Capital and small letters are variables of the same semantic category. Leśniewski made use of two kinds of variables in order to give the theorems of ontology a higher degree of clarity. Capital letters appear in such theorems of ontology whose intuitive meaning becomes clearest after substituting general names for the variables. Small letters, on the other hand, appear in theorems, if on intuitive grounds we would rather substitute singular names. Thus, in theorems of Aristotelian syllogistics the variables will be capital letters, while in expressions of the type 'x is S' the subject will always be denoted by a small letter.[13] As both capital and small letters are variables of the same semantic category, small letters may be substituted for capital ones, and vice versa. Outside the scope of elementary ontology, however, arbitrary names may be substituted for variables of either kind. Let me add that I will not write general quantifiers at the head of theorems. Hence, free variables may appear in theorems of ontology, contrary to the way in which Leśniewski used to note the theorems of his systems. This is not my only deviation from Leśniewski's symbolism. One unessential point of difference between Leśniewski's original symbolism[14] and that employed in this paper is the inversion of the roles of capital and small letters (nominal variables) in the two symbolisms. Of greater importance is the fact that the functors in Leśniewski's symbolism always precede the arguments included in parentheses the shape of which together with the number of arguments they include determines the semantic category of the functor. However, since the number of semantic categories in elementary ontology is restricted, this rule need not be observed. Thus, e.g., in accordance with the symbolism used in traditional logic and in mathematics I put functors of two arguments between the arguments, instead of before them. Moreover the shape of functors I make use of in noting theorems is often different from that of the corresponding functors in Leśniewski's

---

[13] By expressions of the same type I mean expressions that differ only as to the form of their variables.

[14] Leśniewski made use of two symbolisms, one of his own invention and one modelled on that of Peano and Russell. By Leśniewski's symbolism I mean here always the former.

symbolism, as it was my intention to employ a symbolism resembling that of traditional logic as closely as possible.

The following two examples may serve to illustrate the above description of the symbolism used in this paper. I write the expressions 'for every x: if x is S, then x is P' thus: '$[x]\{x \varepsilon S \supset x \varepsilon P\}$'; and the expressions: 'for a certain x: it is not true that x is S' thus: '$[\exists x]\{\sim(x \varepsilon S)\}$'.

§ 3. The Primitive Term of Ontology

If we assume, in accordance with Leśniewski's point of view, that the general quantifier is not a primitive term of ontology although it cannot be defined in this system, then 'is' remains the sole primitive term of ontology.* 'Is' appears always in theorems of elementary ontology as part of expressions of the following type:

(1) $x \varepsilon S$.

* The task of rendering in English the intuitive sense of the symbol 'ε' which is a primitive term of Leśniewski's ontology, meets with some difficulties. Leśniewski 'was thinking in Polish' when constructing his ontology, the symbols of his ontology correspond to meanings in Polish and the syntax of his ontology is based upon syntactical principles of the Polish language. One of the fundamental differences between the Polish and the English language is that there are no articles in the former (just as in Latin), while they occur in the latter where the use or nonuse of an article before a noun, the use of a definite or indefinite article affects the meaning of the verb which refers to that noun. The English verbal form 'is' has three different meanings in the following sentences: 'Socrates is a man', 'The dog is an animal', 'Socrates is the husband of Xantippe'. One of the following three relations corresponds to the word 'is' in these three sentences, the membership-relation, the inclusion-relation and the identity-relation. In Polish as in Latin the copula 'jest' or 'est', respectively, joins two nouns without any articles. In Latin it does make sense to say: 'Socrates est homo', 'Canis est animal', 'Socrates est coniunx Xantippae'. There are two possible viewpoints as to the meaning of the word 'est' in these sentences. Firstly, we may hold that the above three Latin sentences are translations of the respective English sentences mentioned before, hence that the Latin 'est' has a different meaning in each of the three sentences. Secondly, it is possible to hold that 'est' has the same meaning in all three Latin sentences. In that case we may either identify the meaning of the Latin word 'est' with one and only one of the above mentioned meanings of the English word 'is', or admit that 'est' has one meaning in all three quoted Latin sentences and that this meaning is different from any of the three meanings of the English word 'is'. This latter view has been adopted by Leśniewski. He holds, namely, that the Latin 'est' has one meaning in all contexts of the type 'A est B' just as the Polish word 'jest' in contexts of the type 'A jest B', but that this meaning is different from any of the meanings of the English 'is' in the sentences: 'Socrates is a man', 'The dog is an animal', 'Socrates is the husband of Xantippe'.

What this meaning is like we may guess from the axiom of Leśniewski's ontology. It is at any rate, different from that associated in Russell's writings with the symbol 'ε', as Russell means by this symbol approximately the same as is meant by the English word 'is' in the context: 'Socrates is a man'. This difference is discussed by the author in his paper.

I shall denote expressions of this type as singular expressions or, in case constant terms have been substituted for either variable in expressions (1), as singular propositions. The variables 'x' and 'S' in expressions (1) are nominal variables. Hence the primitive term of ontology is a proposition-formative functor of two nominal arguments.

The following propositions exemplify singular propositions:

Uranus is a planet,

John the III Sobieski is the deliverer of Vienna.[15]

Usually the subject of a singular proposition is a singular name. The two propositions given above are indeed propositions of this kind. However, according to the fundamental property of the subdivision of expressions into semantic categories as discussed in § 1, and keeping in mind that all names belong to the same semantic category, expressions we obtain from (1) by substituting an empty or general name for the variable 'x' and an arbitrary name for 'S', are propositions. Thus the following expressions may be obtained by substitution from (1):

Hamlet is the hero of a tragedy by Shakespeare.

The whale is a mammal.

No doubt these expressions are currently considered to be propositions.

However, Leśniewski understands a singular proposition in such a manner that it can be true only if its subject is a singular name. In other words, any singular proposition, the subject of which is an empty name or a general name, is false.

One may raise the question whether the meaning of the word 'is' in ontology is the same as that currently attributed to it. It is commonplace to say that Leśniewski was not able to fix the meaning of the primitive term of ontology so as to be in conformity with all current meanings of the word 'is', the more so as propositions of the type 'x is S' are used with an evident lack of consequence in current speech.[16]

The difference which in Leśniewski's view exists between any of the meanings attributed to the English word 'is' and the meaning of the Polish 'jest' or Latin 'est' and thus also of the meaning of the primitive term 'ε' in Leśniewski's ontology makes it impossible to illustrate the schema '$A \varepsilon B$' by means of examples taken from the English language. Neither the sentences 'Uranus is a planet', 'John the III Sobieski is the deliverer of Vienna' nor the sentence: 'The whale is a mammal' are good examples of the schema '$A \varepsilon B$' as Leśniewski meant it to be understood, but the Polish or Latin translations of the same sentences may serve as good examples: 'Uranus est planeta', 'Johannes III Sobieski est liberator Viennae', 'Cetus est mammale'. For the English reader, wishing to grasp Leśniewski's intentions with regard to the meaning of the symbol 'ε', it is advisable to replace the English examples given in the text by their Latin translations, and the symbol 'ε' of Leśniewski's ontology should be interpreted rather as the Latin 'est' than the English 'is'.

[15] Both these propositions, as well as almost all other examples of singular propositions are taken from Kotarbiński [1929].

[16] Leśniewski [1927–1931, Ch. XI, p. 166].

As I am unable to enumerate all the meanings the word 'is' may take in current speech, I shall mention only the most important ones. To a large extent my remarks will be identical with those of Kotarbiński [1929].[17] These remarks of Prof. Kotarbiński are also quoted by Leśniewski [1927-1931, Ch. XI].

The extent to which the meaning of the word 'is' may vary in current speech can be seen from the fact that it has there no definite semantic category. In the Latin equivalent of the proposition 'there is justice' the word 'is' (its Latin equivalent) acts as a functor of one argument (est iustitia). In the proposition 'seniority is a transitive relation' the subject is a word classified by Prof. Kotarbiński as an apparent name; apparent names are not names, just as 'false money is not money'.[18] The gap between the current meaning of the word 'is' and that defined in ontology is probably most pronounced in singular propositions the subjects of which are empty or general names. The propositions 'Hamlet is the hero of a tragedy by Shakespeare' and 'The whale is a mammal' are currently considered to be true, whereas the meaning attached to the word 'is' in ontology is such as to make them false.

One more example of a proposition in which the word 'est' has a different meaning than in ontology: 'Wrocław is a city of three hundred thousand inhabitants'. Here the word 'is' means 'is now'. Its meaning contains the time factor, which is neither included in the meaning of the primitive term of ontology, nor in that of the symbol used in the theory of sets for denoting that an individual belongs to a set.

I shall try to show that each of the above propositions which contains the word 'is' in a meaning different from that established in ontology can be replaced by a proposition not containing this word or containing it in its ontological meaning.

In propositions in which the word 'is' acts as a functor of one argument (e.g. in 'est iustitia') this word can be replaced by the word 'exists'. The proposition 'justice exists' reflects with exactitude the thought we wish to express when we say in Latin 'est iustitia'.

Prof. Kotarbiński proposes to replace the proposition 'seniority is a transitive relation' by the proposition 'If an object is older than another object and if that latter object is older than some other object, then the first is older than the third'.[19] In this proposition the meaning of the word 'is' no longer differs from that attached to it in ontology. In view of the fact that the decision as to whether any given word is or is not an apparent name is beyond the scope of logic, it is difficult to analyze the meaning of singular propositions whose subjects are apparent names. For this

---

[17] Kotarbiński [1929, pp. 8-10]. (*Editorial Note*: Also Kotarbiński [1966, pp.7–9]).
[18] Kotarbiński [1929, p. 9]. (*Editorial Note*: Also Kotarbiński [1966, p. 9]).
[19] Kotarbiński [1929, p. 9]. (*Editorial Note*: Also Kotarbiński [1966, p. 9]).

reason I shall not discuss in more detail singular propositions the subjects of which are names of this type, and I refer the reader to the works of Prof. Kotarbiński.[20]

Propositions with an empty subject classified as true by current intuition, i.e. propositions like the one about Hamlet or the proposition 'Zeus is a god of Olympus' may be replaced by intuitively equivalent propositions the subject of which are singular names. To this effect it suffices to replace the subjects of such a singular proposition by the same subject placed in inverted commas and to amplify the predicate by adding the term 'name' or any other term having a similar meaning. We thus obtain a proposition such as: "'Hamlet' is the name of the hero of a tragedy by Shakespeare", the subject of which is the term 'Hamlet' employed in *suppositio materialis*, as such it ceases to be an empty name.

A singular proposition of which the subject is a general name may be replaced by a general-affirmative proposition of Aristotelian syllogistics. Thus, the information conveyed by the proposition 'The whale is a mammal' is also conveyed, and possibly with even more precision, by the proposition 'Every whale is a mammal'. In this proposition the word 'is' has no meaning of its own, it is only a part of the functor 'every .... is .....'. The definition of this functor is given in II, § 1.

It remains to consider those propositions in which the meaning of the word 'is' implies the time factor. It is evident that the time factor may be transferred to the subject or predicate. For instance, taking the proposition 'Wrocław is a city of three hundred thousand inhabitants' and replacing the subject 'Wrocław' by the composite name 'Wrocław at such and such a period of time' or 'Wrocław at such and such a date' we obtain propositions in which the word 'is' no longer involves the time factor. An analysis of propositions in which the subject or predicate implies the time factor is contained in II, § 1.

The preceding remarks on the current meaning of the word 'is', though possibly altogether too lengthy and detailed are justified on the one hand by the need for preserving the intuitive values of the system (a characteristic feature of Leśniewski's entire work), and, on the other hand by the fact that the primitive term of ontology is often misunderstood. These misunderstandings are exposed in the following fragment taken from XI of "Foundations of Mathematics":[21]

"Conversations I held on various occasions made me believe that the afore-said 'singular' propositions were not always understood as I intended them to be. It was often the case that the person I spoke to failed to realize the meaning in which I employed the symbol 'ε' in my ontology, and he had interpretational difficulties when trying to bring this symbol into some semantic relation to the traditional symbols of 'mathe-

---

[20] Problems connected with apparent names are dealt with, among others, in Kotarbiński [1929 and 1952]. (*Editorial Note*: Also Kotarbiński [1966]).

[21] Leśniewski [1927–1931, Ch. XI, pp. 163–164].

matical logic' and 'the theory of sets'. From a theoretical point of view this state of affairs put me in an awkward position and I found myself sharing the difficulties of all whom circumstances force to express 'in their own words' the meaning of primitive terms in deductive theories of their making. I brought such situations to a close with or without success by employing various methods of approach to the intuitions of the person I spoke to. Thus if one person might be convinced by the fact that I use the term 'ε' in such a meaning as to satisfy the above axiom of ontology, another would feel more satisfied on hearing that propositions of the type '$A \varepsilon B$' are equivalent to corresponding propositions of the type 'every A is B and one object at most is A'; if a third person was satisfied with the fact that in my current language propositions of the type 'A is B' are equivalent to corresponding propositions of the type 'A is one of the objects B' so understood that they remain valid also when A is the only object which is B, – yet another might begin to grasp the semantic situation on being told that I employ the symbol 'ε' in propositions of the type '$A \varepsilon B$' in that meaning in which I use the word 'is' in propositions belonging to current language such as 'this man is long-lived', 'Rome is older than Warsaw', 'the point of intersection of the lines P and R is the centre of the circle K'."

How the axiom of ontology determines the meaning of the term 'is' will be explained in the next paragraph.

I shall now briefly compare the primitive term of ontology with the symbol '$\in$' denoting the relation of an object to the set of which it is a member. In the theory of sets not based on the theory of types the meaning of the symbol '$\in$' is determined by axioms. By axioms of the theory of sets I shall understand the system of axioms given in the "Theory of Sets" of Kuratowski and Mostowski,[22] this system being more intuitive than other systems. The other primitive term of the Kuratowski-Mostowski's[22] system, beside '$\in$', is the term '0' which is the symbol of the empty set. Its equivalent in ontology is the empty name.[23] To the symbol of identity appearing in the axioms of the theory of sets I shall coordinate the symbol of equality of extensions of two names.[24] All the axioms of the theory of sets, except the axiom of the exponential set and the axiom of infinity, remain true propositions on replacing the terms of the theory of sets by the corresponding terms of ontology. However, the intuitive meaning of most of the axioms of the theory of sets undergoes thereby a profound modification.

I will show this for the axiom of summation. In the Kuratowski-Mostowski theory of sets we find the following two definitions:

1. $Z(x) \equiv (x = 0) + \sum_{v} (v \in x)$ (x is a set),

2. $R(x) \equiv Z(x) \cdot \prod_{y} [(y \in x) \rightarrow Z(y)]$ (x is a set of sets).

---

[22] Kuratowski and Mostowski [1952, pp. 252–253].

[23] The definition of the empty name is given in II, § 4.

[24] The definition of the symbol of the equality of extensions is contained in II, § 3.

Employing the terms defined above the axiom of summation may be noted as follows:

$$3.\ \prod_{x}\Big\{R(X)\rightarrow\sum_{y}\Big[Z(Y)\cdot\prod_{x}\Big((x\in Y)\equiv\sum_{z}[(x\in z)\cdot(z\in X)]\Big)\Big]\Big\}.$$

To propositions 1–3 correspond in Leśniewski's system the propositions:

$$1'.\ Z/X/\equiv(X\mathrel{\overline{\overline{Z}}}\Lambda\vee[\exists v]\{v\,\varepsilon\,X\}),$$

$$2'.\ R/X/\equiv Z/X/\wedge[y]\{y\,\varepsilon\,X\supset Z/y/\},$$

$$3'.\ [X]\{R/X/\supset[\exists Y]\{Z/Y/\wedge[x]\{x\,\varepsilon\,Y\equiv[\exists z]\{x\,\varepsilon\,z\wedge z\,\varepsilon\,Y\}\}\}\}.$$

It is easy to see that proposition '$X\mathrel{\overline{\overline{Z}}}\Lambda\vee[\exists v]\{v\,\varepsilon\,X\}$' is true. Its proof in ontology does not involve any difficulties. The expression '$Z/X/$' is thus always true. Hence, it is easily proved that also the expression '$R/X/$' is always true. The proposition 3′ is therefore equivalent to

$$3''.\ [X,\exists Y,x]\{x\,\varepsilon\,Y\equiv[\exists z]\{x\,\varepsilon\,z\wedge z\,\varepsilon\,X\}\}.$$

This proposition is true, since the propositions

$$x\,\varepsilon\,X\equiv x\,\varepsilon\,x\wedge x\,\varepsilon\,X,$$
$$x\,\varepsilon\,z\wedge z\,\varepsilon\,X\supset x\,\varepsilon\,X$$

are true.

However, the intuitive meaning of proposition 3″ differs significantly from that of proposition 3.

I shall also point out that the meaning of the term '$\in$' as fixed in the theory of sets not based on the theory of types allows the following propositions to be simultaneously true:

$$x\in z,\ y\in z,\ \sim(x\in y),\ z\in t.$$

On the other hand, in ontology the corresponding propositions are not allowed to be simultaneously true, for if the propositions '$x\,\varepsilon\,z$', '$y\,\varepsilon\,z$' and '$\sim(x\,\varepsilon\,y)$' are all of them true, then 'z' is a general name and hence the proposition '$z\,\varepsilon\,t$' is false for any 't'.[25]

The problem of similarity or dissimilarity of the meanings of the primitive term of ontology and the symbol '$\in$' of the theory of sets based on the theory of types, requires a few explanatory words concerning the relation the theory of the division of objects into logical types bears to the theory of semantic categories.[26] Notwithstanding essential differences these theo-

---

[25] In explanations concerning the intuitive meaning of certain expressions, for instance of expressions which are theorems of ontology, I shall usually assume that the variables contained therein are arbitrary names; this will enable me to classify these expressions as propositions.

[26] The simple theory of types is described, for instance, in Mostowski [1948, pp. 213–221].

ries are formally alike. The analogy between them stands out clearly if we assume that to every constant symbol of logic, excepting quantifiers and the constants of the calculus of propositions, corresponds an object the name of which is this symbol, and that two objects are of the same logical type if and only if their names belong to the same semantic category.

The primitive term of ontology as regards its meaning stands nearest to the symbol '$\in$', when used to denote the relation of membership holding between an individual and a set of individuals. But even then the meanings of these symbols are essentially different, as the symbol '$\in$' serves to denote a relation holding between objects of different logical types, while the symbol 'ε' is a functor whose arguments belong to the same semantic category.

Singular propositions have a simple interpretation in the theory of sets based on the theory of types.[27] However, this interpretation deviates markedly from the intuitions which Leśniewski connected with the primitive term of his system. It consists in coordinating to the singular proposition the relation which holds between two sets of individuals if and only if the former is a unit set contained in the latter. This is of course a relation between objects of the same type. Under this interpretation the axiom of ontology is a theorem of the theory of sets.[28]

Notwithstanding the essential difference of meaning between the primitive term of ontology and the symbol denoting the relation of being a member of a set these two terms may often replace one another in propositions without bringing about a noticeable change of meaning. This is best shown by the following table which contains the axioms of Peano's arithmetic written first in terms of ontology and then in those of the theory of sets.[29]

1) $1 \varepsilon N$,	1′) $1 \in N$,
2) $x \varepsilon N \supset seqx \varepsilon N$,	2′) $x \in N \supset seqx \in N$,
3) $x \varepsilon N \supset seqx \neq 1$,	3′) $x \in N \supset seqx \neq 1$,
4) $x \varepsilon N \wedge y \varepsilon N \wedge seqx = seqy \supset x = y$,	4′) $x \in N \wedge y \in N \wedge seqx = seqy \supset x = y$,
5) $1 \varepsilon X \wedge [x]\{x \varepsilon X \supset seqx \varepsilon X\} \supset NaX$.	5′) $1 \in X \wedge [x]\{x \in X \supset seqx \in X\} \supset N \subset X$.

In the axioms to the left the letter 'N' is a general name having the same meaning as the name 'natural number' while the letters 'x', 'y' and 'X' are nominal variables. In those to the right the letter 'N' denotes the set of natural numbers, and the letters 'x', 'y' and 'X' are variables of

27 This interpretation has been the theme of a lecture held by Prof. J. Łoś in a Seminar on Mathematical Logic, he and myself conducted jointly in 1950.

28 The proof of this theorem will be given in III, § 3.

29 The axioms of Peano's arithmetic terms of ontology are given on the basis of notes from the course of lectures on Inductive Definitions, held by Leśniewski in 1933/34.

which the first two represent individuals and the third sets. Thus axiom (1) means: one is natural number; and axiom (1′): one is an element of the set of natural numbers. The expression 'NaX' which appears in axiom (5) is a general affirmative proposition; the counterpart of this expression in axiom (5′) is the expression '$N \subset X$' stating the inclusion of set N in set X.

§ 4. The Axiom of Ontology

The only axiom of ontology, formulated by Leśniewski for the first time in 1920, has been subsequently simplified in various ways by Leśniewski as well as by his students. However, the system of elementary ontology is based here on the axiom from 1920, for I am convinced that this axiom determines the intuitive meaning of the primitive term of the system better than its simpler variants. I shall discuss the latter in IV § 2. Here now is the axiom from 1920:

(1) $x \varepsilon X \equiv [\exists y]\{y \varepsilon x\} \wedge [y, z]\{y \varepsilon x \wedge z \varepsilon x \supset y \varepsilon z\} \wedge [y]\{y \varepsilon x \supset y \varepsilon X\}$.

Just as the only axioms of the system of protothetics in which the sign of equivalence acts as primitive term and as other variants of the only axiom of ontology, axiom (1) is an equivalence the left side of which contains only variables and one occurrence of the primitive term of the system. The axiom has thus a structure resembling that of the definitions of Leśniewski's systems; it cannot be considered, however, to form a definition of the term 'ε', as the latter appears also to the right of the sign of equivalence. The expression standing to the right of the sign of equivalence in axiom (1) is a conjunction of the following three expressions:

(2) $[\exists y]\{y \varepsilon x\}$,
(3) $[y, z]\{y \varepsilon x \wedge z \varepsilon x \supset y \varepsilon z\}$,
(4) $[y]\{y \varepsilon x \supset y \varepsilon X\}$.

I shall try to explain how these expressions determine the meaning of the expression '$x \varepsilon X$', which is equivalent to their conjunction. I hope the aim I pursue will excuse some lack of precision.

In what follows I assume the letters 'x' and 'X' to be arbitrary though fixed names. This assumption transforms the expressions '$x \varepsilon X$' and (2)–(4) into propositions.

The proposition

(3′) $[z, y]\{z \varepsilon x \wedge y \varepsilon x \supset z \varepsilon y\}$

differs from (3) only by the form of the variables it contains. The rules of inference of elementary ontology[30] make it possible to state that each of these propositions is a consequence of the other. These propositions are therefore inferentially equivalent. From this and also from theorems of the calculus of propositions we conclude that proposition (3) is inferentially equivalent to the following proposition:

$$[y, x]\{y \varepsilon x \wedge z \varepsilon x \supset y \varepsilon z \wedge z \varepsilon y\}.$$

From this and from the fact that in ontology the conjunction '$y \varepsilon z \wedge z \varepsilon y$' is the definiens of the expression '$y = z$',[31] follows the inferential equivalence of propositions (3) and

(3″) $[y, z]\{y \varepsilon x \wedge z \varepsilon x \supset y = z\}$.

Propositions (2), (3″) and (4), on being translated freely into current speech, take the forms:

a certain object is x,
any two objects that are x are identical,
anything that is x is also X.

The first and second of these propositions imply that x is a singular name. These propositions thus ensure that the necessary condition of the truth of the singular proposition '$x \varepsilon X$' (discussed in the previous paragraph) is fulfilled. The conjunction of all three propositions constitutes not only a necessary but also a sufficient condition for the proposition '$x \varepsilon X$' to be true.

I will show that if the term 'ε' is given the meaning ascribed to it in ontology, axiom (1) is true if either 'x' or 'X' is an empty name.

If 'x' is an empty name, then proposition (2) is obviously false. On the other hand, if 'x' is not empty while 'X' is empty, then proposition (4) is false. If therefore one or both of these names are empty, the proposition to the right of the sign of equivalence in axiom (1) is false. Then, however, the proposition to the left of the symbol of equivalence in this proposition is also false. Hence the axiom of ontology is true if either 'x' or 'X' is an empty name. It follows from this that the theorems of ontology, as distinct from those of Aristotelian syllogistics, may be applied in reasonings in which empty names appear.

§ 5. Rules of Inference of Elementary Ontology

Of the three systems of Leśniewski the first to develop was mereology, ontology came next and protothetics followed. The logical order of these systems is the inverse: the most elementary system is protothetics, the

---

[30] These rules will be dealt with in detail in the next paragraph.

[31] The definition of the term '=' is contained in II, § 3.

theorems of which were used by Leśniewski for proving those of ontology. Mereology is based on both these systems. However, in proving the theorems of ontology I shall be making use of the theorems of that fragment of protothetics only which coincides with the ordinary calculus of propositions. This does not apply to the methodological reasonings of II, § 4, in which I employ theorems of the calculus of propositions enlarged to include the terms '1' and '0'.

The deduction of the theorems of elementary ontology does not involve all the rules of inference of Leśniewski's entire system. For instance, the rule of extensionality by which certain propositional expressions of a clearly defined structure may be included into the system cannot be applied in elementary ontology as the expressions included according to this rule into the system contain terms not belonging to the language of elementary ontology.[32] Besides, the rule for adding definitions to the system is of a simpler form in elementary ontology than that valid in the entire system.

I shall discuss this rule briefly. The definitions are theorems of the system; they have the form of equivalences. I shall denote the propositional expressions appearing in these definitions to the left and to the right of the sign of equivalence as definiendum and definiens, respectively. The only constant term of the definiendum, apart from the term 'ε', is the term to be defined, which appears only once in this expression. The definiendum does not contain equiform variables. Every variable of the definiendum is equiform to a free variable of the definiens, and inversely: every free variable of the definiens is equiform to a variable of the definiendum. The definiendum contains, of course, only free variables.

In the definitions of elementary ontology the definiens is always noted only by means of the primitive term of ontology, the functors of the calculus of propositions, nominal variables, quantifiers and parentheses. In elementary ontology the symbol which is being defined is not equiform to any symbol previously introduced into the system.

The language of elementary ontology limits the scope of terms attached to the system by means of definitions to nominal constants and to name-formative or proposition-formative functors of nominal arguments.

The schema for definitions of proposition-formative functors is as follows:

$$\text{(I)} \qquad \varphi / x_1, x_2, \ldots, x_n / \equiv \alpha.$$

The term defined in (I) is φ, the letters 'x_1', 'x_2', ..., 'x_n' are nominal variables, and α is a propositional expression of elementary ontology.[33]

[32] The rules of extensionality will be dealt with in detail in III, § 1.

[33] A definition of this type is D 1.2. The definitions and theorems appearing in II are given in a double numeration in Arabic numerals. The first numeral serves to denote the place occupied by an expression or theorem in the paragraph in which it appears, the second is the number of that paragraph.

The following is the schema for the definition of a name-formative functor:

(II) $x_0 \varepsilon \psi | x_1, \ldots, x_n | \equiv x_0 \varepsilon x_i \wedge \alpha \quad 0 \leqslant i \leqslant n.$

Here, the term defined is ψ, the letters 'x_0', 'x_1', ..., 'x_n' are nominal variables and α is a propositional expression of elementary ontology.[34]

Finally, the schema for defining a nominal constant is as follows:

(III) $x \varepsilon X \equiv x \varepsilon x \wedge \alpha.$

Here, the term defined is X, the letter 'x' is a nominal variable, and α, as above, is a propositional expression of elementary ontology.[35]

The first factor of the definiens of a name-formative functor or of a nominal constant plays an important role, that of preserving the system from contradictions. Thus, for example, the definition of the empty name:

$$x \varepsilon \Lambda \equiv x \varepsilon x \wedge \sim(x \varepsilon x)$$

on omitting the factor '$x \varepsilon x$' takes the form:

$$x \varepsilon \Lambda \equiv \sim(x \varepsilon x).$$

This last expression, however, leads to contradiction, since after substituting in it the symbol 'Λ' for the variable 'x' we obtain a proposition from which two contradictory propositions follow immediately.[36] The reason why the factor '$x \varepsilon y$' forms a necessary part of the definiens in definitions of name-formative functors and of nominal constants is the following: in these definitions the definiendum implies the existence of the object x, and hence the definiens also should guarantee the existence of this object, which the expression '$x \varepsilon y$' does in effect.

When proving the theorems of elementary ontology I shall be applying the rule of substitution. This rule in Leśniewski's system does not differ from analogous rules in other systems of logic. I shall therefore not discuss it here. The rule allowing us to reletter the bound variables is related to that of substitution. By this rule we may put in place of any variable enclosed in the square brackets of a quantifier as well as of all the variables equiform to it and bound by this quantifier, any variable differing in from all the variables occurring in the expression covered by that quantifier. I should point out that the rule of relettering the variables is secon-

---

[34] Definitions of this type are D 2.4 and D 3.4.

[35] A definition of this type is D 4.1. I note that the definition of the name 'object' (D 4.1) does not fall under schema (III). The definiens of this definition is, however, equivalent to the expression '$x \varepsilon x \wedge [\exists x]\{x \varepsilon X\}$'. Thus, it is possible to replace definition D 4.1 by a definition falling under schema (III).

[36] The way in which the contradiction arises here is reminiscent, at least formally, of Russell's antinomy. The condition which definitions of ontology should fulfil reminds us, perhaps again only formally, of the axiom of sub-sets. (Russell's antinomy and the axiom of sub-sets are discussed in Kuratowski and Mostowski [1952, pp. 44–47].)

dary in relation to the rules holding in elementary ontology; the proof of this I omit. I denote the propositional-expression obtained from the expression of elementary ontology α by applying to it the rule of substitution or by reletterring some of its variables, by the symbol α^*. I shall make use of this symbol in defining the notion of direct inference.

The expression α follows by direct inference from expressions $\alpha_1, \alpha_2, \ldots$ $\ldots, \alpha_n$, if and only if the expression $\ulcorner \alpha_1^* \wedge \alpha_2^* \wedge \ldots \wedge \alpha_n^* \supset \alpha \urcorner$ is a result of substitution applied to a theorem of the calculus of propositions. After Prof. Ajdukiewicz I denote this theorem as the principle of inference.[37]

When a certain theorem follows from other theorems by direct inference I limit the notation of its proof to writing next to it in square brackets the numbers of those theorems from which it follows. The principle of inference, and the substitutions and reletterings applied in the proof are not given.

The proofs of theorems not following by direct inference from earlier theorems will be given in the form of suppositional proofs. This was the form Leśniewski chose for them. However, the method of noting suppositional proofs applied in this paper differs from that used by Leśniewski.

A suppositional proof is a column of expressions alongside which stand abbreviatory explanations in square brackets justifying the inclusion of these expressions in the proof. If a conditional is being proved, the first expression of the proof is its antecedent. If this antecedent is a conjunction, then its factors are the initial expressions of the proof. I shall denote these expressions as premisses of the proof.

Expressions which are not premisses of a proof may be subjoined to it, if the subjoining is justified by the laws of the functional calculus or if the expressions subjoined result by direct inference from earlier theorems of the system and from expressions already subjoined to the proof or, lastly, if these expressions may be obtained by substitution from theorems of the calculus of propositions. The method of applying in the proofs the laws of the functional calculus needs explanation. I omit remarks about expressions following from theorems already attached to the system as well as from expressions occurring earlier in the proof for the reason that they are in no way different from remarks about theorems following by direct inference from other theorems. When an expression forming a part of the proof may by obtained by substitution from a theorem of the calculus of propositions, I write this theorem in square brackets alongside the expression subjoined to the proof. A suppositional proof is completed when an expression equiform to the theorem to be proved or, should this theorem be an implication, an expression equiform to its consequent, has been subjoined.

I shall be applying the laws of the functional calculus exclusively as rules for subjoining new expression to the proof. I shall note these rules in

---

[37] I make use of pseudo-inverted commas in the same way as Quine [1940].

the form of schemata consisting of a horizontal line, of expressions written above the line (and called premisses) and of expressions written beneath the line (and called conclusions). Premisses are expressions already subjoined to the proof or earlier theorems of the system, conclusions are expressions following from premisses on the strength of the rule defined by the schema in which they occur.

The schemata of rules for using quantifiers contain the following expressions:

(1) $\alpha(x)$, $\alpha(y_1/x)$, $\alpha(y/x)$, $[x]\{\alpha(x)\}$, $[\exists x]\{\alpha(x)\}$.

The letter 'x' in these expressions denotes an arbitrary nominal variable as admitted in the theorems of elementary ontology. The letter 'y_1' stands for a variable not appearing in theorems of ontology but only in such expressions of the proof as are neither premisses nor last expressions thereof. Variables of this kind will always bear indices. Intuitively, they are arbitrary though fixed names. In attaching new expressions to a proof, I shall also be making use of expressions obtained from earlier theorems of the system through replacing their variables by variables bearing indices. However, it should be emphasized that no expressions may be substituted for variables appearing in expressions already subjoined to a proof.

The letter 'y' denotes either a constant name, or it is the name of an arbitrary variable appearing in the expressions of the proof. In particular, it may denote a variable bearing an index. $\alpha(x)$ is an arbitrary expression of the system containing the free variable x. $\alpha(y_1/x)$ and $\alpha(y/x)$ are expressions obtained from $\alpha(x)$ by substituting the variable y_1 or y for x. The two remaining expressions in (1) denote expressions obtained from $\alpha(x)$ by placing before $\alpha(x)$ either the general, or the existential quantifier binding the variable x.

The schema of the rule for omitting the general quantifier is as follows:

$$\text{O}\Pi.\ \frac{[x]\{\alpha(x)\}}{\alpha(y/x)}.$$

The following is the schema of the rule for omitting the existential quantifier:

$$\text{O}\Sigma.\ \frac{[\exists x]\{\alpha(x)\}}{\alpha(y_1/x)}.$$

The variable y_1 which we substitute in $\alpha(x)$ for the variable x cannot be equiform to any variable appearing in earlier expressions of a proof. This rule is founded on the following intuition: if the premise of this rule is true, then certain names if substituted for the variable x transform this premise into a true proposition. When y_1 stands for any of these names, then the proposition $\alpha(y_1/x)$ is obviously true. The condition which the variable y_1 has to fulfil is necessary in as much as the name that transforms $\alpha(x)$ into a true proposition may differ from all names appearing in earlier expressions of a proof.

The schema of the rule for adding the general quantifier is as follows:

$$\text{D}\Pi.\ \frac{\alpha(x)}{[x]\{\alpha(x)\}}.$$

This rule is limited by the condition that the variable x should not appear in the premisses of the proof. The intuitive justification for this restriction is the fact that variables appearing in the premisses do not stand for arbitrary names, but for such only as transform these expressions into true propositions. The applicability of the rule for adding the general quantifier is restricted by a second condition, namely that the variable x must not appear in any earlier expression of a proof containing, apart from x, a variable bearing an index, for example y_1. It is possible to give the following intuitive reason for this restriction: in an expression containing y_1, the variable x does not stand for arbitrary names, but for such names only as transform, together with the name y_1, a given expression of the proof into a true proposition. I wish to point out that in none of the proofs given in this paper will it be necessary to take this restriction into account.

The following is the schema for the rule of adding the existential quantifier:

$$\text{D}\Sigma.\ \frac{\alpha(y/x)}{[\exists x]\{\alpha(x)\}}.$$

No conditions limit the application of this rule.

It can easily be proved that the rules we have given are equivalent to those for attaching quantifiers to or omitting them from the antecedent or the consequent of a conditional (together with the rule of substitution for nominal variables), holding in systems of the functional calculus.[38] The proof of this I omit.

Apart from primitive rules I shall be using in proofs secondary rules defined by the following schemata:

$$\Pi\text{N}.\ \frac{[x]\{\sim(\alpha(x))\}}{\sim([\exists x]\{\alpha(x)\})},$$

$$\Sigma\text{N}.\ \frac{[\exists x]\{\sim(\alpha(x))\}}{\sim([x]\{\alpha(x)\})},$$

$$\text{N}\Pi.\ \frac{\sim([x]\{\alpha(x)\})}{[\exists x]\{\sim(\alpha(x))\}},$$

$$\text{N}\Sigma.\ \frac{\sim([\exists x]\{\alpha(x)\})}{[x]\{\sim(\alpha(x))\}}.$$

[38] These rules are given in Mostowski [1948, pp. 53–56].

I shall not prove here that these rules are secondary with regard to the accepted primitive rules and the theorems of the calculus of propositions. Their meaning is clearly connected with de Morgan's laws of the functional calculus. I may add that the restricted number of secondary rules lengthens the proofs of some theorems of ontology; as this lengthening is quite insignificant, I refrain from increasing the number of these rules.

When applying in a proof one of the rules referring to quantifiers, I shall be writing next to the expression subjoined to the proof the abbreviated name of the rule applied in that case (the abbreviated names of rules are given alongside their schemata) and, following a colon, the number of the premise, which may be an earlier theorem of the system or an earlier expression of the proof.

I shall be omitting proofs of the theorems, when they present no difficulties, those of theorems appearing in II, § 1 and of the first nine in II, § 2 forming an exception. I give the proofs of these theorems because of the very special interest they present in relation to the system; at the same time these proofs will serve as examples of proofs of elementary ontology. When a theorem is a direct consequence of earlier theorems, I give their numbers.

In omitting the proofs of the greater part of the theorems of elementary ontology I act contrary to Leśniewski's tradition; he always gave the proofs of all theorems adduced in his papers.

II

§ 1. Basic Theorems of Elementary Ontology

The only axiom of ontology is the expression:

T.1.1. $x \varepsilon X \equiv [\exists y]\{y \varepsilon x\} \wedge [y, z]\{y \varepsilon x \wedge z \varepsilon x \supset y \varepsilon z\}$
$\wedge [y]\{y \varepsilon x \supset y \varepsilon X\}$.

The intuitive meaning of this axiom has been discussed in I, § 4.

I will now give three definitions which will make it possible to shorten the notation of the axiom of ontology. They also throw a new light on the intuitive meaning of that axiom.

D1.1. $ex\,/X/ \equiv [\exists x]\{x \varepsilon X\}$.

We read the expression '$ex\,/X/$' as follows: 'there exists at least one X'. A special case of this expression is for example the proposition 'there exists at least one aquatic mammal'. According to D1.1 this proposition is true if for a certain x it is true that x is an aquatic mammal. Further remarks on the intuitive meaning of the defined term are given in § 4 of this chapter.

D2.1. $\rightharpoonup /x/ \equiv [y, z]\{y \varepsilon x \wedge z \varepsilon x \supset y \varepsilon z\}$.

We read the expression '$\rightharpoondown |x|$' as follows: 'there exists one x at the most'. The intuitive meaning of the definiens of D2.1 has been discussed in I, § 3.

D3.1. $SaP \equiv [x]\{x \varepsilon S \supset x \varepsilon P\}$.

We read the definiendum of D3.1 as follows: every 'S is P'. Thus, the term defined is one of the terms of Aristotelian syllogistics. If the name 'S' is empty the proposition 'SaP' is true since the expression '$x \varepsilon S$', which is the antecedent of the conditional covered by the quantifier in the definiens of D3.1 is false for every x. In the following paragraph we shall give a different definition of the general-affirmative proposition, according to which this proposition is false if its subject is an empty name. The relation defined in D3.1 corresponds also to the inclusion of sets.

The following theorem is equisignificant to the axiom of ontology:

T2.1. $x \varepsilon X \equiv ex |x| \wedge \rightharpoondown |x| \wedge xaX$. {T1.1, D1.1, D2.1, D3.1}

According to T2.1 x is X if and only if there exists at least one and at the most one x, and if every x is X.

The following four theorems are direct conclusions from T2.1.

T3.1. $x \varepsilon X \supset ex |x|$.
T4.1. $x \varepsilon X \supset \rightharpoondown |x|$.
T5.1. $x \varepsilon X \supset xaX$.
T6.1. $ex |x| \wedge \rightharpoondown |x| \wedge xaX \supset x \varepsilon X$.
{T2.1}

According to T3.1 and T4.1, if the name 'x' appears in the subject of an arbitrary true singular proposition, there exists at least one and at most one x. T5.1 states that from a singular proposition follows the general-affirmative proposition in the subject and predicate of which appear the same names as in the subject and predicate of the singular proposition. T6.1 sets out a sufficient condition for the truth of a singular proposition.

The following theorem is not a direct consequence of earlier theorems. It should be remembered that the rules of inference of ontology as well as the method of noting suppositional proofs have been discussed in I, § 5.

T7.1. $x \varepsilon X \wedge y \varepsilon x \supset y \varepsilon X$.

Proof. (1) $x \varepsilon X$ {premise}
(2) $y \varepsilon x$ {premise}
(3) $[y]\{y \varepsilon x \supset y \varepsilon X\}$ {T1.1, 1}
(4) $y \varepsilon x \supset y \varepsilon X$ {OΠ: 3}
$y \varepsilon X$ {2, 4}

T7.1. is the Aristotelian syllogism Barbara expressed in terms of ontology.

T8.1. SaS.

Proof.	(1) $x \varepsilon S \supset x \varepsilon S$	$\{p \supset p\}$
	(2) $[x]\{x \varepsilon S \supset x \varepsilon S\}$	$\{D\Pi: 1\}$
	SaS	$\{D3.1, 2\}$

This theorem is one of the four axioms of Aristotelian syllogistics, set up for that system by Prof. J. Łukasiewicz.[39] Let me point out that in the proof of T8.1 a theorem of the calculus of propositions, a definition and a rule for the use of a quantifier is applied. I have therefore included this theorem in the system without referring to the axiom. In further paragraphs of this paper we shall encounter more theorems possessing the same property.

T9.1. $x \varepsilon X \supset x \varepsilon x$.

Proof.	(1) $x \varepsilon X$	{premise}
	(2) $ex\,/x/$	{T3.1, 1}
	(3) $\rightharpoonup /x/$	{T4.1, 1}
	$x \varepsilon x$	{T6.1, 2, 3, T8.1}

T9.1. is the law of identity of elementary ontology. It is not possible to omit therein the antecedent '$x \varepsilon X$', as the expression '$x \varepsilon x$' becomes a false proposition on substituting an empty or a general name for the variable 'x'.

T10.1. $x \varepsilon X \wedge y \varepsilon x \supset x \varepsilon y$.

Proof.	(1) $x \varepsilon X$	{premise}
	(2) $y \varepsilon x$	{premise}
	(3) $x \varepsilon x$	{T9.1, 1}
	(4) $[y, z]\{y \varepsilon x \wedge z \varepsilon x \supset y \varepsilon z\}$	{T1.1, 1}
	(5) $x \varepsilon x \wedge y \varepsilon x \supset x \varepsilon y$	$\{O\Pi : 4\}$
	$x \varepsilon y$	{5, 3, 2}

This theorem will be called the fundamental law of ontology. It enables us to convert any singular proposition if its predicate is the subject of an arbitrary true singular proposition, that is to say, it enables us to convert any singular proposition if its predicate is the subject of an arbitrary true singular proposition, that is to say, if it is a singular name.

The following theorem is easily inferred from D1.1.

T11.1. $x \varepsilon X \supset ex\,/X/$.

Proof.	(1) $x \varepsilon X$	{premise}

[39] This system is discussed in Łukasiewicz [1951].

(2) $[\exists x]\{x \varepsilon X\}$ {DΣ : 1}
$ex\,|X|$ {D1.1, 2}

T12.1. $ex\,|x| \wedge \neg\,|x| \supset x \varepsilon x$ {T6.1, T8.1}

This theorem gives a condition for the truth of the proposition '$x \varepsilon x$', different from that given by the identity law of elementary ontology.

T13.1. $\neg\,|x| \wedge y \varepsilon x \supset x \varepsilon y$.

Proof. (1) $\neg\,|x|$
(2) $y \varepsilon x$ } {premise}
(3) $ex\,|x|$ {T11.1, 2}
(4) $x \varepsilon x$ {T12.1, 3, 1}
$x \varepsilon y$ {T10.1, 4, 2}

According to T13.1. the singular proposition '$x \varepsilon y$' may be converted if at least one x exists.

D4.1. $x \varepsilon \mathrm{V} \equiv [\exists X]\{x \varepsilon X\}$.

We read the definiendum of this definition as follows: 'x is an object'. The term 'V', denoted by Leśniewski as full name, corresponds to the term denoting the universal class of individuals.

The intuitive meaning of theorems T3.1–T5.1, T9.1 and T10.1 becomes clearer on substituting the symbol 'V' for the variable 'X' therein.

T14.1. $x \varepsilon X \supset x \varepsilon \mathrm{V}$.

Proof. (1) $x \varepsilon X$ {premise}
(2) $[\exists X]\{x \varepsilon X\}$ {DΣ : 1}
$x \varepsilon \mathrm{V}$ {D4.1, 2}

This theorem may be translated freely into current speech as follows: if x is anything at all, then x is an object.

T15.1. $x \varepsilon \mathrm{V} \equiv x \varepsilon x$. {T9.1, T14.1}

This theorem might replace D4.1.

T16.1. $Xa\mathrm{V}$.

Proof. (1) $[x]\{x \varepsilon X \supset x \varepsilon \mathrm{V}\}$ {DΠ : T14.1}
$Xa\mathrm{V}$ {D3.1, 1}

This theorem corresponds to the theorem of the algebra of sets, stating that every set of individuals is contained in the universal class.

The following four theorems serve to establish certain simple relations between the primitive term of ontology and the terms defined in the course of this paragraph.

T17.1. $x \varepsilon \mathrm{V} \equiv ex\,/x/ \wedge \rightharpoonup /x/$. {T2.1, T16.1}

T18.1. $x \varepsilon X \equiv x \varepsilon \mathrm{V} \wedge x a X$. {T2.1, T17.1}

T19.1. $\sim(ex\,/X/) \supset \rightharpoonup /X/$.

Proof. (1) $\sim(ex\,/X/)$ {premise}
(2) $\sim([\exists x]\{x \varepsilon X\})$ {D1.1, 1}
(3) $[x]\{\sim(x \varepsilon X)\}$ {NΣ : 2}
(4) $\sim(x \varepsilon X)$ {OΠ : 3}
(5) $x \varepsilon X \wedge y \varepsilon X \supset x \varepsilon y$ {4}
(6) $[x, y]\{x \varepsilon X \wedge y \varepsilon X \supset x \varepsilon y\}$ {DΠ : 5}
$\rightharpoonup /X/$ {D2.1, 6}

T20.1. $\rightharpoonup /x/ \equiv x \varepsilon \mathrm{V} \vee \sim(ex\,/x/)$. {T17.1, T19.1}

Two auxiliary theorems now follow.

T21.1. $x \varepsilon X \supset [\exists y]\{x \varepsilon y \wedge y \varepsilon X\}$.

Proof. (1) $x \varepsilon X$ {premise}
(2) $x \varepsilon x$ {T9.1, 1}
(3) $x \varepsilon x \wedge x \varepsilon X$ {1, 2}
$[\exists y]\{x \varepsilon y \wedge y \varepsilon X\}$ {DΣ : 3}

T22.1. $[\exists y]\{x \varepsilon y \wedge y \varepsilon X\} \supset x \varepsilon X$.

Proof. (1) $[\exists y]\{x \varepsilon y \wedge y \varepsilon X\}$ {premise}
(2) $x \varepsilon y_1 \wedge y_1 \varepsilon X$ {OΣ : 1}
$x \varepsilon X$ {T7.1, 2}

T23.1. $x \varepsilon X \equiv [\exists y]\{x \varepsilon y \wedge y \varepsilon X\}$. {T21.1, T22.1}

T23.1. does not contain any defined terms. In IV, § 2 I shall prove that this theorem can act as the only axiom of ontology.

The theorems of elementary ontology proved so far enable us to analyze the meaning of such singular propositions in which the subject or the predicate implies the time factor. The difficulties we encounter when using such propositions are exemplified by the fact that the following reasoning though seemingly correct leads to a strikingly false conclusion. Leśniewski quotes this reasoning in Leśniewski [1927-1931, Ch. XI].

In the following propositions I assume the letters 'x', 'y', 'z', 'X' and 'Y' to be fixed names:

(a) $x \varepsilon X \wedge y \varepsilon x \supset x \varepsilon y$,
(b) $x \varepsilon y \wedge z \varepsilon x \supset z \varepsilon y$,
(c) $y \varepsilon Y \wedge z \varepsilon y \supset z \varepsilon Y$.

The first of these propositions is a result of substitution applied to T10.1. The remaining two follow through substitution from T7.1. I further assume that the propositions

(d) $x \varepsilon X$,
(e) $y \varepsilon x$,
(f) $z \varepsilon x$,
(g) $y \varepsilon Y$

are true.

Propositions (a), (d) and (e) lead to the proposition:

$$x \varepsilon y,$$

which together with propositions (b) and (f) leads to

$$z \varepsilon y.$$

This latter proposition together with propositions (c) and (g) results in

(h) $z \varepsilon Y$.

Let us now substitute the names 'Warsaw', 'Warsaw in the year 1830', 'Warsaw in the year 1930', 'a city older than Łódź', and 'a city smaller than Warsaw in the year 1930' for the names 'x', 'y', 'z', 'X', 'Y', respectively. The propositions (d)–(g) are then transformed into the following propositions:

(d′) Warsaw is a city older than Łódź,
(e′) Warsaw in the year 1830 is Warsaw,
(f′) Warsaw in the year 1930 is Warsaw,
(g′) Warsaw in the year 1830 is a city smaller than Warsaw in the year 1930.

We would not hesitate to classify these propositions as true. They imply, however, the following propositions obtained from (h) by replacing the letters 'z' and 'Y' by names substituted for these letters in the propositions (f) and (g).

(h′) Warsaw in the year 1930 is a city smaller than Warsaw in the year 1930.

According to Leśniewski, the source of this paradox lies in the fact that the meaning of the name 'Warsaw' in propositions (d′)–(h′) is not precise. If we classify this name as a general name the designata of which are: Warsaw in the year 1830 and Warsaw in the year 1930, or in general: Warsaw in any moment or period of time, then proposition (d′) is false, as its subject is then not a singular name. On the other hand, if we consider the name 'Warsaw' to be equivalent to the name 'Warsaw in the entire period of its existence', then propositions (e′) and (f′) are false.

§ 2. Aristotelian Syllogistics

In § 1 I have defined the general-affirmative proposition of Aristotelian syllogistics.

The following expressions define the remaining terms of syllogistics:

D1.2. $SiP \equiv [\exists x]\{x \varepsilon S \wedge x \varepsilon P\}$,
D2.2. $SeP \equiv [x]\{x \varepsilon S \supset \sim(x \varepsilon P)\}$,
D3.2. $SoP \equiv [\exists x]\{x \varepsilon S \wedge \sim(x \varepsilon P)\}$.

In order to elucidate the relation of Aristotelian syllogistics to the set of theorems of elementary ontology in which the only functors of nominal arguments are the terms defined in D3.1 and D1.2–D3.2, I am giving a concise description of the system of syllogistics as constructed by Prof. J. Łukasiewicz.

Apart from the moods of the syllogism this system contains the laws of the square of opposition and of conversion; its primitive terms are the functors '*a*' and '*i*'. The remaining terms of syllogistics are defined as follows:

(a) $SeP \equiv \sim(SiP)$,
(b) $SoP \equiv \sim(SaP)$.

The following expressions are axioms of the system of Łukasiewicz:

(c) SaS,
(d) SiS,
(e) $MaP \wedge SaM \supset SaP$,
(f) $MaP \wedge MiS \supset SiP$.

In proving the theorems of syllogistics Łukasiewicz makes use of theorems of the calculus of propositions, of the rule of substitution and of the rule allowing detachment of the consequent from a conditional already subjoined to the system.[40]

Some theorems of the system of Łukasiewicz, as for instance (d), turn into false propositions on substituting empty names for their variables. Thus the system of Łukasiewicz may be applied only in reasonings containing no empty names. This restriction does not hold for Leśniewski's system, (cf. I, § 4). It should be kept in mind that D1.3 fixes the meaning of the general-affirmative proposition in such a way that this proposition is always true if its subject is an empty name.

[40] In Łukasiewicz's system definitions are not theorems of the system. Hence apart from the rules of substitution and detachment, the rule of definitional replacement holds in that system. The description I have of Łukasiewicz's system is thus not exact. However, the modifications I introduced into that system in order to bring it nearer to Leśniewski's system in no way change the set of theorems following from axioms (c)–(f).

Theorem T8.1 of § 1 is equiform with the first axiom of Łukasiewicz. Let us remember that this theorem was attached to the system without having recourse to the axiom of ontology. This property will be also shared by all such theorems appearing in this paragraph in which all functors of nominal arguments are terms of syllogistics.

I shall now prove that the expressions (a), (b), (e) and (f) are also theorems of elementary ontology.

T1.2. $SeP \supset \sim(SiP)$.

Proof.	(1) SeP	{premise}
	(2) $[x]\{x \varepsilon S \supset \sim(x \varepsilon P)\}$	{D2.2, 1}
	(3) $x \varepsilon S \supset \sim(x \varepsilon P)$	{OΠ : 2}
	(4) $\sim(x \varepsilon S \wedge x \varepsilon P)$	{3}
	(5) $[x]\{\sim(x \varepsilon S \wedge x \varepsilon P)\}$	{DΠ : 4}
	(6) $\sim([\exists x]\{x \varepsilon S \wedge x \varepsilon P\})$	{ΠN : 5}
	$\sim(SiP)$	{D1.2, 6}

T2.2. $\sim(SiP) \supset SeP$.

Proof.	(1) $\sim(SiP)$	{premise}
	(2) $\sim([\exists x]\{x \varepsilon S \wedge x \varepsilon P\})$	{D1.2, 1}
	(3) $[x]\{\sim(x \varepsilon S \wedge x \varepsilon P)\}$	{NΣ : 2}
	(4) $\sim(x \varepsilon S \wedge x \varepsilon P)$	{OΠ : 3}
	(5) $x \varepsilon S \supset \sim(x \varepsilon P)$	{4}
	(6) $[x]\{x \varepsilon S \supset \sim(x \varepsilon P)\}$	{DΠ : 5}
	SeP	{D2.2, 6}

T3.2. $SeP \equiv \sim(SiP)$. {T1.2, T2.2}

The theorem proved is equiform with expression (a).

T4.2. $SoP \supset \sim(SaP)$.

Proof.	(1) SoP	{premise}
	(2) $[\exists x]\{x \varepsilon S \wedge \sim(x \varepsilon P)\}$	{D3.2, 1}
	(3) $x_1 \varepsilon S \wedge \sim(x_1 \varepsilon P)$	{OΣ : 2}
	(4) $\sim(x_1 \varepsilon S \supset x_1 \varepsilon P)$	{3}
	(5) $[\exists x]\{\sim(x \varepsilon S \supset x \varepsilon P)\}$	{DΣ : 4}
	(6) $\sim([x]\{x \varepsilon S \supset x \varepsilon P)\}$	{ΣN : 5}
	$\sim(SaP)$	{D3.1, 6}

T5.2. $\sim(SaP) \supset SoP$.

Proof.	(1) $\sim(SaP)$	{premise}
	(2) $\sim([x]\{x \varepsilon S \supset x \varepsilon P\})$	{D3.1, 1}
	(3) $[\exists x]\{\sim(x \varepsilon S \supset x \varepsilon P)\}$	{NΠ : 2)
	(4) $\sim(x_1 \varepsilon S \supset x_1 \varepsilon P)$	{OΣ : 3}
	(5) $x_1 \varepsilon S \wedge \sim(x_1 \varepsilon P)$	{4}
	(6) $[\exists x]\{x \varepsilon S \wedge \sim(x \varepsilon P)\}$	{DΣ : 5}
	SoP	{D3.2, 6}

T6.2. $SoP \equiv \sim(SaP)$. {T4.2, T5.2}

This theorem is equiform with expression (b).

T7.2. $MaP \wedge SaM \supset SaP$.

Proof. (1) MaP } {premise}
(2) SaM }
(3) $[x]\{x \varepsilon M \supset x \varepsilon P\}$ {D3.1, 1}
(4) $[x]\{x \varepsilon S \supset x \varepsilon M\}$ {D3.1, 2}
(5) $x \varepsilon M \supset x \varepsilon P$ {OΠ : 3}
(6) $x \varepsilon S \supset x \varepsilon M$ {OΠ : 4}
(7) $x \varepsilon S \supset x \varepsilon P$ {6, 5}
(8) $[x]\{x \varepsilon S \supset x \varepsilon P\}$ {DΠ : 7}
SaP {D3.1, 8}

This theorem is equiform with expression (e).

T8.2. $MaP \wedge MiS \supset SiP$.

Proof. (1) MaP } {premise}
(2) MiS }
(3) $[x]\{x \varepsilon M \supset x \varepsilon P\}$ {D3.1, 1}
(4) $[\exists x]\{x \varepsilon M \wedge x \varepsilon S\}$ {D1.2, 2}
(5) $x_1 \varepsilon M \wedge x_1 \varepsilon S$ {OΣ : 4}
(6) $x_1 \varepsilon M \supset x_1 \varepsilon P$ {OΠ : 3}
(7) $x_1 \varepsilon S \wedge x_1 \varepsilon P$ {5, 6}
(8) $[\exists x]\{x \varepsilon S \wedge x \varepsilon P\}$ {DΣ : 7}
SiP {D1.2, 8}

T8.2. is equiform with expression (f).

Expressions (a), (b), (c), (e) and (f) are therefore theorems of elementary ontology. However, expression (d) is not a theorem of this system and moreover, a theorem of ontology is the negation of the proposition obtained from (d) on substituting the empty name (to be defined in II, § 4) for the variable 'S'. Notwithstanding, the following proposition is a theorem of ontology:

T9.2. $ex\,|S| \supset SiS$.

Proof. (1) $ex\,|S|$ {premise}
(2) $[\exists x]\{x \varepsilon S\}$ {D1.1, 1}
(3) $x_1 \varepsilon S$ {OΣ : 2}
(4) $x_1 \varepsilon S \wedge x_1 \varepsilon S$ {3}
(5) $[\exists x]\{x \varepsilon S \wedge x \varepsilon S\}$ {DΣ : 4}
SiS {D1.2, 5}

Let α be an arbitrary theorem of the system of Łukasiewicz and let α^* be an expression or a conjunction of expressions of the type '$ex\,|S|$'.

I assume that for every variable in α there exists an equiform variable in α^*, and inversely: for every variable in α^* there exists an equiform variable in α. I also assume that no two variables in α^* are equiform. Hence, if α is one of the expressions (e), (f), α^* takes the form

$$ex\,/M/ \wedge ex\,/P/ \wedge ex\,/S/.$$

I shall use this notation in formulating and proving

Theorem I. *If* α *is a theorem of the system of Łukasiewicz, then the expression:* $\ulcorner \alpha^* \supset \alpha \urcorner$ *is a theorem of elementary ontology.*

Proof. I shall prove this theorem in two parts:
(a) proving that if α is equiform with one of the expressions (a)–(f), then $\ulcorner \alpha^* \supset \alpha \urcorner$ is a theorem of ontology;
(b) proving that if α follows from an expression β or from expressions β and γ, then $\ulcorner \alpha^* \supset \alpha \urcorner$ follows from $\ulcorner \beta^* \supset \beta \urcorner$ or from the expressions $\ulcorner \beta^* \supset \beta \urcorner$ and $\ulcorner \gamma^* \supset \gamma \urcorner$.
(a). If α is equiform with one of the expressions (a), (b), (c), (e) or (f), then, as previously stated, α is a theorem of ontology. We then subjoin $\ulcorner \alpha^* \supset \alpha \urcorner$ to ontology on the basis of the following theorem of the calculus of propositions:

(I) $p \supset (q \supset p)$.

If, however, α is equiform with (d), then the expression $\ulcorner \alpha^* \supset \alpha \urcorner$ is equiform to T9.2.
(b). If α may be obtained by substitution from a certain theorem of the calculus of propositions, then $\ulcorner \alpha^* \supset \alpha \urcorner$ may be obtained by substitution from the theorem $\ulcorner t \supset \varphi \urcorner$, where t is a propositional variable differing in form from every variable in φ.[41]

I assume $\ulcorner \alpha' \supset \alpha \urcorner$ to be an arbitrary result of substitution applied to $\ulcorner \beta^* \supset \beta \urcorner$. Hence α is a result of substitution applied to β. The expression α' may differ from α^* only in that some of its factors may be equiform. However, the commutative and the associative laws of conjunction, together with the law

$$p \wedge p \wedge q \equiv p \wedge q$$

yield the equivalence of α' and α^*. Hence, if α is a result of substitution applied to β, then $\ulcorner \alpha^* \supset \alpha \urcorner$ is a consequence of the expression $\ulcorner \beta^* \supset \beta \urcorner$.

I now assume that α may be obtained from the expressions β and γ by applying the rule of detachment, the expression γ being equiform to $\ulcorner \beta \supset \alpha \urcorner$.

In the theorem of the calculus of propositions:

$$(p \supset (q \supset r)) \supset ((s \supset q) \supset (p \wedge s \supset r))$$

[41] The conclusion arrived at can be expressed as follows: if the expression a of the system of Łukasiewicz follows from an empty set of expressions of that system, then the expression $\ulcorner \alpha^* \supset \alpha \urcorner$ of elementary ontology follows from an empty combination of expressions of that system.

I substitute the expressions γ^*, β, α and β^* for the variables 'p', 'q', 'r' and 's', respectively. Hence, the expression $\ulcorner\gamma^* \wedge \beta^* \supset \alpha\urcorner$ is a consequence of the expressions $\ulcorner\gamma^* \supset \gamma\urcorner$ and $\ulcorner\beta^* \supset \beta\urcorner$. If the expression $\ulcorner\gamma^* \wedge \beta^*\urcorner$ contains variables differing in form from all variables of the expression α, I substitute for them a variable equiform to an arbitrary variable of that expression. It is obvious that the expression α' obtained in this way may differ from α^* only in that it may contain equiform factors. The expressions α' and α^* are thus equivalent. It follows from this that the expression $\ulcorner\alpha^* \supset \alpha\urcorner$ is a consequence of the expressions $\ulcorner\beta^* \supset \beta\urcorner$ and $\ulcorner\gamma^* \supset \gamma\urcorner$.

The above proof, together with the remark that the rules of substitution and detachment are the only rules of inference in the system of Łukasiewicz, yields theorem I.

As already mentioned in § 1 above, Leśniewski defines the general-affirmative proposition of syllogistics in two ways. One of his definitions is D3.1 and the other is as follows:

D4.2. $Sa^*P \equiv [\exists x]\{x \varepsilon S\} \wedge [x]\{x \varepsilon S \supset x \varepsilon P\}$.[42]

According to this definition the general-affirmative proposition is false if its subject is an empty name, for then the proposition '$[\exists x]\{x \varepsilon S\}$' is false.

The following theorem might act as the definition of the term 'a^*'; however, its definiens would contain a defined term:

T10.2. $Sa^*P \equiv ex\,/S/ \wedge SaP$. {D4.2, D1.1, D3.1}

In order to explain how the set of theorems of elementary ontology in which the only functors of nominal arguments are the terms defined in D1.2–D4.2 is related to Aristotelian syllogistics I will describe briefly a certain system of Aristotelian syllogistics based on different axioms than Łukasiewicz's system.[43] In noting the theorems of this system I shall be using the symbol 'a^*' in place of 'a'.

The primitive terms of the system under discussion are the functors 'a^*' and 'i'. The definitions of the remaining functors of syllogistics are analogous to the corresponding definitions in Łukasiewicz's system.

(a′) $SeP \equiv \sim(SiP)$,
(b′) $SoP \equiv \sim(Sa^*P)$.

The axioms of the system are the following expressions:

(c′) $Sa^*P \supset SiP$,
(d′) $SiP \supset PiS$,
(e′) $Ma^*P \wedge Sa^*M \supset Sa^*P$,
(f′) $Ma^*P \wedge SiM \supset SiP$.

[42] Prof. Kotarbiński proposes to read the proposition 'SaP' as 'Any S is P', and the proposition 'Sa^*P' as 'All S is P'. (Kotarbiński [1929, p. 232 and 233]). (*Editorial Note*: Also Kotarbiński [1966, pp. 186–187].)

[43] This system is discussed in Słupecki [1946].

The rules of this system coincide with those of the system of Łukasiewicz. These two systems are thus clearly based on similar assumptions. All moods of the syllogism as well as all the laws of the square of opposition and of conversion are theorems of the described system; however, expressions (c) and (d) are not theorems of that system.

Expression (a′) is equiform with T3.2. It is easily proved that expressions (c′)–(f′) are also theorems of elementary ontology. All moods of the syllogism and all laws of the square of opposition and of conversion in which the symbol 'o' does not appear follow from these expressions. Hence, all these laws are theorems of elementary ontology provided the general-affirmative propositions appearing therein are given meanings in accordance with D4.2.

Expression (b′) is not a theorem of elementary ontology. However, the expression

$$\text{(b'')}\qquad ex\,/S/ \supset (SoP \equiv {\sim}(Sa^*P))$$

is a theorem of this system.

The fact that expressions (a′), (b″) and (c′)–(f′) are theorems of elementary ontology yields a theorem analogous to I which treats expressions of elementary ontology differing from those treated in theorem I insofar only, as they contain the symbol 'a'* in place of 'a'.

The remarks presented hitherto in this paragraph lead to the conclusion that it is not possible to include all theorems of Aristotelian syllogistics in elementary ontology if the general-affirmative proposition is to have either the meaning corresponding to D3.1 or that corresponding to D4.2. But whichever of these two meanings we choose, all theorems of syllogistics, when supplemented with antecedents of the type '$ex\,/S/$' or with a conjunction of expressions of this type, will be theorems of elementary ontology.

The next two theorems state the conditions for the equivalence of expressions 'SaP' and 'Sa^*P'.

T11.2. $ex\,/S/ \supset (SaP \equiv Sa^*P)$. {T10, 2}
T12.2. $x \varepsilon \mathrm{V} \supset (xaS \equiv xa^*S)$. {T17.1, T11.2}

The ensuing theorems treat of the relation holding between the singular proposition and the affirmative propositions of syllogistics and that between the negation of a singular proposition and negative propositions.

T13.2. $x \varepsilon \mathrm{V} \supset (x \varepsilon S \equiv xaS)$. {T5.1, T17.1, T6.1}
T14.2. $x \varepsilon \mathrm{V} \supset (x \varepsilon S \equiv xa^*S)$. {T13.2, T12.2}

I omit the proofs of subsequent theorems of this paragraph.

T15.2. $x \varepsilon \mathrm{V} \supset (x \varepsilon S \equiv xiS)$.
T16.2. $x \varepsilon \mathrm{V} \supset ({\sim}(x \varepsilon S) \equiv xeS)$.
T17.2. $x \varepsilon \mathrm{V} \supset ({\sim}(x \varepsilon S) \equiv xoS)$.

On the basis of the above theorems it is easy to prove the following two theorems which exemplify theorems following from the moods of syllogism by replacing in the latter affirmative propositions by singular propositions, and negative propositions by negations of singular propositions.

T18.2. $SaP \wedge x \varepsilon S \supset x \varepsilon P$.
T19.2. $x \varepsilon S \wedge PeS \supset \sim(x \varepsilon P)$.

The ensuing three theorems establish relations between terms of syllogistics and terms defined in I, § 1.

T20.2. $SiP \supset ex\,|S|$.
T21.2. $ex\,|S| \equiv SiS$.
T22.2. $x \varepsilon X \equiv xiX \wedge \rightharpoonup |x|$.

I shall make use of T22.2 in IV, § 3, when proving that the term '*i*' can act as the only primitive term of ontology.

In IV, §§ 2 and 3, I shall refer to the following theorems:

T23.2. $SiP \supset PiS$,
T24.2. $[x]\{xaS \equiv xaP\} \supset [x]\{x \varepsilon S \equiv x \varepsilon P\}$.

T23.2. is equiform to expression (d′).

§ 3. Equality of Extensions and Identity

The ensuing two expressions are definitions of terms the properties of which are to be investigated in this paragraph:

D1.3. $X \underset{Z}{=} Y \equiv [x]\{x \varepsilon X \equiv x \varepsilon Y\}$.
D2.3. $x = y \equiv x \varepsilon y \wedge y \varepsilon x$.

We read the definiendum in D1.3 as follows: the extension of X is equal to that of Y. For example: The extension of 'equilateral rectangle' is equal to that of 'square'. The intuitions from which this definition is derived coincide with those on which the definition of the equality of extensions of sets in the theory of sets is based. The expression '$X \underset{Z}{=} Y$' may be transformed into a true proposition on substituting general, empty or singular names for the variables.

We read the definiendum in D2.3 as follows: 'x and y are identical'' According to this definition the capital of Poland and Warsaw are identical if and only if the capital of Poland is Warsaw and Warsaw is the capital of Poland. The definition of identity in Leśniewski's system is based on intuitions that differ distinctly from those on which the definition of this term is based in other systems of logic. The expression '$x = y$' can become a true proposition only if singular names are substituted for the variables. The two relations defined above are symmetrical and transitive; the relation of equality of extensions is also reflexive.

The expression '$x = x$' is not a theorem of elementary ontology since after substituting in it an empty or general name for the variable 'x' it turns into a false proposition. However, the following expression is a theorem of ontology:

T1.3. $x \varepsilon \mathrm{V} \supset x = x$.

I omit the proofs of this and of subsequent theorems in this paragraph, since they present no serious difficulties.

The ensuing theorem establishes the relation between the equality of extensions and the general-affirmative proposition of Aristotelian syllogistics.

T2.3. $XaY \wedge YaX \supset X \underset{Z}{=} Y$.

The ensuing two theorems establish the relation between equality of extensions and identity.

T3.3. $x = y \equiv x \varepsilon \mathrm{V} \wedge x \underset{Z}{=} y$.
T4.3. $x \varepsilon \mathrm{V} \supset (x = y \equiv x \underset{Z}{=} y)$. {T3.3.}

Let α_t denote an arbitrary propositional expression of elementary ontology not containing any defined terms of this system, and containing the free variable t, but neither x nor y. Further, let α_x and α_y denote expressions obtained from α_t by substituting for $t: x$ and y, respectively. I shall make use of this notation formulating and proving the following

Theorem II. *The expression*

(I) $x = y \wedge \alpha_x \supset \alpha_y$

is a theorem of elementary ontology.

Proof. I assume the following definition to have been attached to ontology:

(II) $t \varepsilon \varphi|z, \ldots, u| \equiv t \varepsilon t \wedge \alpha_t$.

Further, I assume that $t, z, \ldots, u$ are the only free variables in α_t.

I shall write the ensuing part of the proof in the same way as the proofs of theorems of the system:

(III)	$x = y$	{premise}
(IV)	α_x	{premise}
(V)	$x \varepsilon y$	{D2.3, III}
(VI)	$x \varepsilon x$	{T9.1, V}
(VII)	$x \varepsilon \varphi\|z, \ldots, u\|$	{II, VI, IV}
(VIII)	$y \varepsilon \varphi\|z, \ldots, u\|$	{D2.3, T7.1, III, VII}
	α_y	{II, VIII}

Thus, expression (I) is a theorem of elementary ontology.

The assumption that α_i does not contain any defined terms of ontology by no means restricts the generality of the theorem proved. As in IV, § 1 I shall prove every expression of ontology to be equivalent to a certain expression not containing defined terms.

A theorem analogous to theorem II, but treating of equality of extensions instead of identity, is also true. Its proof involves induction and is much longer than that of theorem II. In III, § 2, I shall prove that a certain expression (not an expression of elementary ontology) is a theorem, which will make it possible to ascertain at once the truth of a generalization of the theorem discussed above. For these reasons I omit the proof of this theorem.

I shall now give two theorems containing the term of identity

T5.3. $x = y \equiv x \varepsilon y \wedge y \varepsilon \mathrm{V}$.
T6.3. $x = y \equiv x \varepsilon y \wedge \neg /y/$.

T5.3. and T6.3 could act as definitions of the term of identity. In both, however, the definiens would contain defined terms.

D3.3. $x \neq y \equiv x \varepsilon x \wedge y \varepsilon y \wedge \sim(x \varepsilon y)$.

We read the definiendum of D3.3 as follows: x and y are not identical.

T7.3. $x \neq y \equiv x \varepsilon \mathrm{V} \wedge y \varepsilon \mathrm{V} \wedge \sim(x = y)$. {D3.3, D2.3, T15.1, T10.1}

In IV I shall repeatedly make use of the following two theorems stating that the relation of identity is symmetrical and transitive.

T8.3. $x = y \supset y = x$.
T9.3. $x = y \wedge y = z \supset x = z$.

Other theorems, containing the terms defined in this paragraph, are given in III, § 2.

§ 4. The Algebra of Names

Among theorems of elementary ontology Leśniewski distinguishes those having analogues in the algebra of sets. I shall denote the class of these theorems as the algebra of names. The difference between the algebra of sets and the algebra of names is the following: the variables of the former represent sets, those of the latter are nominal variables. In building up the algebra of names, I shall use only terms analogous to the terms denoting the equality of sets, the inclusion of sets, the universal and the empty set, the sum, the product and the complement of sets.

These will be the only constant terms (apart from the functors of the calculus of propositions) to appear in the theorems of the algebra of names.

The ontological analogues of the terms denoting the equality of sets, the inclusion of sets and the universal set are: the term denoting the equality of extensions, the term 'a' of Aristotelian syllogistics and the universal name, respectively. These terms have been defined in preceding sections.[44] Further definitions of the algebra of names are:

D1.4. $x \varepsilon \Lambda \equiv x \varepsilon x \wedge \sim(x \varepsilon x)$,
D2.4. $x \varepsilon X+Y \equiv x \varepsilon x \wedge (x \varepsilon X \vee x \varepsilon Y)$,
D3.4. $x \varepsilon X \cdot Y \equiv x \varepsilon X \wedge x \varepsilon Y$,
D4.4. $x \varepsilon X' \equiv x \varepsilon x \wedge \sim(x \varepsilon X)$.

We read the definienda of these definitions as follows:

x is a non-object,
x is an X or an Y,
x is an X and an Y,
x is a non-X.

The terms defined in D1.4–D4.4. are: the empty name, the sign of the alternative of names, of the conjunction of names and of the negation of names, respectively.

Here now are four theorems closely connected with these definitions:

T1.4. $x \varepsilon \Lambda \equiv x \varepsilon \mathbf{V} \wedge \sim(x \varepsilon x)$. {D1.4, T15.1}
T2.4. $x \varepsilon X' \equiv x \varepsilon \mathbf{V} \wedge \sim(x \varepsilon X)$. {D4.4, T15.1}
T3.4. $x \varepsilon \mathbf{V} \supset (x \varepsilon X' \equiv \sim(x \varepsilon X))$. {T2.4}
T4.4. $x \varepsilon X+Y \equiv (x \in X \vee x \varepsilon Y)$.

T4.4 indicates that the factor '$x \varepsilon x$', attached to the definiens in D2.4 according to general rules of forming definitions in elementary ontology, is in this case irrelevant. I omit the proof of T4.4 and of all subsequent theorems in this paragraph.

Let small Greek letters denote expressions of the calculus of propositions enlarged so as to include the terms '1' and '0', abbreviations that stand for an arbitrary true and an arbitrary false proposition, respectively. I assume that in these expressions only the functors of alternative, conjunction and negation may appear. I further assume that the propositional variables in the expressions to be discussed below are the symbols:

$$p_1, p_2, p_3, \ldots$$

and the nominal variables in these expressions are the symbols:

$$x, X_1, X_2, X_3, \ldots$$

Let α^* denote the result of replacing the symbols '1', '0' and 'p_i' in the expression α by '$x \varepsilon \mathbf{V}$', '$x \varepsilon \Lambda$' and '$x \varepsilon X_i$', respectively; let X_α denote

[44] These definitions are D1.3, D3.1 and D4.1.

the nominal expression obtained on substituting the symbols '1', '0', '$\vee$', '$\wedge$', '$\sim$' and 'p_i' in the expression α by 'V', 'Λ', '$+$', '$\cdot$', '$'$', and 'X_i', respectively. For example, the expression

$$\sim(p_1 \wedge 1) \vee p_2$$

yields the following expressions α^* and X_α:

$$\sim(x \varepsilon X_1 \wedge x \varepsilon \mathrm{V}) \vee x \varepsilon X_2$$
$$|X_1 \cdot \mathrm{V}|' + X_2.$$

I shall be employing these symbols in this section.

I am now proceeding to prove theorems connecting the calculus of propositions with the algebra of names. The connection established thus is a perfect analogy of that existing between the calculus of propositions and the algebra of sets. The proofs of lemmas and theorems I give, however, though closely resembling analogous proofs in the methodology of the algebra of sets, differ from them essentially in several points.[45]

Lemma I. *The following formulae hold*:

$$\ulcorner \sim(\alpha) \urcorner^* = \ulcorner \sim(\alpha^*) \urcorner,$$
$$\ulcorner \alpha \wedge \beta \urcorner^* = \ulcorner \alpha^* \wedge \beta^* \urcorner,$$
$$\ulcorner \alpha \vee \beta \urcorner^* = \ulcorner \alpha^* \vee \beta^* \urcorner.$$

I omit the easy proof of this lemma.

Lemma II. *The following formulae hold*:

$$X_{\sim(\alpha)} = X'_\alpha,$$
$$X_{\alpha \vee \beta} = X_\alpha + X_\beta,$$
$$X_{\alpha \wedge \beta} = X_\alpha \cdot X_\beta.$$

I omit the equally easy proof of this lemma.

Lemma III. *For any expression* α, *the expression*

(I) $x \varepsilon \mathrm{V} \supset (\alpha^* \equiv x \varepsilon X_\alpha)$

is a theorem of elementary ontology.

Proof. The lemma is proved by induction on the number of proposition-formative functors appearing in α. I shall first prove that lemma III holds when α does not contain any proposition-formative functors. I will discuss three cases:

(a) the expression α is a propositional variable,
(b) ,, ,, α is the constant '1',
(c) ,, ,, α is the constant '0'.

I note the proof of lemma III in the same way as the proofs of theorems.

(II) $x \varepsilon \mathrm{V}$ {premise}

---

[45] The theorems of the methodology of the algebra of sets discussed here are proved in Mostowski [1948, pp. 97–100].

I prove lemma III for case (a):

(III) $\alpha = `p_i'$ {premise of the proof}

(IV) $\alpha^* = `x \varepsilon X_i'$ {III}

(V) $X_\alpha = X_i$ {III}

(VI) $x \varepsilon X_\alpha \equiv x \varepsilon X_i$ {V}

$\alpha^* \equiv x \varepsilon X_\alpha$ {IV, VI}

Hence lemma III is true for case (a). I now prove the lemma for case (b):

(VII) $\alpha = `1'$ {premise of the proof}

(VIII) $\alpha^* = `x \varepsilon V'$ {VII}

(IX) $X_\alpha = V$ {VII}

(X) $x \varepsilon X_\alpha \equiv x \varepsilon V$ {IX}

$\alpha^* \equiv x \varepsilon X_\alpha$ {VIII, X}

Hence lemma III is also true for case (b). The proof for case (c) is quite similar to that for (b). The first part of the proof by induction has thus been completed.

Let k be any natural number. I assume that α contains k proposition-formative functors, and that lemma III is true for all expressions a containing less than k proposition-formative functors. Here again I have three cases to discuss:

(A) there is an expression β fulfilling the equation
$\alpha = \ulcorner \sim(\beta) \urcorner$

(B) there are expressions β and γ fulfilling the equation
$\alpha = \ulcorner \beta \vee \gamma \urcorner$

(C) there are expressions β and γ fulfilling the equation
$\alpha = \ulcorner \beta \wedge \gamma \urcorner$.

It is evident that the expressions β and γ contain less than k proposition-formative functors.

I now proceed to prove lemma III for case (A):

(XI) $\alpha = \ulcorner \sim(\beta) \urcorner$ {premise of the proof}

(XII) $\beta^* \equiv x \varepsilon X_\beta$ {premise of induction}

(XIII) $\alpha^* = \ulcorner \sim(\beta) \urcorner^*$ {XI}

(XIV) $\alpha^* = \ulcorner \sim(\beta^*) \urcorner$ {XIII, lemma I}

(XV) $X_\alpha = X_{\sim(\beta)}$ {XI}

(XVI) $X_\alpha = X'_\beta$ {XV, lemma II}

(XVII) $\sim(\beta^*) \equiv \sim(x \varepsilon X_\beta)$ {XII}

(XVIII) $\sim(x \varepsilon X_\beta) \equiv x \varepsilon X'_\beta$ {T3.4, II}

$\alpha^* \equiv x \varepsilon X_\alpha$ {XIV, XVII, XVIII, XVI}

Thus, lemma III is true in case (A). I shall now prove it for (B):

(XIX) $\alpha = \ulcorner \beta \vee \gamma \urcorner$ {premise of the proof}

(XX) $\gamma^* \equiv x \varepsilon X_\gamma$ {premise of induction}

(XXI) $\alpha^* = \ulcorner \beta \vee \gamma \urcorner^*$ {XIX}

(XXII)	$\alpha^* = \ulcorner\beta^* \vee \gamma^*\urcorner$	{XXI, lemma I}
(XXIII)	$X_\alpha = X_{\beta\vee\gamma}$	{XIX}
(XXIV)	$X_\alpha = X_\beta + X_\gamma$	{XXIII, lemma II}
(XXV)	$(\beta^* \vee \gamma^*) \equiv (x \varepsilon X_\beta \vee x \varepsilon X_\gamma)$	{XII, XX}
(XXVI)	$(x \varepsilon X_\beta \vee x \varepsilon X_\gamma) \equiv x \varepsilon X_\beta + X_\gamma$	{T4.4}
	$\alpha^* \equiv x \varepsilon X_\alpha$	{XXII, XXV, XXVI, XXIV}

Thus, lemma III is also true in case (B). The proof of lemma III for case (C) is quite similar to that for case (B). The second part of the proof by induction has therefore also been completed; hence expression (I) is a theorem of elementary ontology.

Theorem III. *If the expression* $\ulcorner\alpha \supset \beta\urcorner$ *is a theorem of the calculus of propositions enlarged so as to include the symbols* '1' *and* '0', *then the expression* $\ulcorner X_\alpha a X_\beta\urcorner$ *is a theorem of elementary ontology.*

Proof. Let $\ulcorner\alpha \supset \beta\urcorner$ be a theorem of the enlarged calculus of propositions. The expression $\ulcorner\alpha^* \supset \beta^*\urcorner$ is a result of substitution applied to $\ulcorner\alpha \supset \beta\urcorner$ and hence a theorem of elementary ontology. In the following theorem of the calculus of propositions:

$$(p \supset (q \equiv r)) \supset \Big((p \supset (s \equiv t)) \supset \big((q \supset s) \supset ((r \supset p) \supset (r \supset t))\big)\Big)$$

let us substitute the expressions '$x \varepsilon$ V', α^*, '$\ulcorner x \varepsilon X_\alpha\urcorner$', β^* and '$\ulcorner x \varepsilon X_\beta\urcorner$' for the variables '$p$', '$q$', '$r$', '$s$' and '$t$', respectively. From lemma III and the fact that the expressions $\ulcorner\alpha^* \supset \beta^*\urcorner$ and T14.1 are theorems of elementary ontology, it follows that the expression $\ulcorner\{x \varepsilon X_\alpha \supset x \varepsilon X_\beta\}\urcorner$ is also a theorem of this system. This together with rule DΠ leads to the conclusion that the expression $\ulcorner[x]\{x \varepsilon X_\alpha \supset x \varepsilon X_\beta\}\urcorner$, a result of substitution applied to the definiens of D3.1, is a theorem of elementary ontology. This leads immediately to theorem III.

Theorem IV. *If the expression* $\ulcorner\alpha \equiv \beta\urcorner$ *is a theorem of the calculus of propositions enlarged so as to include the symbols* '1' *and* '0', *then the expression* $\ulcorner X_\alpha \underset{Z}{=} X_\beta\urcorner$ *is a theorem of elementary ontology.*

Proof. I assume the expression $\ulcorner\alpha \equiv \beta\urcorner$ to be a theorem of the enlarged calculus of propositions. In that case the expressions $\ulcorner\alpha \supset \beta\urcorner$ and $\ulcorner\beta \supset \alpha\urcorner$ are also theorems of this system. This, together with theorem III, leads to the conclusion that expressions $\ulcorner X_\alpha a X_\beta\urcorner$ and $\ulcorner X_\beta a X_\alpha\urcorner$ are theorems of elementary ontology. This together with the fact that expression T2.3 is a theorem of this system, proves the truth of theorem IV.

By way of example, I give certain theorems of the algebra of names, placing alongside each of them in braces the corresponding theorem of the calculus of propositions enlarged to include the terms '1' and '0'.

T5.4.	XaV.	$\{p \supset 1\}$
T6.4.	$\Lambda a X$.	$\{0 \supset p\}$

T7.4. $XaX + Y$. $\{p \supset (p \vee q)\}$
T8.4. $X \cdot YaX$. $\{p \wedge q \supset p\}$
T9.4. $V' \underset{Z}{=} \Lambda$. $\{1' \equiv 0\}$
T10.4. $X + X' \underset{Z}{=} V$. $\{(p \vee \sim(p)) \equiv 1\}$
T11.4. $X \cdot X' \underset{Z}{=} \Lambda$. $\{(p \wedge \sim(p)) \equiv 0\}$
T12.4. $/X'/' \underset{Z}{=} X$. $\{\sim(\sim p) \equiv p\}$

The ensuing theorems do not belong to the calculus of names. They contain, however, terms the definitions of which have been given in this paragraph. First a few theorems characteristic of the sign of negation.

T13.4. $\sim(x \varepsilon x')$.

This theorem is a variant of the law of double negation.

T14.4. $x \varepsilon X' \supset \sim(x \varepsilon X)$. {D4.4}
T15.4. $\sim(x \varepsilon X \wedge x \varepsilon X')$. {T14.4}
T16.4. $x \varepsilon V \supset (x \varepsilon X \vee x \varepsilon X')$.

The last two theorems are variants of the laws of contradiction and of excluded middle. I wish to point out that on omitting the antecedent '$x \varepsilon V$', T16.4 ceases to be a theorem of ontology, as its consequent turns into a false proposition if an empty or general name is substituted in it for the variable 'x'.

The following theorems serve to connect the functor '*ex*' with the terms defined in this paragraph.[46]

T17.4. $ex\,/X/ \equiv \sim(X \underset{Z}{=} \Lambda)$.

According to T17.4. there exists at least one X if and only if X and the empty name are not extensionally equal.

T18.4. $\sim(ex\,/\Lambda/)$.

We read this theorem as follows: the non-object does not exist.

The following three theorems serve to connect the general-affirmative proposition of Aristotelian syllogistics with the product of names appearing in that proposition.

T19.4. $SiP \equiv ex\,/S \cdot P/$.
T20.4. $SiP \equiv \sim(S \cdot P \underset{Z}{=} \Lambda)$.

The last theorems of elementary ontology I include here establish relations between the functor '$\neq$'[47] and the sign of nominal negation.

T21.4. $x \neq y \equiv x \varepsilon y' \wedge y \varepsilon y$. {D3.3, D4.4}
T22.4. $x \neq y \equiv [\exists X]\{x \varepsilon X' \wedge y \varepsilon X\}$.

This theorem could act as definition of the sign '$\neq$', the definiens, however, would contain a defined term.

---

[46] The definition of the term '*ex*' is D 1.1.
[47] The definition of this functor is D 3.3.

III

§ 1. General Remarks on Non-Elementary Ontology

In this chapter I shall present definitions and theorems lying outside the scope of elementary ontology. As mentioned in the Introduction, this paper is not meant to contain a systematic presentation of non-elementary ontology. In consequence, most of the definitions and theorems given below serve only as examples characterizing Leśniewski's entire system.

The language of ontology has been discused in I, § 1. However, the remarks I made in I, § 2 on the symbols employed in this paper need to be supplemented. In the notation of theorems of non-elementary ontology I shall be following Leśniewski's original symbolism more closely than I have done in the preceding chapter. This will enable the reader to make himself acquainted with the principles of Leśniewski's symbolism which are of a high degree of precision. However, the symbolism used above in the notation of theorems of elementary ontology will be kept unchanged. In particular functors of two nominal arguments as well as functors of two arguments of the calculus of proposition will be written between the arguments. But all functors not appearing in theorems of elementary ontology will precede their arguments and the latter will be enclosed in parentheses the shape of which together with the number of arguments determines the semantic category of the functor. This will enable us to use equiform symbols for functors having analogous properties though belonging to different semantic categories. I wish to point out that functors (constants as well as variables) which are not functors of the calculus of propositions and the semantic category of which differs from that of the functors appearing in the axiom of ontology, may appear in theorems of ontology only if the definition of a functor of that semantic category has already been attached to the system. In the definition of a functor, the semantic category of which is not represented in previous theorems of the system, the arguments should be enclosed in parentheses differing in shape from all those used in the notation of earlier theorems of the system. This, however, does not hold where all the arguments of the functor to be defined belong to the same semantic category and the system already includes functors of a different number of arguments that belong without exception to the same semantic category as the arguments of the functor to be defined. Arguments of functors belonging to the same semantic category are always enclosed in parentheses printed in ordinary type. I shall also be using parentheses in bold type considering them as differing in form from the former.

In enlarging elementary ontology to form a system, the theorems of which may include functors of any order[48], we are neither obliged to introduce new axioms nor to attach new primitive terms to the system. It is, however, necessary to strengthen the rules of inference of elementary ontology. First of all, the rule of substitution is generalized. Namely the scope of expressions which may be substituted for the variables, as well as that of the variables for which we perform the substitutions, is enlarged. The rule of substitution valid in Leśniewski's systems of logic allows to substitute for variables of any semantic category variables or constants of the same semantic category. In particular, we may substitute variable or constant functors for variables representing functors. However, in Leśniewski's systems it is not possible to effect substitution for composite expressions, as exemplified by expressions formed of a variable functor and all its arguments. Neither is it permissible in Leśniewski's systems to substitute composite expressions not belonging to any semantic category, even if the expression obtained by such a substitution were to be a meaningful expression of the system. Thus, it is not permissible to substitute expressions formed of a functor and only some of its arguments.

In proving theorems of non-elementary ontology I shall also be applying the rule of relettering the variables; this rule is however secondary with regard to the primitive rules of the system (in the entire system just as in elementary ontology). I may also mention that the rules referring to the use of quantifiers are analogous to the corresponding rules in elementary ontology.

The rule of attaching definitions to the system differs essentially from the corresponding rule of elementary ontology. In definitions other than those of elementary ontology only functors are being defined. As follows from the description of the language of Leśniewski's system in I, § 1, these functors are either proposition-formative or name-formative. The schemata of the definitions of these functors do not differ essentially from schemata (I) and (II) of the definitions of the functors of elementary ontology as given in I, § 5. However, the functors of non-elementary ontology, in contradistinction to those of elementary ontology, may depend on parameters. Such is, for instance, the functor defined in D4. II of the next paragraph. Parameters of functors are always placed inside parentheses differing in shape from those enclosing arguments of functors. The shape of the parentheses used to enclose parameters depends upon the semantic categories of these parameters, just as the shape of parentheses used to enclose arguments is chosen according to the semantic categories of the latter. I further note that for every variable acting as parameters of a functor, the definiens of the definition of this functor includes a free variable equiform to the former, and that every variable acting as parameter differs in

[48] The term 'order of a functor' has been defined in I, § 1.

form from all other variables appearing in the definiendum. We effect substitution for variables acting as parameters in the same way as for all other variables appearing in expressions of ontology.

The part the definitions of functors depending on parameters play in the proofs of theorems of the system will be explained in ensuing paragraphs of this chapter as well as in IV, § 1.

In order to facilitate the reading of definitions of non-elementary ontology I shall be using also defined terms in their definienses. Such definitions should be regarded as abbreviations of the proper definitions.

A rule valid in ontology, which however I did not apply in proving theorems of elementary ontology, is the rule of extensionality. According to this rule expressions of a specified form may be subjoined to the system irrespective of what theorems were subjoined thereto before. The rule of extensionality is thus an axiomatic rule.

I shall denote expressions subjoined to the system on the basis of the rule of extensionality as laws of extensionality. I proceed to give a description of these expressions. Let the variable Φ represent proposition-formative functors of one argument and let the not equiform variables φ and ψ be arguments of the functor. The laws of extensionality are implications, the consequents of which have the following structure:

(I) $[\Phi]\{\Phi\,/\varphi/ \equiv \Phi\,/\psi/\}$.

The parentheses I used here do not, of course, determine the semantic category of the functor Φ. The same observation holds for all expressions of the meta-language appearing in the text below.

The form of the antecedents in the laws of extensionality depends on the semantic category of the variables φ and ψ. If they are nominal variables not having the form of the variable x, the antecedent in the law of extensionality has the following structure:

$[x]\{x\,\varepsilon\,\varphi \equiv x\,\varepsilon\,\psi\}$.[49]

I now assume φ and ψ to be functors of k arguments. Let $\alpha_1, \alpha_2, \ldots, \alpha_k$ be the arguments of both φ and ψ. $\alpha_1, \alpha_2, \ldots, \alpha_k$ are variables of which no two are equiform, and none has the form of the variable x. If φ and ψ are name-formative functors, the antecedent in the law of extensionality has the following structure:

$[x, \alpha_1, \alpha_2, \ldots, \alpha_k]\{x\,\varepsilon\,\varphi|\alpha_1, \alpha_2, \ldots, \alpha_k| \equiv x\,\varepsilon\,\psi|\alpha_1, \alpha_2, \ldots, \alpha_k|\}$.

If φ and ψ are proposition-formative functors, the antecedent in the law of extensionality has the following structure:

$[\alpha_1, \alpha_2, \ldots, \alpha_k]\{\varphi\,/\alpha_1, \alpha_2, \ldots, \alpha_k/ \equiv \psi\,/\alpha_1, \alpha_2, \ldots, \alpha_k/\}$.

I write the proofs of theorems of non-elementary ontology in the same way as those of theorems of elementary ontology. It should be noticed that

[49] T 9.II is a theorem attached to the system according to a rule of extensionality, the antecedent of which possesses this structure.

all theorems and definitions in this chapter, just as those in the preceding chapter, are numbered doubly. The Arabic numeral denotes the successive number of the theorem or definition in the paragraph, while the Roman numeral stands for the number of that paragraph.

I may add that the proofs of the theorems of non-elementary ontology are obviously more complicated than those of the theorems of elementary ontology. For this reason in this section the proofs will only exceptionally be omitted.

§ 2. Theorems of Non-Elementary Ontology, in which the Symbols of Identity and Extensional Equality Appear

The rules of inference of elementary ontology, strengthened as explained in the preceding paragraph, enable us to prove several theorems containing the symbols of identity and extensional equality[50] which we could not prove by the rules of elementary ontology. These theorems are essential to the system of ontology.

I shall first prove several theorems in which the symbol of identity appears. For this purpose the following definition is required:

D1.II $x \varepsilon \mathit{stsf} [\![\varphi]\!] \equiv x \varepsilon \mathbf{V} \wedge \varphi /x/$.

We read the definiendum of this definition as follows: the object x fulfils the property φ. The name-formative functor defined in D 1.II does not belong to the language of elementary ontology since its argument is a proposition-formative functor of a nominal argument.

T1.II $x = y \wedge \varphi /x/ \supset \varphi /y/$.

Proof.	(1) $x = y$	{premise}
	(2) $\varphi /x/$	{premise}
	(3) $x \varepsilon \mathbf{V}$	{D2.3, T14.1, 1}
	(4) $x \varepsilon \mathit{stsf} [\![\varphi]\!]$	{D1.II, 3, 2}
	(5) $y \varepsilon x$	{D2.3, 1}
	(6) $y \varepsilon \mathit{stsf} [\![\varphi]\!]$	{T7.1, 4, 5}
	$\varphi /y/$	{D1.II, 6}

T2.II. $x = y \supset [\varphi]\{\varphi /x/ \equiv \varphi /y/\}$.

Proof.	(1) $x = y$	{premise}
	(2) $\varphi /x/ \supset \varphi /y/$	{T1.II, 1}
	(3) $y = x$	{T8.3, 1}

[50] The definitions of these terms have been given in II, § 3.

(4) $\varphi\,/y/ \supset \varphi\,/x/$ {T1.II, 3}
(5) $\varphi\,/x/ \equiv \varphi\,/y/$ {2, 4}
$[\varphi]\{\varphi\,/x/ \equiv \varphi\,/y/\}$ {DΠ : 5}

Leśniewski called this theorem the law of extensionality for identity. From the intuitive and formal point of view, this theorem is closely akin to theorems attached to the system on the basis of the rule of extensionality. It is, however, proved independently of this rule. The fact that T2.II is independent of the rule of extensionality impelled Leśniewski to take up semantic investigations on intentional functions, with the aim of eliminating these functions from current language without essentially impoverishing the latter.[51]

D2.II. $\varepsilon\,\langle X\rangle\,/x/ \equiv x\,\varepsilon\,X$.

The functor defined depends on a nominal parameter. The definitions D1.II and D2.II indicate that in ontology a name is coordinated to every proposition-formative functor of one nominal argument, and inversely: a proposition-formative functor of nominal argument is coordinated to every name.

T3.II. $x\,\varepsilon\,\mathrm{V} \wedge [\varphi]\{\varphi\,/x/ \equiv \varphi\,/y/\} \supset x = y$.

Proof. (1) $x\,\varepsilon\,\mathrm{V}$ } {premise}
(2) $[\varphi]\{\varphi\,/x/ \equiv \varphi\,/y/\}$ }
(3) $\varepsilon\,\langle x\rangle\,/x/ \equiv \varepsilon\,\langle x\rangle\,/y/$ {OΠ : 2}
(4) $x\,\varepsilon\,x \equiv y\,\varepsilon\,x$ {D2.II, 3}
(5) $x\,\varepsilon\,x$ {T9.1, 1}
(6) $y\,\varepsilon\,x$ {4, 5}
(7) $x\,\varepsilon\,y$ {T10.1, 1, 6}
$x = y$ {D2.3, 7, 6}

The above proof shows clearly the part played by definitions of functors depending on parameters in proofs of theorems of ontology. Expression (4), which in certain systems of logic[52] may be obtained by substitution from the expression derived from expression (2) by omitting the general quantifier, has here been obtained by applying definition D2.II to expression (3). Expression (3) is a result of substitution applied to the expression derived from expression (2) by omitting the general quantifier. The procedure of substitution consists here in replacing a variable functor by a functor depending on parameters.

T4.II. $x = y \equiv x\,\varepsilon\,\mathrm{V} \wedge [\varphi]\{\varphi\,/x/ \equiv \varphi\,/y/\}$ {D2.3, T14.1, T2.II, T3.II}

---

[51] Leśniewski dedicated a special course of lectures (under the heading Intentional Functions in Deductive Languages) to problems concerning intentional functions.

[52] Such a system is, for instance, the functional calculus given in Hilbert and Bernays [1934].

This theorem might act as definition of the symbol of identity. It indicates that the meaning of the relation of identity, as established in ontology, is in accordance with the Leibnizian notion of indiscernible objects.[53]

In the next three theorems appears the symbol '$\neq$'; its definition has been given in II, § 3.

T5.II. $x \neq y \supset [\exists\varphi]\{\sim(\varphi\,/x/) \wedge \varphi\,/y/\}$.

Proof.	(1) $x \neq y$	{premise}
	(2) $[\exists X]\{x \,\varepsilon\, X' \wedge y \,\varepsilon\, X\}$	{T22.4, 1}
	(3) $x \,\varepsilon\, X_1' \wedge y \,\varepsilon\, X_1$	{OΣ : 2}
	(4) $\sim(x \,\varepsilon\, X_1) \wedge y \,\varepsilon\, X_1$	{3, T14.4}
	(5) $\sim(\varepsilon\,\langle X_1\rangle\,/x/) \wedge \varepsilon\,\langle X_1\rangle\,/y/$	{4, D2.II}
	$[\exists\varphi]\{\sim(\varphi\,/x/) \wedge \varphi\,/y/\}$	{DΣ : 5}

The part played by D2.II in the proof of T5.II is different from that played in the proof of T3.II. In either case, however, this definition enables us to obtain from certain expressions other equivalent expressions, containing terms of a semantic category different from that of the terms appearing in the expressions from which the expressions obtained follow.

T6.II. $x \,\varepsilon\, \mathrm{V} \wedge y \,\varepsilon\, \mathrm{V} \wedge [\exists\varphi]\{\sim(\varphi\,/x/) \wedge \varphi\,/y/\} \supset x \neq y$.

Proof.	(1) $x \,\varepsilon\, \mathrm{V} \wedge y \,\varepsilon\, \mathrm{V}$	{premise}
	(2) $[\exists\varphi]\{\sim(\varphi\,/x/) \wedge \varphi\,/y/\}$	{premise}
	(3) $\sim(\varphi_1\,/x/) \wedge \varphi_1\,/y/$	{OΣ : 2}
	(4) $x = y \wedge \varphi_1\,/y/ \supset \varphi_1\,/x/$	{T1.II, T8.3}
	(5) $\sim(x = y)$	{3, 4}
	$x \neq y$	{T7.3, 1, 5}

T7.II. $x \neq y \equiv x \,\varepsilon\, \mathrm{V} \wedge y \,\varepsilon\, \mathrm{V} \wedge [\exists\varphi]\{\sim(\varphi\,/x/) \wedge \varphi\,/y/\}$.

{T7.3, T5.II, T6.II.}

This theorem may act as definition of the term '$\neq$'.

D3.II. $\varepsilon^*\langle x\rangle\,/X/ \equiv x \,\varepsilon\, X$.

This definition is analogous to D2.II. The ensuing theorems will contain the symbol of equality of extension.

T8.II. $[\varphi]\{\varphi\,/X/ \equiv \varphi\,/Y/\} \supset X \underset{Z}{=} Y$.

Proof.	(1) $[\varphi]\{\varphi\,/X/ \equiv \varphi\,/Y/\}$	{premise}
	(2) $\varepsilon^*\langle x\rangle\,/X/ \equiv \varepsilon^*\langle x\rangle\,/Y/$	{OΠ : 1}
	(3) $x \,\varepsilon\, X \equiv x \,\varepsilon\, Y$	{2, D3.II}
	(4) $[x]\{x \,\varepsilon\, X \equiv x \,\varepsilon\, Y\}$	{DΠ : 3}
	$X \underset{Z}{=} Y$	{D1.3, 4}

---

[53] This notion is mentioned in Mostowski [1948, p. 10].

The rules I applied in this paragraph in proving theorems differed from those of elementary ontology chiefly in so far as they made it possible to attach to the system definitions not belonging to the language of elementary ontology. The next theorem is attached to the system by applying the rule of extensionality.

T9.II. $[x]\{x \varepsilon X \equiv x \varepsilon Y\} \supset [\varphi]\{\varphi /X/ \equiv \varphi /Y/\}$. {rule of extensionality}

T10.II. $X \underset{Z}{=} Y \supset [\varphi]\{\varphi /X/ \equiv \varphi /Y/\}$. {T9.II, D1.3}

T11.II. $X \underset{Z}{=} Y \equiv [\varphi]\{\varphi /X/ \equiv \varphi /Y/\}$. {T10.II, T8.II}

This theorem, which is analogous to T4.II, may act as definition of the symbol of equality of extensions.

T12.II. $X \underset{Z}{=} Y \wedge \varphi /X/ \supset \varphi /Y/$.

Proof. (1) $X \underset{Z}{=} Y$ }{premise}
(2) $\varphi /X/$ }
(3) $[\varphi]\{\varphi /X/ \equiv \varphi /Y/\}$ {T10.II, 1}
(4) $\varphi /X/ \equiv \varphi /Y/$ {OΠ : 3}
$\varphi /Y/$ {4, 2}

Once theorem T12.II has been attached to the system the theorem on expressions of ontology in which the symbol of equality of extension appears may be proved (see II, § 3).

Let α_Z denote an arbitrary propositional expression of ontology, containing the free nominal variable Z, but not the variables X and Y. Moreover, let α_X and α_Y denote expressions obtained from α_Z by substituting in α_Z for Z the variables X and Y, respectively. I shall use the above notation in formulating and proving the following

Theorem V. *The expression*

(I) $X \underset{Z}{=} Y \wedge \alpha_X \supset \alpha_Y$

is a theorem of ontology.

Proof. I assume that the variables $\beta_1, \beta_2, \ldots, \beta_k$ are all the free variables in α_Z non-equiform to Z. I also assume that the following definition has been attached to the system:

(II) $\varphi\langle\beta_1, \beta_2, \ldots, \beta_k\rangle /Z/ \equiv \alpha_Z$.

The remaining part of the proof will be noted in the same way as the proofs of theorems of the system are noted.

(III) $X \underset{Z}{=} Y$ }{premise}
(IV) α_X }
(V) $\varphi\langle\beta_1, \beta_2, \ldots, \beta_k\rangle /X/$ {II, IV}
(VI) $\varphi\langle\beta_1, \beta_2, \ldots, \beta_k\rangle /Y/$ {T12.II, III, V}
α_Y {II, VI}

Expression (I) is thus a theorem of ontology.

It is also easily proved that expressions of a structure analogous to that of (I), the antecedents of which include, however, the symbol of identity in place of that of extensional equality, are theorems of ontology. The theorem stating this — a generalization of theorem II the proof of which was given in II § 3 — may be proved now owing to the fact that T1.II has been attached to the system.

I now proceed to give further definitions and theorems of non-elementary ontology.

D4.II. $*\langle f, x\rangle /X/ \equiv x \varepsilon f|X|$.

T13.II. $X \underset{Z}{=} Y \supset [f]\{f|X| \underset{Z}{=} f|X|\}$.

Proof.	(1) $X \underset{Z}{=} Y$	{premise}
	(2) $[\varphi]\{\varphi /X/ \equiv \varphi /Y/\}$	{T10.II, 1}
	(3) $*\langle f, x\rangle /X/ \equiv *\langle f, x\rangle /Y/$	{OΠ : 2}
	(4) $x \varepsilon f|X| \equiv x \varepsilon f|Y|$	{3, D4.II}
	(5) $[x]\{x \varepsilon f|X| \equiv x \varepsilon f|Y|\}$	{DΠ : 4}
	(6) $f|X| \underset{Z}{=} f|Y|$	{D1.3, 5}
	$[f]\{f|X| \underset{Z}{=} f|Y|\}$	{DΠ : 7}

D5.II. $x \varepsilon *|X| \equiv x \varepsilon X$.

T14.II. $[f]\{f|X| \underset{Z}{=} f|Y|\} \supset X \underset{Z}{=} Y$.

Proof.	(1) $[f]\{f|X| \underset{Z}{=} f|Y|\}$	{premise}
	(2) $*|X| \underset{Z}{=} *|Y|$	{OΠ : 1}
	(3) $[x]\{x \varepsilon *|X| \equiv x \varepsilon *|Y|\}$	{D1.3, 2}
	(4) $x \varepsilon *|X| \equiv x \varepsilon *|Y|$	{OΠ : 3}
	(5) $x \varepsilon X \equiv x \varepsilon Y$	{4, D5.II}
	(6) $[x]\{x \varepsilon X \equiv x \varepsilon Y\}$	{DΠ : 5}
	$X \underset{Z}{=} Y$	{D1.3, 6}

T15.II. $X \underset{Z}{=} Y \equiv [f]\{f|X| \underset{Z}{=} f|Y|\}$ {T13.II, T14.II}

T16.II. $x = y \equiv x \varepsilon V \wedge [f]\{f|x| \underset{Z}{=} f|y|\}$ {T3.3, T15.II}

The last two theorems are analogous to T11.II and T4.II. They contain, however, name-formative functors in place of proposition-formative functors.

The following exemplifies a proof based on T16.II.

T17.II. $x = y \supset x' \underset{Z}{=} y'$.

Proof.	(1) $x = y$	{premise}
	(2) $[f]\{f|x| \underset{Z}{=} f|y|\}$	{T16.II, 1}
	$x' \underset{Z}{=} y'$	{OΠ : 2}

I shall now prove a theorem analogous to T9.II, containing proposition-formative functors of two nominal arguments. The proof requires the following two auxiliary definitions:

D6.II. $*\langle\varphi, Y\rangle /X/ \equiv \varphi /X, Y/$.
D7.II. $**\langle\varphi, X\rangle /Y/ \equiv \varphi /X, Y/$.

T17.II. $[x]\{x \varepsilon X \equiv x \varepsilon Z\} \wedge [x]\{x \varepsilon Y \equiv x \varepsilon U\} \supset [\varphi]\{\varphi /X, Y/ \equiv \varphi /Z, U/\}$.

Proof.	(1) $[x]\{x \varepsilon X \equiv x \varepsilon Z\}$	{premise}
	(2) $[x]\{x \varepsilon Y \equiv x \varepsilon U\}$	{premise}
	(3) $[\varphi]\{\varphi /X/ \equiv \varphi /Z/\}$	{T9.II, 1}
	(4) $*\langle\varphi, Y\rangle /X/ \equiv *\langle\varphi, Y\rangle /Z/$	{OП : 3}
	(5) $\varphi /X, Y/ \equiv \varphi /Z, Y/$	{4, D6.II}
	(6) $[\varphi]\{\varphi /Y/ \equiv \varphi /U/\}$	{T9.II, 2}
	(7) $**\langle\varphi, Z\rangle /Y/ \equiv **\langle\varphi, Z\rangle /U/$	{OП : 6}
	(8) $\varphi /Z, Y/ \equiv \varphi /Z, U/$	{7, D7.II}
	(9) $\varphi /X, Y/ \equiv \varphi /Z, U/$	{5, 8}
	$[\varphi]\{\varphi /X, Y/ \equiv \varphi /Z, U/\}$	{DП : 9}

By analogy with the proofs of T13.II and T17.II it is easily proven that all expressions differing, on the basis of the rule of extensionality, from expressions attached to the system with respect to the number of their antecedents or because their consequents contain functors of an arbitrary semantic order and an arbitrary number of arguments in place of proposition-formative functors of one argument are theorems of ontology.[54] (By functors I understand here of course functors that are not arguments of other functors in these expressions.)

Let us also note that the role played by definitions of functors depending on parameters is most evident as we direct our attention to the proofs of T13.II and T17.II; such definitions make it possible to subjoin to the proof expressions equivalent to those already subjoined thereto but different as to the semantic categories of the terms employed in their notation.

I shall now give five theorems without proof. I shall apply them in proving theorems in the next paragraph.

T18.II. $\varepsilon \langle X\rangle /x/ \supset x \varepsilon \mathrm{V}$.
T19.II. $x \varepsilon \mathrm{V} \wedge \varepsilon \langle x\rangle /y/ \wedge \varepsilon \langle x\rangle /z/ \supset y = z$.
T20.II. $x \varepsilon \mathrm{V} \wedge y \varepsilon \mathrm{V} \wedge (\varepsilon \langle x\rangle /x/ \supset \varepsilon \langle y\rangle /x/) \supset x = y$.
T21.II. $x \varepsilon \mathrm{V} \supset [\exists y]\{\varepsilon \langle x\rangle /y/\}$.
T22.II. $\varepsilon \langle x\rangle /y/ \wedge x \varepsilon \mathrm{V} \supset x = y$.

---

[54] The detailed proof of an analogous theorem referring to theses of the system of protothetics in which the rule of extensionality is valid, is given in Słupecki [1953].

§ 3. Further Definitions and Theorems of Non-Elementary Ontology

I now proceed to define a proposition-formative functor, the argument of which is itself a proposition-formative functor.

D1.III. $prpr\boldsymbol{/}\Phi\boldsymbol{/} \equiv [x]\{\Phi\,/x/ \supset x\ \varepsilon\ \mathrm{V}\}$.

We read the definiendum of the above definition as follows: Φ is a property. Thus, according to this definition, the expression 'Φ is a property' is equivalent to the expression: '$\Phi(x)$' is true only if x is an object. We may also read the expression '$prpr\boldsymbol{/}\Phi\boldsymbol{/}$' as follows: 'Φ is a set of individuals.'

D2.III. $\varepsilon\boldsymbol{/}\varphi, \Phi\boldsymbol{/} \equiv prpr\boldsymbol{/}\varphi\boldsymbol{/} \wedge prpr\boldsymbol{/}\Phi\boldsymbol{/} \wedge [\exists x]\{\varphi\,/x/\} \wedge [x, y]\{\varphi\,/x/ \wedge \varphi\,/y/ \supset x = y\} \wedge [x]\{\varphi\,/x/ \supset \Phi\,/x/\}$.

Having in view the second way of reading the definiendum of D1.III, we may read that of D2.III as follows: the unit set of individuals φ is included in the set of individuals Φ.

My aim is now to prove a theorem of ontology differing from the axiom of ontology (beside the form of the variables) only in that at every place at which in the axiom appears the primitive term of ontology the theorem will contain the term defined in D3.II. Let me note that some of the proofs of theorems in this paragraph will be given in an abbreviated form.

T1.III. $\varepsilon\boldsymbol{/}\varphi, \Phi\boldsymbol{/} \supset [\exists\psi]\{\varepsilon\boldsymbol{/}\psi, \varphi\boldsymbol{/}\}$.

Proof.
(1) $\varepsilon\boldsymbol{/}\varphi, \Phi\boldsymbol{/}$ {premise}
(2) $prpr\boldsymbol{/}\varphi\boldsymbol{/} \wedge [\exists x]\{\varphi\,/x/\} \wedge [x, y]\{\varphi\,/x/ \wedge \varphi\,/y/ \supset x = y\}$ {D2.III, 1}
(3) $\varphi\,/x/ \supset \varphi\,/x/$ {$\mathrm{p} \supset \mathrm{p}$}
(4) $[x]\{\varphi\,/x/ \supset \varphi\,/x/\}$ {$\mathrm{D}\Pi : 3$}
(5) $\varepsilon\boldsymbol{/}\varphi, \varphi\boldsymbol{/}$ {D2.III.2, 4}
$[\exists\psi]\{\varepsilon\boldsymbol{/}\psi, \varphi\boldsymbol{/}\}$ {$\mathrm{D}\Sigma : 5$}

T2.III. $\varepsilon\boldsymbol{/}\varphi, \Phi\boldsymbol{/} \wedge \varepsilon\boldsymbol{/}\psi, \varphi\boldsymbol{/} \supset \varepsilon\boldsymbol{/}\psi, \Phi\boldsymbol{/}$.

Proof.
(1) $\varepsilon\boldsymbol{/}\varphi, \Phi\boldsymbol{/}$
(2) $\varepsilon\boldsymbol{/}\psi, \varphi\boldsymbol{/}$ {premise}
(3) $prpr\boldsymbol{/}\Phi\boldsymbol{/}$
(4) $[x]\{\varphi\,/x/ \supset \Phi\,/x/\}$ {D2.III, φ}
(5) $prpr\boldsymbol{/}\psi\boldsymbol{/} \wedge [\exists x]\{\psi\,/x/\} \wedge [x, y]\{\psi\,/x/ \wedge \psi\,/y/ \supset x = y\}$
(6) $[x]\{\psi\,/x/ \supset \varphi\,/x/\}$ {D2.III, 2}
(7) $[x]\{\psi\,/x/ \supset \Phi\,/x/\}$ {4, 6}
$\varepsilon\boldsymbol{/}\psi, \Phi\boldsymbol{/}$ {D2.III, 3, 5, 7}

This theorem is analogous to T7.1.

T3.III. $\varepsilon\mathbf{I}\varphi, \Phi\mathbf{I} \supset [\psi]\{\varepsilon\mathbf{I}\psi, \varphi\mathbf{I} \supset \varepsilon\mathbf{I}\psi, \Phi\mathbf{I}\}$ {T2.III}

T1.III and T3.III are analogous to theorems of elementary ontology which are implications, the antecedent of which is formed by the expression appearing in axiom T1.1, to the left of the symbol of equivalence while the consequence is the first or the third factor of the conjunction appearing in this axiom to the right of the symbol of equivalence.

T4.III. $\varepsilon\mathbf{I}\varphi, \Phi\mathbf{I} \wedge \varepsilon\mathbf{I}\psi, \varphi\mathbf{I} \wedge \varphi\,/x/ \supset \psi\,/x/$.

Proof.	(1)	$\varepsilon\mathbf{I}\varphi, \Phi\mathbf{I}$	{premise}
	(2)	$\varepsilon\mathbf{I}\psi, \varphi\mathbf{I}$	{premise}
	(3)	$\varphi\,/x/$	{premise}
	(4)	$[\exists x]\{\varphi\,/x/\}$	{D2.III, 1}
	(5)	$[x, y]\{\varphi\,/x/ \wedge \varphi\,/y/ \supset x = y\}$	{D2.III, 1}
	(6)	$\varphi\,/x_1/$	{OΣ : 4}
	(7)	$\varphi\,/x_1/ \wedge \varphi\,/x/ \supset x_1 = x$	{OΠ : 5}
	(8)	$x_1 = x$	{7, 6, 3}
	(9)	$[\exists x]\{\psi\,/x/\}$	{D2.III, 2}
	(10)	$[x]\{\psi\,/x/ \supset \varphi\,/x/\}$	{D2.III, 2}
	(11)	$\psi\,/x_2/$	{OΣ : 9}
	(12)	$\psi\,/x_2/ \supset \varphi\,/x_2/$	{OΠ : 10}
	(13)	$\varphi\,/x_2/$	{12, 11}
	(14)	$\varphi\,/x_2/ \wedge \varphi\,/x_1/ \supset x_2 = x_1$	{OΠ : 5}
	(15)	$x_2 = x_1$	{14, 13, 6}
	(16)	$x_2 = x$	{T9.3, 15, 8}
		$\psi\,/x/$	{T1.II, 16, 11}

T5.III. $\varepsilon\mathbf{I}\varphi, \Phi\mathbf{I} \wedge \varepsilon\mathbf{I}\psi, \varphi\mathbf{I} \supset \varepsilon\mathbf{I}\varphi, \psi\mathbf{I}$.

Proof.	(1)	$\varepsilon\mathbf{I}\varphi, \Phi\mathbf{I}$	{premise}
	(2)	$\varepsilon\mathbf{I}\psi, \varphi\mathbf{I}$	
	(3)	$prpr\mathbf{I}\varphi\mathbf{I} \wedge [\exists x]\{\varphi\,/x/\} \wedge [x, y]$ $\{\varphi\,/x/ \wedge \varphi\,/y/ \supset x = y\}$	{D2.III, 1}
	(4)	$prpr\mathbf{I}\psi\mathbf{I}$	{D2.III, 2}
	(5)	$\varphi\,/x/ \supset \psi\,/x/$	{T4.III, 1, 2}
	(6)	$[x]\{\varphi\,/x/ \supset \psi\,/x/\}$	{DΠ : 5}
		$\varepsilon\mathbf{I}\varphi, \psi\mathbf{I}$	{D2.III, 3, 4, 6}

This theorem is analogous to T10.1.

T6.III. $\varepsilon\mathbf{I}\varphi, \Phi\mathbf{I} \wedge \varepsilon\mathbf{I}\psi, \varphi\mathbf{I} \wedge \varepsilon\mathbf{I}\chi, \varphi\mathbf{I} \supset \varepsilon\mathbf{I}\psi, \chi\mathbf{I}$.

Proof.	(1)	$\varepsilon\mathbf{I}\varphi, \Phi\mathbf{I}$	{premise}
	(2)	$\varepsilon\mathbf{I}\psi, \varphi\mathbf{I}$	{premise}
	(3)	$\varepsilon\mathbf{I}\chi, \varphi\mathbf{I}$	{premise}

(4) $\varepsilon\mathbf{/}\varphi, \chi\mathbf{/}$ {T5.III, 1, 3}
$\varepsilon\mathbf{/}\psi, \chi\mathbf{/}$ {T2.III, 4, 2}

T7.III. $\varepsilon\mathbf{/}\varphi, \Phi\mathbf{/} \supset [\psi, \chi]\{\varepsilon\mathbf{/}\psi, \varphi\mathbf{/} \wedge \varepsilon\mathbf{/}\chi, \varphi\mathbf{/} \supset \varepsilon\mathbf{/}\psi, \chi\mathbf{/}\}$ {T6.III}

This theorem is analogous to the theorem of elementary ontology which is an implication the antecedent of which is formed by the expression appearing in axiom T1.1 to the left of the symbol of equivalence and the consequent is the second factor of the conjunction standing in this axiom to the right of the symbol of equivalence.

T8.III. $prpr\mathbf{/}\varepsilon\langle x\rangle\mathbf{/}$ {D1.III, T18.II}

T9.III. $\varepsilon\mathbf{/}\psi, \varphi\mathbf{/} \wedge \varphi\,/x/ \supset \varepsilon\mathbf{/}\varepsilon\langle x\rangle, \varphi\mathbf{/}.$

Proof. (1) $\varepsilon\mathbf{/}\psi, \varphi\mathbf{/}$ {premise}
(2) $\varphi\,/x/$ {premise}
(3) $prpr\mathbf{/}\varphi\mathbf{/}$ {D2.III, 1}
(4) $x\,\varepsilon\,\mathrm{V}$ {D1.III, 3, 2}
(5) $\varepsilon\langle x\rangle\,/y/ \wedge \varepsilon\langle x\rangle\,/z/ \supset y = z$ {T19.II, 4}
(6) $[y, z]\{\varepsilon\langle x\rangle\,/y/ \wedge \varepsilon\langle x\rangle\,/z/ \supset y = z\}$ {DΠ : 5}
(7) $[\exists y]\{\varepsilon\langle x\rangle\,/y/\}$ {T21.II, 4}
(8) $\varepsilon\langle x\rangle\,/y/ \supset x = y$ {T22.II, 4}
(9) $x = y \supset \varphi\,/y/$ {T1.II, 2}
(10) $\varepsilon\langle x\rangle\,/y/ \supset \varphi\,/y/$ {8, 9}
(11) $[y]\{\varepsilon\langle x\rangle\,/y/ \supset \varphi\,/y/\}$ {DΠ : 10}
$\varepsilon\mathbf{/}\varepsilon\langle x\rangle, \varphi\mathbf{/}$ {D2.III, T8.III, 3, 7, 6, 11}

T10.III. $\varepsilon\mathbf{/}\psi, \varphi\mathbf{/} \wedge [\psi, \chi]\{\varepsilon\mathbf{/}\psi, \varphi\mathbf{/} \wedge \varepsilon\mathbf{/}\chi, \varphi\mathbf{/} \supset \varepsilon\mathbf{/}\psi, \chi\mathbf{/}\} \wedge \varphi\,/x/ \wedge \varphi\,/y/ \supset x = y.$

Proof. (1) $\varepsilon\mathbf{/}\psi, \varphi\mathbf{/}$ {premise}
(2) $[\psi, \chi]\{\varepsilon\mathbf{/}\psi, \varphi\mathbf{/} \wedge \varepsilon\mathbf{/}\chi, \varphi\mathbf{/} \supset \varepsilon\mathbf{/}\psi, \chi\mathbf{/}\}$ {premise}
(3) $\varphi\,/x/$ {premise}
(4) $\varphi\,/y/$ {premise}
(5) $\varepsilon\mathbf{/}\varepsilon\langle x\rangle, \varphi\mathbf{/}$ {T9.III, 1, 3}
(6) $\varepsilon\mathbf{/}\varepsilon\langle y\rangle, \varphi\mathbf{/}$ {T9.III, 1, 4}
(7) $\varepsilon\mathbf{/}\varepsilon\langle x\rangle, \varphi\mathbf{/} \wedge \varepsilon\mathbf{/}\varepsilon\langle y\rangle, \varphi\mathbf{/} \supset \varepsilon\mathbf{/}\varepsilon\langle x\rangle, \varepsilon\langle y\rangle\mathbf{/}$ {OΠ : 2}
(8) $\varepsilon\mathbf{/}\varepsilon\langle x\rangle, \varepsilon\langle y\rangle\mathbf{/}$ {7, 5, 6}
(9) $[z]\{\varepsilon\langle x\rangle\,/z/ \supset \varepsilon\langle y\rangle\,/z/\}$ {D2.III, 8}
(10) $\varepsilon\langle x\rangle\,/x/ \supset \varepsilon\langle y\rangle\,/x/$ {OΠ : 9}
(11) $prpr\mathbf{/}\varphi\mathbf{/}$ {D2.III, 1}
(12) $x\,\varepsilon\,\mathrm{V}$ {D1.III, 11, 3}
(13) $y\,\varepsilon\,\mathrm{V}$ {D1.III, 11, 4}
$x = y$ {T20.II, 12, 13, 10}

T11.III. $[\exists\psi]\{\varepsilon\mathbf{/}\psi, \varphi\mathbf{/}\} \wedge [\psi, \chi]\{\varepsilon\mathbf{/}\psi, \varphi\mathbf{/} \wedge \varepsilon\mathbf{/}\chi, \varphi\mathbf{/} \supset \varepsilon\mathbf{/}\psi, \chi\mathbf{/}\} \wedge$
$[\psi]\{\varepsilon\mathbf{/}\psi, \varphi\mathbf{/} \supset \varepsilon\mathbf{/}\psi, \Phi\mathbf{/}\} \supset \varepsilon\mathbf{/}\varphi, \Phi\mathbf{/}$.

Proof. (1)	$[\exists\psi]\{\varepsilon\mathbf{/}\psi, \varphi\mathbf{/}\}$	{premise}
(2)	$[\psi, \chi]\{\varepsilon\mathbf{/}\psi, \varphi\mathbf{/} \wedge \varepsilon\mathbf{/}\chi, \varphi\mathbf{/} \supset \varepsilon\mathbf{/}\psi, \chi\mathbf{/}\}$	
(3)	$[\psi]\{\varepsilon\mathbf{/}\psi, \varphi\mathbf{/} \supset \varepsilon\mathbf{/}\psi, \Phi\mathbf{/}\}$	
(4)	$\varepsilon\mathbf{/}\psi_1, \varphi\mathbf{/}$	{OΣ : 1}
(5)	$prpr\mathbf{/}\varphi\mathbf{/}$	{D2.III, 4}
(6)	$[\exists x]\{\psi_1 /x/\} \wedge [x]\{\psi_1 /x/ \supset \varphi /x/\}$	
(7)	$[\exists x]\{\varphi /x/\}$	{6}
(8)	$\varphi /x/ \wedge \varphi /y/ \supset x = y$	{T10 III, 4, 2}
(9)	$[x, y]\{\varphi /x/ \wedge \varphi /y/ \supset x = y\}$	{DΠ : 8}
(10)	$\varphi /x/ \supset \varphi /x/$	{p ⊃ p}
(11)	$[x]\{\varphi /x/ \supset \varphi /x/\}$	{DΠ : 10}
(12)	$\varepsilon\mathbf{/}\varphi, \varphi\mathbf{/}$	{D2.III, 5, 7, 9, 11}
(13)	$\varepsilon\mathbf{/}\varphi, \varphi\mathbf{/} \supset \varepsilon\mathbf{/}\varphi, \Phi\mathbf{/}$	{OΠ : 3}
	$\varepsilon\mathbf{/}\varphi, \Phi\mathbf{/}$	{13, 12}

This theorem is analogous to the theorem of elementary ontology which is an implication the antecedent and consequent of which are the expressions appearing in axiom T1.1 to the right and to the left of the symbol of equivalence.

T12.III. $\varepsilon\mathbf{/}\varphi, \Phi\mathbf{/} \equiv [\exists\psi]\{\varepsilon\mathbf{/}\psi, \varphi\mathbf{/}\} \wedge [\psi, \chi]\{\varepsilon\mathbf{/}\psi, \varphi\mathbf{/} \wedge \varepsilon\mathbf{/}\chi, \varphi\mathbf{/} \supset$
$\varepsilon\mathbf{/}\psi, \chi\mathbf{/}\} \wedge [\psi]\{\varepsilon\mathbf{/}\psi, \varphi\mathbf{/} \supset \varepsilon\mathbf{/}\psi, \Phi\mathbf{/}\}$. {T1.III, T7.III, T3.III, T11.III}

T12.III is analogous to the axiom of ontology. This shows that the theorems of ontology remain true if we interpret the primitive term of the system as the symbol of inclusion of a unit set of individuals in the set of individuals. Mention has been made of this interpretation of theorems of ontology in I, § 3.

Let φ and ψ be variables representing functors and belonging to an arbitrary, but both to the same semantic category. It is a very important feature of non-elementary ontology that it is possible to define in this system a proposition-formative functor with ψ and φ as its only arguments, and to prove a theorem containing this functor, analogous to axiom T1.1 in the meaning in which T12.III is analogous to that axiom. Thus, for every semantic category of the language of ontology it is possible to construct a fragment of the system, the theorems of which are analogous to the theorems of elementary ontology. It is even possible to do this in more than one way. The term defined in the following definition belongs to the same semantic category as that defined in D2.III. For this term also, just as for the term defined in D2.III it is possible to prove a theorem analogous to the axiom of ontology.

D3.III. $\varepsilon^{*}\langle\varphi, \Phi\rangle \equiv [X, Y]\{\varphi\langle X\rangle \wedge \varphi\langle Y\rangle \supset X \underset{Z}{=} Y\} \wedge$
$\wedge [X]\{\varphi\langle X\rangle \supset \Phi\langle X\rangle\}.$

The meaning of expression '$\varepsilon^{*}\langle\varphi, \Phi\rangle$' differs from that of expression '$\varepsilon\langle\varphi, \Phi\rangle$' in that the truth of propositions '$\varepsilon^{*}\langle\varphi, \Phi\rangle$' and '$\varphi\langle x\rangle$' or '$\Phi\langle x\rangle$' does not imply that x is an object.

I shall define now the relation 'equinumerous' understood as holding between names; it is analogous to the relation 'equinumerous' understood as holding between sets.

The following are four auxiliary definitions.

D4.III. $rel\,\langle\langle\varphi\rangle\rangle \equiv [x, y]\{\varphi\langle x, y\rangle \supset x\, \varepsilon\, \mathrm{V} \wedge y\, \varepsilon\, \mathrm{V}\}.$

This definition is analogous to D1.III.

D5.III. $x\, \varepsilon\, \mathrm{D}\langle\langle\varphi\rangle\rangle \equiv x\, \varepsilon\, \mathrm{V} \wedge rel\,\langle\langle\varphi\rangle\rangle \wedge [\exists y]\{\varphi\langle x, y\rangle\}$
D6.III. $y\, \varepsilon\, \mathrm{ᗡ}\langle\langle\varphi\rangle\rangle \equiv y\, \varepsilon\, \mathrm{V} \wedge rel\,\langle\langle\varphi\rangle\rangle \wedge [\exists x]\{\varphi\langle x, y\rangle\}.$
D7.III. $1\text{–}1\,\langle\langle\varphi\rangle\rangle \equiv rel\,\langle\langle\varphi\rangle\rangle \wedge [x, y, z]\{(\varphi\langle x, z\rangle \wedge \varphi\langle y, z\rangle \supset x = y)$
$\wedge (\varphi\langle z, x\rangle \wedge \varphi\langle z, y\rangle \supset x = y)\}.$

We read the definienda in D5.III–D7.III as follows:

φ is a relation,
x is an element of the domain of relation φ,
x is an element of the converse domain of relation φ,
φ is a one-one relation.

The following is the definition of the relation 'equinumerous':

D8.III. $X \sim Y \equiv [\exists\varphi]\{1\text{–}1\,\langle\langle\varphi\rangle\rangle \wedge X \underset{Z}{=} \mathrm{D}\langle\langle\varphi\rangle\rangle \wedge Y \underset{Z}{=} \mathrm{ᗡ}\langle\langle\varphi\rangle\rangle\}.$

IV

§ 1. The Properties of Definitions. Consistency of the System

This chapter deals with problems of methodology. I shall restrict my remarks to the discussion of those properties of ontology which I believe to be most important or most characteristic for Leśniewski's system. In view of the fact that definitions play a special part in ontology and in Leśniewski's other systems, I shall discuss them in more detail.

I begin by discussing the problem of translatability of definitions.[55] The definitions of a given system fulfil the condition of translatability if for every propositional expression φ of that system, in the notation of which

[55] The condition of translatability of definitions is discussed in Ajdukiewicz [1934, p. 33].

defined terms appear, there exists an equivalent expression ψ containing no such terms; the expressions φ and ψ are equivalent if and only if the expression $\varphi \equiv \psi$ is a theorem of the system.

Theorem VI. *The definitions of ontology fulfil the condition of translatability.*

Proof. Let φ_ψ be an arbitrary propositional expression containing the term ψ defined in one of the definitions of ontology, and let α be a variable not appearing in φ_ψ and belonging to the same semantic category as ψ. Next, let φ_α denote the expression obtained from φ_ψ by substituting the variables α for all the symbols ψ. Thus, φ_α is a result of substitution applied to the expression φ_ψ. Obviously φ_α contains fewer defined terms than the expression φ_ψ. I also assume that the variables $\varepsilon_1, \varepsilon_2, \ldots, \varepsilon_k$ are all the free variables of φ_ψ, and that the following definition has been attached to the system:

(I) $\gamma\langle\varepsilon_1, \varepsilon_2, \ldots, \varepsilon_k\rangle /\alpha/ \equiv \varphi_\alpha.$

I further assume that δ is a variable of the same semantic category as the functor $\gamma\langle\varepsilon_1, \varepsilon_2, \ldots, \varepsilon_k\rangle$.

In proving the theorem I shall classify the nominal constant as a name-formative functor, considering it to be a functor of zero arguments. This reduces the discussion in the proof to two cases only:

(a) ψ is a proposition-formative functor, and

(b) ψ is a name-formative functor.

I shall be proving the theorem for a functor ψ that does not depend on parameters. However, the theorem is proved in a similar manner if the functor depends on parameters.

(a) Let the definition of ψ be as follows:

(II) $\psi /\beta_1, \beta_2, \ldots, \beta_l/ \equiv \Phi.$

I assume that the definiens Φ in this definition does not contain any defined terms,[56] and that the variables $\beta_1, \beta_2, \ldots, \beta_l$ do not appear in expression φ_α. Thus α is not a variable of the expression Φ.

I shall prove that the expression:

(III) $\varphi_\psi \equiv [\exists\alpha]\{[\beta_1, \beta_2, \ldots, \beta_l]\{\alpha /\beta_1, \beta_2, \ldots, \beta_l/ \equiv \Phi\} \wedge \varphi_\alpha\}$

is a theorem of ontology. In order to prove this, I shall first prove that the implications

(IV) $\varphi_\psi \supset [\exists\alpha]\{[\beta_1, \beta_2, \ldots, \beta_l]\{\alpha /\beta_1, \beta_2, \ldots, \beta_l/ \equiv \Phi\} \wedge \varphi_\alpha\},$

(V) $[\exists\alpha]\{[\beta_1, \beta_2, \ldots, \beta_l]\{\alpha /\beta_1, \beta_2, \ldots, \beta_l/ \equiv \Phi\} \wedge \varphi_\alpha\} \supset \varphi_\psi,$

are theorems of the system.

56 It should be remembered that the only reason why I employed defined terms in the notation of the definienda of the definitions in the preceding chapter was to render them more clear.

I now proceed to prove that expression (IV) is a theorem of ontology. The way of noting the proof will be the same as that used in the proofs of theorems of the system.

(VI)	φ_ψ	{premise}
(VII)	$[\beta_1, \beta_2, \ldots, \beta_l]\{\psi/\beta_1, \beta_2, \ldots, \beta_l/ \equiv \Phi\}$	{DΠ : II}
(VIII)	$[\beta_1, \beta_2, \ldots, \beta_l]\{\psi/\beta_1, \beta_2, \ldots, \beta_l/ \equiv \Phi\} \wedge \varphi_\psi$	{VI, VII}
	$[\exists\alpha]\{[\beta_1, \beta_2, \ldots, \beta_l]\{\psi/\beta_1, \beta_2, \ldots, \beta_l/ \equiv \Phi\} \wedge \varphi_\alpha\}$	{DΣ : VIII}

In proving expression (V) to be a theorem of ontology I shall make use of the rule of extensionality.[57] Let me also mention that the symbol φ_{α_1} appearing in the expressions of this and of the ensuing proof denotes an expression following from φ_α by the substitution of the variable α_1 for α.

(IX)	$[\exists\alpha]\{[\beta_1, \beta_2, \ldots, \beta_l]\{\alpha/\beta_1, \beta_2, \ldots, \beta_l/ \equiv \Phi\} \wedge \varphi_\alpha\}$	{premise}
(X)	$[\beta_1, \beta_2, \ldots, \beta_l]\{\alpha_1/\beta_1, \beta_2, \ldots, \beta_l/ \equiv \Phi\} \wedge \varphi_{\alpha_1}$	{OΣ : IX}
(XI)	$\gamma\langle\varepsilon_1, \varepsilon_2, \ldots, \varepsilon_k\rangle/\alpha_1/$	{I, X}
(XII)	$\alpha_1/\beta_1, \beta_2, \ldots, \beta_l/ \equiv \Phi$	{OΠ : X}
(XIII)	$\alpha_1/\beta_1, \beta_2, \ldots, \beta_l/ \equiv \psi/\beta_1, \beta_2, \ldots, \beta_l/$	{XII, II}
(XIV)	$[\beta_1, \beta_2, \ldots, \beta_l]\{\alpha_1/\beta_1, \beta_2, \ldots, \beta_l/ \equiv \psi/\beta_1, \beta_2, \ldots, \beta_l/\}$	{DΠ : XIII}
(XV)	$[\beta_1, \beta_2, \ldots, \beta_l]\{\alpha_1/\beta_1, \beta_2, \ldots, \beta_l/ \equiv \psi/\beta_1, \beta_2, \ldots, \beta_l/\} \supset [\delta]\{\delta/\alpha_1/ \equiv \delta/\psi/\}$	{rule of extensionality}
(XVI)	$[\delta]\{\delta/\alpha_1/ \equiv \delta/\psi/\}$	{XV, XIV}
(XVII)	$\gamma\langle\varepsilon_1, \varepsilon_2, \ldots, \varepsilon_k\rangle/\alpha_1/ \equiv \gamma\langle\varepsilon_1, \varepsilon_2, \ldots, \varepsilon_k\rangle/\psi/$	{OΠ : XVI}
(XVIII)	$\gamma\langle\varepsilon_1, \varepsilon_2, \ldots, \varepsilon_k\rangle/\psi/$	{XVII, XI}
	φ_ψ	{I, XVIII}

The fact that expressions (IV) and (V) are theorems of ontology implies directly that expression (III) is also a theorem of this system. This, however, together with the assumption stating that Φ contains no defined terms, and the fact that φ_α contains fewer defined terms than φ_ψ, leads to the conclusion that in the case under discussion the expression φ_ψ is equivalent to an expression containing a smaller number of defined terms.
(b) I assume that the following expression defines the name-formative functor:

$$\text{(XIX)} \quad x \varepsilon \psi|\beta_1, \beta_2, \ldots, \beta_l| \equiv x \varepsilon x \wedge \Psi.$$

I also assume that Ψ does not contain any defined terms and that the variables $x, \beta_1, \beta_2, \ldots, \beta_l$ are not variables of the expression φ_α. Thus α is not a variable of the expression Ψ.

---

[57] This rule has been discussed in III, § 1.

I shall first prove that the following two implications are theorems of ontology:

(XX) $\varphi_\psi \supset [\exists\alpha]\{[x, \beta_1, \beta_2, \ldots, \beta_l]\{x \,\varepsilon\, \alpha|\beta_1, \beta_2, \ldots, \beta_l| \equiv x \,\varepsilon\, x \wedge \Psi\} \wedge \varphi_a\}$,

(XXI) $[\exists\alpha]\{[x, \beta_1, \beta_2, \ldots, \beta_l]\{x \,\varepsilon\, \alpha|\beta_1, \beta_2, \ldots, \beta_l| \equiv x \,\varepsilon\, x \wedge \Psi\} \wedge \varphi_a\} \supset \varphi_\psi$.

I now prove that expression (XX) is a theorem of ontology.

(XXII) φ_ψ {premise}

(XXIII) $[x, \beta_1, \beta_2, \ldots, \beta_l]\{x \,\varepsilon\, \psi|\beta_1, \beta_2, \ldots, \beta_l| \equiv x \,\varepsilon\, x \wedge \Psi\}$ {DΠ : XIX}

(XXIV) $[x, \beta_1, \beta_2, \ldots, \beta_l]\{x \,\varepsilon\, \psi|\beta_1, \beta_2, \ldots, \beta_l| \equiv x \,\varepsilon\, x \wedge \Psi\} \wedge \varphi_\psi$ {XXII, XXIII}

$[\exists\alpha]\{[x, \beta_1, \beta_2, \ldots, \beta_l]\{x \,\varepsilon\, \alpha|\beta_1, \beta_2, \ldots, \beta_l| \equiv x \,\varepsilon\, x \wedge \Psi\} \wedge \varphi_a\}$ {DΣ : XXIV}

I further prove that expression (XXI) is a theorem of ontology.

(XXV) $[\exists\alpha]\{[x, \beta_1, \beta_2, \ldots, \beta_l]\{x \,\varepsilon\, \alpha|\beta_1, \beta_2, \ldots, \beta_l| \equiv x \,\varepsilon\, x \wedge \Psi\} \wedge \varphi_a\}$ {premise}

(XXVI) $[x, \beta_1, \beta_2, \ldots, \beta_l]\{x \,\varepsilon\, \alpha_1|\beta_1, \beta_2, \ldots, \beta_l| \equiv x \,\varepsilon\, x \wedge \Psi\} \wedge \varphi_{a_1}$ {OΣ : XXV}

(XXVII) $\gamma\langle\varepsilon_1, \varepsilon_2, \ldots, \varepsilon_k\rangle /\alpha_1/$ {I, XXVI}

(XXVIII) $x \,\varepsilon\, \alpha_1|\beta_1, \beta_2, \ldots, \beta_l| \equiv x \,\varepsilon\, x \wedge \Psi$ {OΠ : XXVI}

(XXIX) $x \,\varepsilon\, \alpha_1|\beta_1, \beta_2, \ldots, \beta_l| \equiv x \,\varepsilon\, \psi|\beta_1, \beta_2, \ldots, \beta_l|$ {XXVIII, XIX}

(XXX) $[x, \beta_1, \beta_2, \ldots, \beta_l]\{x \,\varepsilon\, \alpha_1|\beta_1, \beta_2, \ldots, \beta_l| \equiv x \,\varepsilon\, \psi|\beta_1, \beta_2, \ldots, \beta_l|\}$ {DΠ : XXIX}

(XXXI) $[x, \beta_1, \beta_2, \ldots, \beta_l]\{x \,\varepsilon\, \alpha_1|\beta_1, \beta_2, \ldots, \beta_l| \equiv x \,\varepsilon\, \psi|\beta_1, \beta_2, \ldots, \beta_l|\} \supset [\delta]\{\delta /\alpha_1| \equiv \delta /\psi/\}$ {rule of extensionality}

(XXXII) $[\delta]\{\delta /\alpha_1/ \equiv \delta /\psi/\}$ {XXXI, XXX}

(XXXIII) $\gamma\langle\varepsilon_1, \varepsilon_2, \ldots, \varepsilon_k\rangle /\alpha_1/ \equiv \gamma\langle\varepsilon_1, \varepsilon_2, \ldots, \varepsilon_k\rangle /\psi/$ {OΠ : XXXII}

(XXXIV) $\gamma\langle\varepsilon_1, \varepsilon_2, \ldots, \varepsilon_k\rangle /\psi/$ {XXXIII, XXVII}

φ_ψ {I, XXXIV}

The fact that expressions (XX) and (XXI) are theorems of ontology implies directly that the following expression:

$$\varphi_\psi \equiv [\exists\alpha]\{[x, \beta_1, \beta_2, \ldots, \beta_l]\{x \,\varepsilon\, \alpha|\beta_1, \beta_2, \ldots, \beta_l| \equiv x \,\varepsilon\, x \wedge \Psi\} \wedge \varphi_a\}$$

is a theorem of this system.

From this, from the assumption stating that there occur no defined terms in expression Ψ and from the fact that φ_a contains fewer defined terms than φ_ψ it follows that also in the second of the possible cases, the expression φ_ψ is equivalent to an expression containing a smaller number of defined terms. This proves theorem VI, if we keep in mind that the relation of equivalence of expressions is transitive.

Let me point out that in case of definitions of elementary ontology it is possible to prove theorem VI without applying the rule of extensionality.

I shall discuss in turn the problem of non-creativity of the definitions of ontology. The definitions of a given system fulfil the condition of non-creativity if for any expression not containing defined terms, if that expression follows from definitions and a particular set of expressions of that system, it follows also from the expressions of that set.[58]

Theorem VII. *The definitions of ontology do not fulfil the condition of non-creativity.*

Proof. I shall prove that from T7.1 and from the following definition:

(I) $x \varepsilon X^*y \equiv x \varepsilon X \wedge y \varepsilon x,$

follows the expression:

(II) $x \varepsilon X \wedge y \varepsilon x \wedge z \varepsilon x \supset y \varepsilon z,$

and that the latter expression does not follow from T7.1.

The following is the proof of expression (II) on the basis of T7.1 together with definition (I):[59]

(III)	$x \varepsilon X \wedge y \varepsilon x$	{premise}
(IV)	$z \varepsilon x$	{premise}
(V)	$x \varepsilon X^*y$	{I, III}
(VI)	$z \varepsilon X^*y$	{T7.1, V, IV}
	$y \varepsilon z$	{I, VI}

In order to prove that expression (II) does not follow from T7.1 on the basis of the rules of inference of ontology (minus the rule of attaching definitions to the system), I let the term 'ε' have the meaning given to the term 'a' by D3.1. Under this interpretation of expressions of ontology T7.1 remains true (as on replacing in it 'ε' by 'a' we obtain an expression which is a result of substitution applied to T7.2). All rules of ontology retain the property of yielding exclusively true expressions when applied to true expressions. The rule of extensionality is no exception here since T24.2 is a theorem of ontology and expression (II) turns into a false ex-

[58] The condition of non-creativity of definitions is mentioned in Mostowski [1948, p. 257].

[59] The proof was given by Prof. A. Tarski. It is contained in Sobociński [1934, p. 149].

pression; hence expression (II) does not follow from T7.1. There exists therefore an expression following from a certain expression together with a certain definition on the basis of the rules of inference of ontology (minus the rule of attaching definitions to the system) and not following from that expression alone. This proves the theorem.

However, the definitions of ontology will be fulfilling the condition of non-creativity if we attach to the system a certain definitional axiom being one of the axioms of the full system of the simple theory of types. This system is discussed in Tarski [1933]* and Mostowski [1948].

I present this axiom in the form of the following two rules:

1. If φ is an arbitrary propositional expression of ontology not containing the free variable 'Y', and if 'X' is an arbitrary nominal variable non-equiform to 'Y', then it is permissible to subjoin to the system the following expression:

(a) $[\exists Y][x]\{x \varepsilon Y \equiv x \varepsilon X \wedge \varphi\}$.

2. If $\varphi_{\alpha_1, \alpha_2, \ldots, \alpha_n}$ is an arbitrary propositional expression containing the free variables $\alpha_1, \alpha_2, \ldots, \alpha_n$ belonging to arbitrary semantic categories, then it is permissible to attach to the system the expression:

$$[\exists \Phi]\{\Phi /\alpha_1, \alpha_2, \ldots, \alpha_n/ \equiv \varphi_{\alpha_1, \alpha_2, \ldots, \alpha_n}\}.$$

By way of example, I give the proof of expression (II), mentioned above when proving theorem VII. In proving expression (II), I shall not be employing definitions.

Rule I and OΣ enable us to subjoin to the system expression

(b) $x \varepsilon Y_1 \equiv x \varepsilon X \wedge y \varepsilon x$,

corresponding to expression (I) appearing in the proof of theorem VII.

Proof of expression (II):

(c)	$x \varepsilon X \wedge y \in x$	{premise}
(d)	$z \varepsilon x$	
(e)	$x \varepsilon Y_1$	{b, c}
(f)	$z \varepsilon Y_1$	{T7, I, e, d}
	$y \varepsilon z$	{b, f}

In order to obtain the last expression of the proof, I have substituted in (b) the variable 'z' for 'x'.

Theorem VIII. *If Leśniewski's protothetics is consistent, then ontology is also consistent.*[60]

Proof. I am giving only an abridged proof of this theorem.

If in the axiom of ontology we agree to consider nominal variables as

* *Editorial Note*: English translation in Tarski [1956].

[60] A detailed exposition of protothetics is given in Słupecki [1953].

propositional variables and the functor 'ε' as the symbol of conjunction, then this axiom becomes a theorem of protothetics. Under this interpretation[61], all the rules of inference of ontology become primitive or secondary rules of protothetics. This proves the theorem.

§ 2. Simplified Axioms of Ontology

In I, § 4, I have drawn attention to the fact that beside T1.1 other theorems of ontology can also act as its only axiom. The simplest theorem known to possess this property is T23.1. Detailed information on the consecutive simplification of the axiom of ontology is found in Sobociński [1934].

In order to prove that a given propositional expression can replace axiom T1.1, it is sufficient to prove that it follows from T1.1 according to rules holding in Leśniewski's system, and that T1.1 follows from this expression. T23.1 has been deduced from T1.1 in II, § 1. I shall now prove that T1.1 follows from T23.1. In this proof I shall apply the rule of extensionality. Thus, in the system the only axiom of which is T23.1, we must make use of this rule when proving certain theses of elementary ontology.

I will append Roman numerals to expressions following from T23.1. I repeat T23.1:

(I) $x \varepsilon X \equiv [\exists y]\{x \varepsilon y \wedge y \varepsilon X\}$.

(II) $x \varepsilon y \wedge y \varepsilon X \supset x \varepsilon X$.

Proof. (1)	$x \varepsilon y \wedge y \varepsilon X$	{premise}
(2)	$[\exists y]\{x \varepsilon y \wedge y \varepsilon X\}$	$\{D\Sigma : 1\}$
	$x \varepsilon X$	{I, 2}

(III) $x \varepsilon X \supset [y]\{y \varepsilon x \supset y \varepsilon X\}$.

Proof. (1)	$x \varepsilon X$	{premise}
(2)	$y \varepsilon x \supset y \varepsilon X$	{II, I}
(3)	$[y]\{y \varepsilon x \supset y \varepsilon X\}$	$\{D\Pi : 2\}$

The antecedent of this theorem is equiform with the expression appearing in T1.1 to the left of the symbol of equivalence; the consequent is equiform with the expression constituting the last factor in the conjunction appearing in T1.1 to the right of the symbol of equivalence.

I repeat the definition given in the preceding paragraph in the proof of theorem VII:

(IV) $x \varepsilon X{*}y \equiv x \varepsilon X \wedge y \varepsilon x$.

[61] In the notes from Leśniewski's course of lectures "An Outline of Elementary Ontology" mention is made that Kruszewski was the author of this interpretation.

(V) $x \varepsilon X \wedge y \varepsilon x \wedge z \varepsilon x \supset y \varepsilon z$.

Proof. (1) $x \varepsilon X \wedge y \varepsilon x$ } {premise}
(2) $z \varepsilon x$ }
(3) $x \varepsilon X{*}y$ {IV, 1}
(4) $z \varepsilon X{*}y$ {II, 2, 3}
$y \varepsilon z$ {IV, 4}

(VI) $x \varepsilon X \supset [y, z]\{y \varepsilon x \wedge z \varepsilon x \supset y \varepsilon z\}$.

Proof. (1) $x \varepsilon X$ {premise}
(2) $y \varepsilon x \wedge z \varepsilon x \supset y \varepsilon z$ {V, 1}
$[y, z]\{y \varepsilon x \wedge z \varepsilon x \supset y \varepsilon z\}$ {DΠ : 2}

The antecedent of this theorem is equiform with the expression appearing in T1.1 to the left of the symbol of equivalence; the consequent is equiform with the expression constituting the second factor of the conjunction appearing in T1.1 to the right of the symbol of equivalence.

(VII) $x \varepsilon X \supset [\exists y]\{y \varepsilon x\}$.

Proof. (1) $x \varepsilon X$ {premise}
(2) $[\exists y]\{x \varepsilon y \wedge y \varepsilon X\}$ {I, 1}
(3) $x \varepsilon y_1 \wedge y_1 \varepsilon X$ {OΣ : 2}
(4) $x \varepsilon x$ {V, 3}
$[\exists y]\{y \varepsilon x\}$ {DΣ : 4}

In order to subjoin expression (4) to the proof, I have substituted in theorem (V) the variable "y", for "x", and "x" for the variables "y" and "z". The antecedent of (VII) is equiform with the expression appearing in T1.1 to the left of the symbol of equivalence; the consequent is equiform with the first factor of the conjunction appearing in T1.1 to the right of the symbol of equivalence.

The ensuing theorem has been attached to the system on the authority of the rule of extensionality.

(VIII) $[x]\{x \varepsilon X \equiv x \varepsilon Y\} \supset [\varphi]\{\varphi /X/ \equiv \varphi /Y/\}$.
{rule of extensionality}

This theorem is equiform with T9.II.

The antecedent and the consequent of the theorem proved below are aquiform, respectively, with the expression appearing in T1.1 to the right end to the left of the symbol of equivalence. The proof involves the following definition which is equiform with D2.II:

(IX) $\varepsilon \langle X \rangle /x/ \equiv x \varepsilon X$.

(X) $[\exists y]\{y \varepsilon x\} \wedge [y, z]\{y \varepsilon x \wedge z \varepsilon x \supset y \varepsilon z\} \wedge [y]\{y \varepsilon x \supset y \varepsilon X\}$
$\supset x \varepsilon X$.

Proof.

(1)	$[\exists y]\{y \varepsilon x\}$	{premise}
(2)	$[y, z]\{y \varepsilon x \wedge z \varepsilon x \supset y \varepsilon z\}$	{premise}
(3)	$[y]\{y \varepsilon x \supset y \varepsilon X\}$	{premise}
(4)	$y_1 \varepsilon x$	{OΣ : 1}
(5)	$y_1 \varepsilon x \supset y_1 \varepsilon X$	{OΠ : 3}
(6)	$y_1 \varepsilon X$	{5, 4}
(7)	$y \varepsilon y_1 \supset y \varepsilon x$	{II, 4}
(8)	$y \varepsilon x \wedge y_1 \varepsilon x \supset y \varepsilon y_1$	{OΠ : 2}
(9)	$y \varepsilon x \supset y \varepsilon y_1$	{8, 4}
(10)	$y \varepsilon y_1 \equiv y \varepsilon x$	{7, 9}
(11)	$[y[\{y \varepsilon y_1 \equiv y \varepsilon x\}$	{DΠ : 10}
(12)	$[\varphi]\{\varphi/y_1/ \equiv \varphi/x/\}$	{VIII, 11}
(13)	$\varepsilon\langle X\rangle/y_1/ \equiv \varepsilon\langle X\rangle/x/$	{OΠ : 12}
(14)	$\varepsilon\langle X\rangle/y_1/$	{IX, 6}
(15)	$\varepsilon\langle X\rangle/x/$	{13, 14}
	$x \varepsilon X$	{IX, 15}

(XI) $x \varepsilon X \equiv [\exists y]\{y \varepsilon x\} \wedge [y, z]\{y \varepsilon x \wedge z \varepsilon x \supset y \varepsilon z\} \wedge [y]\{y \varepsilon x \supset y \varepsilon X\}$. {VII, VI, III, X}

This theorem is equiform with T1.1. Thus, expression (I) can act as the only axiom of ontology.

The following expression can also act as the only axiom of ontology:

(α) $x \varepsilon X \equiv [\exists y]\{y \varepsilon x \wedge y \varepsilon X\} \wedge [y, z]\{y \varepsilon x \wedge z \varepsilon x \supset y \varepsilon z\}$.

This follows immediately from T22.2, D1.2 and D2.1. The proof of T1.1[62] based on (α) does not involve the rule of extensionality It is also worth noting that the system of ontology as given in the notes from the course of lectures: "An Outline of Elementary Ontology" is based on expression (α) as its only axiom.

§ 3. Terms which Can Act as the Only Primitive Term of Ontology

Just as it is possible to replace axiom T1.1 by other axioms, so can the term 'ε' be replaced by other primitive terms. For instance, the term '*i*' and also '*a*' or '*a**' can act as the only primitive term of ontology. '*i*', '*a*' and '*a**' are terms of Aristotelian syllogistics; they are defined in D1.2, D3.1 and D4.2.

In order to prove that a given term can act as the only term of ontology, it is sufficient to prove that an expression, formed of the symbol of equivalence having on one side the expression '$x \varepsilon X$', and on the other an ex-

---

[62] The proof is given in Sobociński [1934].

pression in which, apart from functors of the calculus of propositions, the only logical constant is the given term, and the only free variables are 'x' and 'X', is a theorem of this system. The following three expressions are formed in this manner:

$$(\alpha)\ x \varepsilon X \equiv xiX \wedge [S, P]\{xiS \wedge xiP \supset SiP\},$$
$$(\beta)\ x \varepsilon X \equiv xaX \wedge [\exists S]\{\sim(xaS)\} \wedge [S, P]\{Sax \wedge \sim(SaP) \supset xaS\},$$
$$(\gamma)\ x \varepsilon X \equiv xa^*X \wedge [S]\{Sa^*x \supset xa^*S\}.$$

Each of these expressions is a theorem of ontology. Hence, they can act as definitions of the term 'ε' in systems of ontology in which the only primitive term is the symbol 'i', 'a', or 'a^*', respectively.

I shall limit myself to proving that expression (α) is a theorem of ontology. I shall denote theorems proved in this paragraph by means of Roman numerals preceded by the letter T.

TI. $[S, P]\{xiS \wedge xiP \supset SiP\} \wedge y \varepsilon x \wedge z \varepsilon x \supset y \varepsilon z.$

Proof. (1)	$[S, P]\{xiS \wedge xiP \supset SiP\}$	{premise}
(2)	$y \varepsilon x$	{premise}
(3)	$z \varepsilon x$	{premise}
(4)	$xiy \wedge xiz \supset yiz$	{OΠ : 1}
(5)	yix	{T22.2, 2}
(6)	zix	{T22.2, 3}
(7)	xiy	{T23.2, 5}
(8)	xiz	{T23.2, 6}
(9)	yiz	{4, 7, 8}
(10)	$y \varepsilon \mathrm{V}$	{T14.1, 2}
	$y \varepsilon z$	{T15.2, 10, 9}

TII. $[S, P]\{xiS \wedge xiP \supset SiP\} \supset \rightharpoonup /x/.$

Proof. (1)	$[S, P]\{xiS \wedge xiP \supset SiP\}$	{premise}
(2)	$y \varepsilon x \wedge z \varepsilon x \supset y \varepsilon z$	{TI, 1}
(3)	$[y, z]\{z \varepsilon x \wedge z \varepsilon x \supset y \varepsilon z\}$	{DΠ : 2}
	$\rightharpoonup /x/$	{D2.1, 3}

TIII. $\rightharpoonup /x/ \wedge xiS \wedge xiP \supset SiP.$

Proof. (1)	$\rightharpoonup /x/$	{premise}
(2)	xiS	{premise}
(3)	xiP	{premise}
(4)	$x \varepsilon S$	{T22.2, 2, 1}
(5)	$x \varepsilon P$	{T22.2, 3, 1}
(6)	$x \varepsilon S \wedge x \varepsilon P$	{4, 5}
(7)	$[\exists x]\{x \varepsilon S \wedge x \varepsilon P\}$	{DΣ : 6}
	SiP	{D1.2, 7}

TIV. $\rightharpoonup /x/ \supset [S, P]\{xiS \wedge xiP \supset SiP\}$.

Proof.	(1) $\rightharpoonup /x/$	{premise}
	(2) $xiS \wedge xiP \supset SiP$	{TIII, 1}
	$[x]\{xiS \wedge xiP \supset StP\}$	{DII : 2}

TV. $\rightharpoonup /x/ \equiv [S, P]\{xiS \wedge xiP \supset SiP\}$. {TIV, TII}

This theorem can act as definition of the symbol '$\rightharpoonup$' in the system in which 'i' acts as primitive term.

TVI. $x \varepsilon X \equiv xiX \wedge [S, P]\{xiS \wedge xiP \supset SiP\}$. {T22.2, TV}

This theorem is equiform with expression (α). Thus, the symbol 'i' can act as the only primitive term of ontology.

Czesław Lejewski

On Leśniewski's Ontology[1]

The study group which before the Second World War became known as the Warsaw School of Logic, was, throughout its only too short history, dominated by philosophers rather than by mathematicians.[2] Although Łukasiewicz and Leśniewski, the two founders of the School, held chairs in the Faculty of Mathematics and Natural Sciences in the University of Warsaw, neither of them had a science degree or a degree in mathematics. Both were Arts scholars or, to be more exact, both had graduated in philosophy under Twardowski in Lwów. They had, of course, studied mathematics but within the province of logic philosophy seems to have inspired their research. Łukasiewicz missed no opportunity to acknowledge his debt to Aristotle and Leśniewski meant his logic to be a true, though very general, description of reality, a kind of πρώτη φιλοσοφία. It is significant that the Warsaw School of Logic influenced the development of philosophy in Poland to an extent to which it never succeeded in influencing the development of mathematics.

Both Łukasiewicz and Leśniewski were masters in logical precision but they did not indulge in 'pure' formalism. This is particularly true of Leśniewski. He used to mock 'mathematical games' which had nothing to offer except the possibility of transforming certain initial formulae into other formulae in accordance with somewhat conventional rules. Contrary to the fashion prevailing among mathematicians Leśniewski insisted that only *true* propositions should be allowed as axioms of a deductive theory and that only those rules of transformation should be

---

[1] An earlier version of this paper was read and discussed at a logical colloquium held in Oxford in July 1956.

I am indebted to Mr. F. P. Chambers, Mr. W. C. Kneale, Professor Julius Kraft, Mr. E. C. Luschei, Professor A. N. Prior, Father Ivo Thomas, and Professor H. J. Woodger for their helpful suggestions which enabled me to improve the composition of my essay.

I should also like to express my gratitude to Dr. B. Sobociński for his kind permission to include some of his own unpublished results in the present paper, and for his generous comments, from which I have benefited a great deal.

[2] For a discussion of the revival of logical and philosophical studies in Poland before the war see e.g. Ajdukiewicz [1935a] and Jordan [1945]. See also Sobociński [1956].

admitted which embodied intuitively *valid* rules of inference. He did not reject formalization as a method of presenting deductive theories. Far from it. But he maintained that a theory which was intuitively sound, did not lose its meaning through being subjected to the process of formalization. He consequently described himself as a confirmed intuitionist who at the same time was a radical formalist, and a better characterization of his standpoint could hardly be given.[3]

Leśniewski's criticism of 'pure' formalism shows that he had never ceased to be a philosopher. There are many 'pure' formalists among logicians and mathematicians but there are few 'pure' formalists among philosophers. For philosophers are, for the most part, preoccupied with the problem of meaning. Whether they deal with expressions of ordinary language or with logical formulae, they are concerned with interpretation rather than with formal elegance alone. The doctrine of 'pure' formalists could perhaps be condensed into the following motto: formalization before interpretation. Leśniewski's principle would read in the reverse.

For the most part Leśniewski's published papers present his theories at the stage of formalization with the problems of interpretation either left out or touched upon in an incidental manner. This makes the reading of these papers extremely difficult. It is the aim of the present contribution to bring the problems of interpretation to the foreground and by so doing serve as an informal introduction to one of the principal theories conceived by Leśniewski.

Leśniewski's System of Logic

Leśniewski's system of logic consists of two theories, which he called Protothetic and Ontology. Of these the former is perhaps the most comprehensive Logic of Propositions which has ever been devised. It goes beyond the classical Calculus of Propositions in several respects. It allows for instance for functorial variables for which constant functors of the Calculus of Propositions can be substituted.[4] It provides for the use of

---

[3] See Leśniewski [1929a, p. 78].

[4] One of the simplest protothetical theses has the form of the following expression:

A. $$[fp]: f(p). \equiv .f(f(f(p))).$$

In this thesis 'p' is a propositional variable while 'f' is a functorial variable of the same semantical category as, for instance, the functor of negation '$\sim$'. Thus from A we get

B. $$[p]: \sim(p). \equiv .\sim(\sim(\sim(p)))$$

by applying the rule of substitution.

For the explanation of the symbolism see pp. 132-133. In the terminology of the Warsaw School of Logic the term 'functor', which is due to Kotarbiński, denotes both propositional connectors, like 'if', 'or', 'and', and operators, like '$+$', '$\times$' in mathematics. We characterize the semantical category of a functor by saying what it forms, whether a proposition or a name, how many arguments it requires, whether one, two or more, and to what semantical category these arguments belong.

the universal quantifier to bind both the propositional and the functorial variables. It has a rule of definitions, which enables us to extend at will the variety of semantical categories within the field of the Logic of Propositions, and, in addition, it has a rule of extensionality; but the most significant point about Protothetic is that with its aid we can derive theses which enable us to dispose of the-usual rules for operating with the universal quantifier in any deductive theory of loser generality.[5] In the edifice of the possible deductive theories Protothetic forms the very base. It requires no more fundamental theory than itself whereas other deductive theories, not included in it, have to be built on it or on a part of it. This is the case with Ontology.

If Protothetic is the most comprehensive Logic of Propositions then Ontology is the most comprehensive Logic of Names. Roughly speaking, it comprises the traditional logic in its modernized form and has counterparts of the Calculus of Predicates, the Calculus of Classes, and the Calculus of Relations including the Theory of Identity.[6]

The title of this paper indicates that Ontology will be our main concern. Protothetic will of course be presupposed but we shall make no explicit use of protothetical theses in any deductions that may be required in the course of this study. This means that the proofs of ontological theorems will be based on 'intuition' rather than on explicit references to various laws of the Logic of Propositions. In this respect our proofs will not differ from the proofs which we ordinarily find in mathematics.

At first sight Leśniewski's use of the term *Ontology* may seem curious and daring, but it should become clearer as we proceed that he was eminently justified. In fact his whole conception of logic was ontological through and through in a truly classical sense.

Although Ontology was the subject of several university courses given by Leśniewski during the twenty years of his academic career in Warsaw until his death in 1939, there have been few papers published on it. The fundamental and most authoritative source is Leśniewski [1930]. It is an extremely condensed and difficult paper as it was meant to be a sort of 'identity card' of Ontology and not its 'lengthy biography'. It gives an axiom of Ontology and, with reference to Leśniewski [1929a, pp. 59-67], it also gives the rules of inference for Ontology stated here with a precision which has not since been improved upon. In addition, the paper contains a brief account of the researches of Leśniewski and his collaborators into the axiomatic foundations of Ontology.[7]

---

[5] In this connection see Leśniewski [1929a, p. 38].

[6] According to Leśniewski himself, Ontology is nearest to Schröder's Calculus of Classes provided the latter includes the theory of individuals. See Leśniewski [1929a, p. 5].

[7] A detailed report on these researches is contained in Sobociński [1934].

There are two more papers by Leśniewski on special problems theoretically belonging to Ontology. They are Leśniewski [1929] and Leśniewski [1929 b]. These papers are also worth mentioning for the fact that they contain some of the neatest examples of Leśniewski's method of setting out his deductions.

Finally, there is Leśniewski [1927-1931, Ch. XI], where he gives his analysis of the meaning of the primitive constant of Ontology as used in Leśniewski [1930]. Naturally enough this analysis is made from the point of view of the Polish language.

This is all that Leśniewski himself ever published on his Ontology. His copious notes and manuscripts, which contained a wealth of new results and which were to have been prepared for publication by Sobociński, were destroyed in 1944 during the war.

There have been some references to Ontology by other authors.

Contrary to the widely accepted practice Leśniewski intended his logic to be an interpreted system. He attached definite meanings to his constants and regarded the theses of his deductive theories as true propositions in the sense in which propositions of empirical sciences are accepted as true. It is with this in mind that we should approach his theories. Ontology has been described as the most comprehensive Logic of Names because its most characteristic expressions belong to the semantical category of names, just as the most characteristic expressions of Protothetic belong to the semantical category of propositions. If, however, we take into account the contents of Ontology then it would be more appropriate to describe it as a theory of what there is. Just as astronomy tells us something about heavenly bodies, the theses of Ontology tell us something about things, or objects if one prefers, or individuals. Since in accordance with Leśniewski's intentions theses of Ontology are to be regarded not as mere 'well formed formulae' but as meaningful propositions which can be examined for their truth or falsity, it is essential that we should understand as clearly as possible the ontological vocabulary. Only when we have mastered this vocabulary can a further step be made, namely the one which consists of a critical study of those ontological theses which are already at our disposal. Finally, we may try to discover new truths, which we can formulate in terms of the ontological vocabulary, and assign them their proper places in the theory by deducing them from the axiom or establishing their independence.

The Ontological Table

Learning the ontological vocabulary, like learning any other vocabulary, is an informal affair. Any method is good as long as it gives the required results. The meaning of ontological constants might be best explained by translating them into English. The ambiguity of ordinary usage, however, renders this method less effective than we might like it to be, and compels us to have recourse to other devices. In the logic of propositions the method of determining the meaning of the various proposition-forming functors for propositional arguments with the aid of truth tables has proved to be very helpful. In what follows we shall try to construct an "Ontological Table" and with its aid indicate the meaning of certain ontological constants in a way that should leave less room for misunderstanding than a direct translation.

The Ontological Table we have in mind is an extension of the well-known Eulerian diagrams. Among the names and name-like expressions of ordinary language we can distinguish (I) unshared names, each of which names one object only, *e.g.* 'Socrates', 'the Moon', *etc.*, (2) shared names, each of which names more than one object, *e.g.* 'satellite of Jupiter', 'heavenly body', *etc.*, and (3) fictitious names, *i.e.*, expressions which as regards their syntax behave like unshared or shared names, but which do not name anything, *e.g.* 'Pegasus', 'centaur', 'object which does not exist'.[8]

The Eulerian diagrams, the purpose of which was to illustrate the meaning of the syllogistic expressions, do not take into account unshared names or fictitious names. Thus the idea calls for an extension. In The Ontological Table below we shall use a shaded circle to represent the only object named by an unshared name, and an unshaded one to represent the many objects each of which is named by a shared name. No circle will be used in the case of a fictitious name. Thus the diagrams I.1, I.2, and I.3 below will serve to illustrate what one might call the semantical status of single names. The diagrams numbered from II.1 to II.16 will illustrate the semantical status of pairs of names. Analogous diagrams could be easily worked out for sets consisting of three or more names.

It should not be difficult to grasp what is meant when we say that diagram I.1 in The Ontological Table illustrates the semantical status of the name 'the Moon', or that diagram I.3 illustrates the semantical status of the name 'centaur'. The shaded circle in I.1 reminds us of the fact that there is exactly one object named by the name 'the Moon'. I.3 has no circle at all because the names whose semantical status it is meant to illustrate, name nothing existing. The name 'centaur' is one of such names. Of the name 'man' we can correctly say that its semantical status is illustrated by diagram I.2.

The case of pairs of names is equally simple. Diagram II.3, for instance, illustrates the semantical status of such pairs as the one consisting of

---

[8] The terminology 'unshared names', 'shared names' and 'fictitious names' is due to J. H. Woodger. See Woodger [1952a, p. 17] and Woodger [1952, p. 196].

The Ontological Table

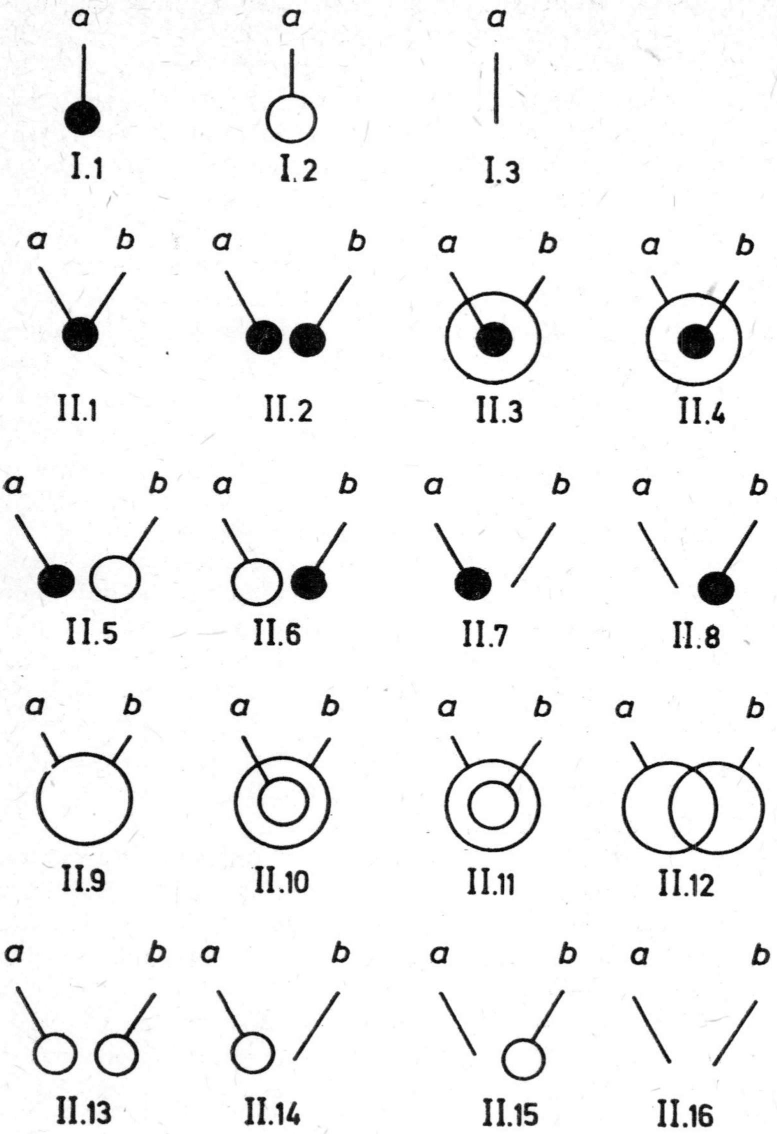

'the Moon' and 'heavenly body'. The shaded circle in II.3 is inside the unshaded one to remind us that the only object named by the name 'the moon' happens to be one of the many objects each of which is named by the name 'heavenly body'. To take another example, diagram II.12 illustrates the semantical status of such pairs of names as the one consisting of 'painter' and 'sculptor'. In this case the circles — which are unshaded to indicate that the names 'painter' and 'sculptor' name several individuals each — intersect to show that there are individuals each of which is named by both the names involved.

Now the usefulness of The Ontological Table lies in this: with its aid we can determine the meaning of certain proposition-forming functors for one or for two arguments which are names. And what is more, we can do so with a fairly high degree of precision. Strictly speaking, the functors whose meaning will be determined with the aid of The Ontological Table do not occur in ordinary usage, just as for instance the word 'infant' does not occur in ordinary language in the sense in which it is used by the Local Education Authorities. In a sense the functors of Ontology are 'technical' terms like the technical terms used by lawyers, stockbrokers, or engineers. They serve the purpose of expressing one's thoughts with more precision than can be attained by means of the ordinary idiom.

Primarily ontological functors are used in their graphical form. However, for the purpose of reading them out I shall equate them with their nearest counterparts from ordinary language, and hope that this will cause no misunderstanding.

The functors selected for characterizing the ontological vocabulary will be introduced under the following headings: functors of inclusion, functors of exclusion, functors of identity, and functors of existence. We begin with

(a) Functors of Inclusion

1. *The functor of singular inclusion* occurs in expressions of the type '$a \varepsilon b$' (to be read: a is b). A proposition of this type will be said to be true if and only if diagram II.1 or diagram II.3 illustrates the semantical status of the two names for which the 'a' and the 'b' stand. Otherwise it will be a false proposition.

2. *The functor of strong inclusion* occurs in expressions of the type '$a < b$' (to be read: every a is b). A proposition of this type will be said to be true if and only if one of the following diagrams, II.1, II.3, II.9, or II.10, illustrates the semantical status of the two names for which the 'a' and the 'b' stand. Otherwise it will be a false proposition.

3. *The functor of weak inclusion* occurs in expressions of the type '$a \subset b$' (to read: all a is b). A proposition of this type will be said to be true if and only if one of the following diagrams, II.1, II.3, II.8, II.9, II.10, II.15, or II.16, illustrates the semantical status of the two names for which the 'a' and the 'b' stand. Otherwise it will be a false proposition.

4. *The functor of partial inclusion* occurs in expressions of the type '$a \triangle b$' (to be read: some a is b). A proposition of this type will be said to be true if and only if one of the following diagrams, II.1, II.3, II.4, II.9, II.10, II.11, or II.12, illustrates the semantical status of the two names for which the 'a' and the 'b' stand. Otherwise it will be a false proposition.

(b) Functors of Exclusion

5. *The functor of singular exclusion* occurs in expressions of the type '$a \bar{\varepsilon} b$' (to be read: a is-not b). A proposition of this type will be said to be true if and only if one of the following diagrams, II.2, II.5, or II.7, illustrates the semantical status of the two names for which the 'a' and the 'b' stand. Otherwise it will be a false proposition.

6. *The functor of strong exclusion* occurs in expressions of the type '$a \not< b$' (to be read: every a is-not b). A proposition of this type will be said to be true if and only if one of the following diagrams, II.2, II.5, II.6, II.7, II.13, or II.14, illustrates the semantical status of the two names for which the 'a' and the 'b' stand. Otherwise it will be a false proposition.

7. *The functor of weak exclusion* occurs in expressions of the type '$a \not\subset b$' (to be read: no a is b). A proposition of this type will be said to be true if and only if one of the following diagrams, II.2, II.5, II.6, II.7, II.8, II.13, II.14, II.15, or II.16, illustrates the semantical status of the two names for which the 'a' and the 'b' stand. Otherwise it will be a false proposition.

8. *The functor of partial exclusion* occurs in expressions of the type '$a \bar{\Delta} b$' (to be read: some a is-not b). A proposition of this type will be said to be true if and only if one of the following diagrams, II.2, II.4, II.5, II.6, II.7, II.11, II.12, II.13, or II.14, illustrates the semantical status of the two names for which the 'a' and the 'b' stand. Otherwise it will be a false proposition.

(c) Functors of Identity

9. *The functor of singular identity* occurs in expressions of the type '$a = b$' (to be read: a is the same object as b). A proposition of this type will be said to be true if and only if diagram II.1 illustrates the semantical status of the two names for which the 'a' and the 'b' stand. Otherwise it will be a false proposition.

10. *The functor of strong identity* occurs in expressions of the type '$a \square b$' (to be read: only every a is b). A proposition of this type will be said to be true if and only if diagram II.1 or diagram II.9 illustrates the semantical status of the two names for which the 'a' and the 'b' stand. Otherwise it will be a false proposition[9].

11. *The functor of weak identity* occurs in expressions of the type '$a \circ b$' (to be read: a is identical with b). A proposition of this type will be said to be true if and only if one of the following diagrams, II.1, II.9, or II.16, illustrates the semantical status of the two names for which the 'a' and the 'b' stand. Otherwise it will be a false proposition.

---

[9] I have included this functor in the list on the suggestion of Dr. Sobociński.

(d) Functors of Existence

12. *The first functor of existence* occurs in expressions of the type 'ex(a)' (to be read: there exists at least one a). A proposition of this type will be said to be true if and only if diagram I.1 or diagram I.2 illustrates the semantical status of the name for which the 'a' stands. Otherwise it will be a false proposition.

13. *The second functor of existence* occurs in expressions of the type 'sol(a)' (to be read: there exists at most one a). A proposition of this type will be said to be true if and only if diagram I.1 or diagram I.3 illustrates the semantical status of the name for which the 'a' stands. Otherwise it will be a false proposition.

14. *The third functor of existence* occurs in expressions of the type 'ob(a)' (to be read: there exists exactly one a). A proposition of this type will be said to be true if and only if diagram I.1 illustrates the semantical status of the name for which the 'a' stands. Otherwise it will be a false proposition.

Here we end our list of ontological functors. It is quite clear that with the aid of The Ontological Table many more functors could be introduced. The ones I have included in the list appear to be of special interest. Thus the functor of singular inclusion plays the part of the only undefined constant in Leśniewski's original system of Ontology. Propositions formulated with the aid of the functor of strong inclusion and those formulated with the aid of the functor of partial inclusion have the same meaning as the meaning of the two affirmative categorical propositions of syllogistic, the universal and the particular respectively, provided we agree to regard the singular affirmative proposition as a special case of either. The functor of weak inclusion has the meaning which we sometimes attribute to the functor of inclusion in Boolean Algebra.

Among the functors of exclusion we do not find functors which could be said to form universal negative or particular negative propositions of syllogistic. A negative proposition of syllogistic has 'existential import' with respect to both its arguments, *i.e.*, with respect to the subject and with respect to the predicate. This means that in a true negative proposition of syllogistic neither the subject nor the predicate can be a fictitious name. Now a true proposition formed with the aid of any of the functors of exclusion above distinguished may have a fictitious name as the second argument, *i.e.*, as the predicate, and the functor of weak exclusion produces a true proposition if the first argument, *i.e.*, the subject, is a fictitious name. It may be of interest to note that a weak exclusion is the contradictory of the corresponding partial inclusion and that a partial exclusion is the contradictory of the corresponding weak inclusion.

The distinction between the various functors of identity appears to be very useful, particularly in connection with the problem whether identity implies existence or not. It seems to be evident from what has been said that if by identity we mean weak identity then there is no ground for asserting that it implies existence. On the other hand the inference from identity to existence is quite legitimate if by identity we mean strong identity or singular identity.

The functors of existence will prove very useful in formulating simple ontological theses which eventually will lead to the formulation of axioms.

Theses Relating Different Functors

Now that we know the meaning of a number of ontological functors, we can proceed to constructing a few propositions about what there is. In doing so we shall have at our disposal the vocabulary of the Logic of Propositions including the vocabulary of the Theory of Quantification. For as it has already been said, Ontology presupposes Protothetic.

We shall be using the Peano-Russellian symbolism for the purpose of formulating ontological theses and it may not be out of place to give a very brief account of it with a view to meeting our present requirements only. Small letters of the initial portion of the Latin alphabet will serve as variables for which any name, unshared, shared, or fictitious, can be substituted. In contradistinction to *propositional* variables, like $p, q, r, \ldots$, they may conveniently be described as *nominal* variables. Together with appropriate functors they form propositional functions of which the following may serve as examples: $a \varepsilon b$, $a < c$, $d \not\subset b$, $b = a$, $c\, \bar{\varepsilon}\, d$, $\mathrm{ex}(a)$, $\mathrm{sol}(a)$, *etc.* Out of these simple propositional functions compound propositional functions can be formed by using proposition-forming functors for propositional arguments. Thus for instance we have '$a = b . \equiv . b = a$' to mean '*a* is the same object as *b* if and only if *b* is the same object as *a*', '$a \varepsilon b . \supset . b \triangle a$' to mean 'if *a* is *b* then some *b* is *a*', $. a \varepsilon b . \vee . a\, \bar{\varepsilon}\, b$' to mean '*a* is *b* or *a* is-not *b*', '$a \varepsilon b . b \varepsilon a$' to mean '*a* is *b* and *b* is *a*', and finally '$\sim(a \varepsilon b)$' to mean 'it is not true that *a* is *b*'. In addition to proposition-forming functors for propositional arguments we shall make use of the quantifiers: the universal quantifier '$[a\, b\, c \ldots]$' to mean 'for all *a*, *b*, *c*, ...' and the particular quantifier '$[\exists\, a\, b\, c \ldots]$' to mean 'for some *a*, *b*, *c*, ...'. Accordingly, the expressions '$[a] . a \circ a$' and '$[\exists a] . \sim(a \varepsilon a)$' mean 'for all *a*, — *a* is identical with *a*' and 'for some *a*, — it is not true that *a* is *a*'.

The use of dots as a 'punctuation' device is one of the characteristic features of the symbolism and calls for explanation[10]. If a dot or a cluster of dots adjoins, *i.e.*, immediately precedes or immediately follows one of

[10] See Whitehead and Russell [1935, p. 94].

the following '≡', '⊃', or '∨', then it only indicates the scope of the functor. Thus if in an expression we find a cluster of say three dots immediately preceding '⊃' then this indicates that the antecedent of the implication involved extends to the left until it reaches a cluster of more than three dots, or failing that until it reaches a cluster of more than three dots, or failing that until it reaches the beginning of the expression under consideration. Analogously, a cluster of three dots immediately following '⊃' indicates that the consequent of the implication involved extends to the right up to the first cluster of more than three dots, or failing that up to the end of the expression under consideration. If in an expression a dot or a cluster of say five dots immediately follows a quantifier then it indicates that the expression 'covered' by the quantifier extends to the right over the clusters of less than five dots, and over the clusters of five dots provided the latter are not immediately followed by '≡', '⊃', or '∨', until it reaches a cluster of five dots immediately followed by '≡' or '⊃' or '∨', or — failing that — until it reaches a cluster of more than five dots. If neither is the case, the quantifier covers the whole of the expression to its right. A dot or a cluster of say four dots which neither adjoins '≡' or '⊃' or '∨' nor immediately follows a quantifier means 'and', *i.e.*, it stands for the functor of conjunction. The first argument of the conjunction extends to the left of the cluster over the clusters of less than four dots until it reaches a cluster of four or more dots. The second argument of the conjunction is indicated in an analogous way, and the procedure is generalized to apply to conjunctions with more than two conjuncts.

In order to leave as few doubts as possible as regards the interpretation of the symbolism, a number of ontological theses will be given below together with a translation into ordinary language.

If we bear in mind the meaning we have attached to the various ontological functors then little reflection will be needed to convince ourselves of the truth of the following theses:

T1. $[a]{:}\mathrm{ex}(a).\equiv.[\exists b].b\,\varepsilon\,a$

i.e., for all *a*, — there exists at least one *a* if and only if for some *b*, — *b* is *a*.

T2. $[a]{:}\mathrm{ex}(a).\equiv.[\exists b].b < a$

i.e., for all *a*, — there exists at least one *a* if and only if for some *b*, — every *b* is *a*.

T3. $[a]{:}\mathrm{ex}(a).\equiv.[\exists b].\sim(a \subset b)$

i.e., for all *a*, — there exists at least one *a* if and only if for some *b*, — it is not true that all *a* is *b*.

T4. $[a]{:}\mathrm{ex}(a).\equiv.[\exists b].a \nless b$

i.e., for all *a*, — there exists at least one *a* if and only if for some *b*, every *a* is-not *b*.

T5. $[a]\therefore \mathrm{sol}(a). \equiv :[b\,c]:b \mathrel{\varepsilon} a . c \mathrel{\varepsilon} a . \supset . b \mathrel{\varepsilon} c$
i.e., for all *a*, — there exists at most one *a* if and only if for all *b* and *c*, — if *b* is *a* and *c* is *a* then *b* is *c*.

T6. $[a]\therefore \mathrm{sol}(a). \equiv :[b]:b \mathrel{\varepsilon} a . \supset . a \mathrel{\varepsilon} b$
i.e., for all *a*, — there exists at most one *a* if and only if for all *b*, — if *b* is *a* then *a* is *b*[11].

T7. $[a]\therefore \mathrm{sol}(a). \equiv :[b\,c]:b < a . c < a . \supset . b < c$
i.e., for all *a*, — there exists at most one *a* if and only if for all *b* and *c*, — if every *b* is *a* and every *c* is *a* then every *b* is *c*.

T8. $[a]\therefore \mathrm{sol}(a). \equiv :[b]:b < a . \supset . a < b$
i.e., for all *a*, — there exists at most one *a* if and only if for all *b*, — if every *b* is *a* then every *a* is *b*[11].

T9. $[a]::\mathrm{sol}(a). \equiv \therefore [b\,c\,d]\therefore b \subset a . c \subset a . \supset : b \subset c . \vee . c \subset d$
i.e., for all *a*, — there exists at most one *a* if and only if for all *b*, *c*, and *d*, — if all *b* is *a* and all *c* is *a* then (all *b* is *c* or all *c* is *d*).

T10. $[a]::\mathrm{sol}(a). \equiv \therefore [b\,c]\therefore b \subset a . \supset : a \subset b . \vee . b \subset c$
i.e., for all *a*, — there exists at most one *a* if and only if for all *b* and *c*, — if all *b* is *a* then (all *a* is *b* or all *b* is *c*)[11].

T11. $[a]\therefore \mathrm{sol}(a). \equiv :[b\,c]:b \mathrel{\triangle} a . c \mathrel{\triangle} a . \supset . b \mathrel{\triangle} c$
i.e., for all *a*, — there exists at most one *a* if and only if for all *b* and *c*, — if some *b* is *a* and some *c* is *a* then some *b* is *c*.

T12. $[a]::\mathrm{sol}(a). \equiv \therefore [b\,c\,d]\therefore b \mathrel{\bar{\varepsilon}} c . c \mathrel{\bar{\varepsilon}} d . \supset : b \mathrel{\bar{\varepsilon}} a . \vee . c \mathrel{\bar{\varepsilon}} a$
i.e., for all *a*, — there exists at most one *a*, if and only if for all *b*, *c*, and *d*, — if *b* is-not *c* and *c* is-not *d* then (*b* is-not *a* or *c* is-not *a*).

T13. $[a]::\mathrm{sol}(a). \equiv \therefore [b\,c\,d]\therefore b \nless c . c \nless d . \supset : b \nless a . \vee . c \nless a$
i.e., for all *a*, — there exists at most one *a* if and only if for all *b*, *c*, and *d*, — if every *b* is-not *c* and every *c* is-not *d* then (every *b* is-not *a* or every *c* is-not *a*).

T14. $[a]::\mathrm{sol}(a). \equiv \therefore [b\,c]\therefore b \not\subset c . \supset : b \not\subset a . \vee . c \not\subset a$
i.e., for all *a*, — there exists at most one *a* if and only if for all *b* and *c*, — if no *b* is *c* then (no *b* is *a* or no *c* is *a*).

T15. $[a]\therefore \mathrm{sol}(a). \equiv :[b\,c]:b \mathrel{\overline{\triangle}} c . a \mathrel{\overline{\triangle}} b . \supset . b \mathrel{\overline{\triangle}} a$
i.e., for all *a*, — there exists at most one *a* if and only if for all *b* and *c*, — if some *b* is-not *c* and some *a* is-not *b* then some *b* is-not *a*.

T16. $[a]:\mathrm{ob}(a). \equiv . [\exists b] . a \mathrel{\varepsilon} b$
i.e., for all *a*, — there exists exactly one *a* if and only if for some *b*, — *a* is *b*.

T17. $[a]:\mathrm{ob}(a). \equiv . \mathrm{ex}(a) . \mathrm{sol}(a)$
i.e., for all *a*, — there exists exactly one *a* if and only if (there exists at least one *a* and there exists at most one *a*).

---

[11] T6, T8, and T10 are due to Sobociński.

The functor of singular inclusion suffices for the purpose of defining all the remaining functors whose meaning has been determined with the aid of The Ontological Table. This is evident from the following theses, which will now be given in symbols only.

T18. $[a\,b]::a < b\,.\equiv\therefore[\exists c].c\,\varepsilon\,a\therefore[c]:c\,\varepsilon\,a\,.\supset.\,c\,\varepsilon\,b$
T19. $[a\,b]\therefore a \subset b\,.\equiv:[c]:c\,\varepsilon\,a\,.\supset.\,c\,\varepsilon\,b$
T20. $[a\,b]:a \triangle b\,.\equiv.[\exists c].c\,\varepsilon\,a.c\,\varepsilon\,b$
T21. $[a\,b]:a\,\bar{\varepsilon}\,b\,.\equiv.a\,\varepsilon\,a.\sim(a\,\varepsilon\,b)$
T22. $[a\,b]::a \nless b\,.\equiv\therefore[\exists c].c\,\varepsilon\,a\therefore[c]:c\,\varepsilon\,a\,.\supset.\sim(c\,\varepsilon\,b)$
T23. $[a\,b]\therefore a \not\subset b\,.\equiv:[c]:c\,\varepsilon\,a\,.\supset.\sim(c\,\varepsilon\,b)$
T24. $[a\,b]:a\,\overline{\triangle}\,b\,.\equiv.[\exists c].c\,\varepsilon\,a.\sim(c\,\varepsilon\,b)$
T25. $[a\,b]:a = b\,.\equiv.a\,\varepsilon\,b.b\,\varepsilon\,a$
T26. $[a\,b]::a \square b\,.\equiv\therefore[\exists c].c\,\varepsilon\,a\therefore[c]:c\,\varepsilon\,a\,.\equiv.c\,\varepsilon\,b$
T27. $[a\,b]\therefore a \circ b\,.\equiv:[c]:c\,\varepsilon\,a\,.\equiv.c\,\varepsilon\,b$

Finally it may be worth noting that the following theses hold:

T28. $[a\,b]:a < b\,.\equiv.\mathrm{ex}(a).a \subset b$
T29. $[a\,b]:a \nless b\,.\equiv.\mathrm{ex}(a).a \not\subset b$
T30. $[a\,b]:a \not\subset b\,.\equiv.\sim(a \triangle b)$
T31. $[a\,b]:a\,\overline{\triangle}\,b\,.\equiv.\sim(a \subset b)$
T32. $[a\,b]:a \square b\,.\equiv.\mathrm{ex}(a).a \circ b$
T33. $[a\,b]:a = b\,.\equiv.\mathrm{sol}(a).a \square b$

Historical Review of Axioms of Ontology

Leśniewski's original system of Ontology constructed in 1920 was based on the following single axiom:

T34. $[a\,b]::a\,\varepsilon\,b\,.\equiv\therefore[\exists c].c\,\varepsilon\,a\therefore[c]:c\,\varepsilon\,a\,.\supset.c\,\varepsilon\,b\therefore[c\,d]:c\,\varepsilon\,a.$ $d\,\varepsilon\,a\,.\supset.c\,\varepsilon\,d$

The meaning of T34 becomes clear as soon as we realize that in virtue of T1, T19, and T5 this thesis is inferentially equivalent to

T35. $[a\,b]:a\,\varepsilon\,b\,.\equiv.\mathrm{ex}(a).a \subset b.\mathrm{sol}(a)$

In 1921 Leśniewski found that T34 could be replaced as a single axiom of Ontology by

T36. $[a\,b]::a\,\varepsilon\,b\,.\equiv\therefore[\exists c].c\,\varepsilon\,a.c\,\varepsilon\,b\therefore[c\,d]:c\,\varepsilon\,a.d\,\varepsilon\,a\,.\supset.c\,\varepsilon\,d$

In virtue of T20 and T5, T36, which is shorter than T34, is inferentially equivalent to

T37. $[a\,b]:a\,\varepsilon\,b\,.\equiv.a \triangle b.\mathrm{sol}(a)$

On the basis of Protothetic alone we can easily show that T34 is inferentially equivalent to the following four theses:

T38. $[a\, b]\!:\! a \varepsilon b . \supset . [\exists c] . c \varepsilon a$

T39. $[a\, b\, c]\!:\! a \varepsilon b . c \varepsilon a . \supset . c \varepsilon b$

T40. $[a\, b\, c\, d]\!:\! a \varepsilon b . c \varepsilon a . d \varepsilon a . \supset . c \varepsilon d$

T41. $[a\, b\, c]\!::\! c \varepsilon a \therefore [d]\!:\! d \varepsilon a . \supset . d \varepsilon b \therefore [d\, e]\!:\! d \varepsilon a . e \varepsilon a . \supset . d \varepsilon e \therefore \supset . a \varepsilon b$

In 1921 Tarski showed that by making use of an appropriate definition we can deduce T40 from T39, and in 1929 Sobociński succeeded in proving that T38 and T39 alone could serve as an axiom system of Ontology and that T41 could be deduced from them by applying, in the course of deductions, the rule of extensionality, which at that time had already been added to Ontology. In the same year Sobociński established that

T42. $[a\, b]\!::\! a \varepsilon b . \equiv \therefore [\exists c] . c \varepsilon a . c \varepsilon b \therefore [c]\!:\! c \varepsilon a . \supset . a \varepsilon c$

which is shorter than T36, could also be used as a single axiom of Ontology. In virtue of T20 and T6, T42 is inferentially equivalent to T37.

In connection with these results of Sobociński a further simplification of the axiomatic foundations of the system of Ontology with the functor of singular inclusion as a single undefined functor was achieved by Leśniewski, who in 1929 noticed that the axiom system consisting of T38 and T39 was inferentially equivalent to

T43. $[a\, b]\!:\! a \varepsilon b . \equiv . [\exists c] . a \varepsilon c . c \varepsilon b$

which thus qualified to the status of a single axiom of the theory[12].

In 1946 I set myself the task of finding single axioms for ontological functors other than the functor of single inclusion. I remembered from Leśniewski's lectures that systems of Ontology could be constructed with the aid of the functor of strong inclusion or with the aid of the functor of partial inclusion, but no theses which could be used as the respective single axioms had ever come to my knowledge.

As a result of my research I found that the following thesis

T44. $[a\, b]\!::\! a \varepsilon b . \equiv \therefore a < b \therefore [c\, d]\!:\! c < a . d < a . \supset . c < d$

which in virtue of T7 is inferentially equivalent to

T45. $[a\, b]\!:\! a \varepsilon b . \equiv . a < b . \mathrm{sol}(a)$

could be used as a definition of singular inclusion in terms of strong nclusion, and that a system of Ontology with the functor of strong in-

[12] See Leśniewski [1930] and Sobociński [1934].

clusion as a single undefined functor could be based on the following axiom:

T46. $[a\,b]::\!:a < b. \equiv :\cdot:[\exists c].c < a:\cdot:[c]::c < a. \supset \therefore [\exists d]\therefore d < c. d < b\therefore [e\,f]:e < d.f < d. \supset .e < f$

It is not difficult to see from T2, T7, and T45 that T46 is inferentially equivalent to

T47. $[a\,b]::a < b. \equiv \therefore \mathrm{ex}(a)\therefore [c]:c < a. \supset .[\exists d].d \varepsilon c.d \varepsilon b$

which thus throws some light on the meaning of T46.

As a further result I established that if we used

T48. $[a\,b]::a \varepsilon b. \equiv \therefore a \triangle b\therefore [c\,d]:c \triangle a.d \triangle a. \supset .c \triangle d$

as a definition of singular inclusion in terms of partial inclusion then

T49. $[a\,b]::a \triangle b. \equiv \therefore [\exists c]\therefore c \triangle a.c \triangle b\therefore [d\,e]:d \triangle c.e \triangle c. \supset .d \triangle e$

could be used as a single axiom for a system of Ontology with the functor of partial inclusion as a single undefined functor.

In virtue of T11, T48 is inferentially equivalent to

T50. $[a\,b]:a \varepsilon b. \equiv .a \triangle b.\mathrm{sol}(a)$

and T49 is inferentially equivalent to T20 in virtue of T11 and T50.

In addition I found that the functor of singular exclusion could be used as a single undefined functor in a system of Ontology. In such a system

T51. $[a\,b]:a \varepsilon b. \equiv .[\exists c].a \bar{\varepsilon} c.\sim(a \bar{\varepsilon} b)$

could be used as a definition of singular inclusion while the single axiom could have the form of the following expression:

T52. $[a\,b]:\cdot:a \bar{\varepsilon} b. \equiv ::[\exists c].c \bar{\varepsilon} b.\sim(c \bar{\varepsilon} a)::[c\,d\,e]\therefore c \bar{\varepsilon} d.d \bar{\varepsilon} e. \supset :c \bar{\varepsilon} a. \vee .d \bar{\varepsilon} a$

The meaning of T52 can be seen from

T53. $[a\,b]\therefore a \bar{\varepsilon} b. \equiv :[\exists c].c \bar{\varepsilon} b.c \varepsilon a:\mathrm{sol}(a)$

which is inferentially equivalent to T52 in virtue of T51 and T12.

I returned to these and similar problems in 1953 when I found that a system of Ontology could be based on the functor of weak inclusion. Singular inclusion could be defined with the aid of the following thesis:

T54. $[a\,b]:\cdot:a \varepsilon b. \equiv ::[\exists c].\sim(a \subset c):a \subset b::[c\,d\,e]\therefore c \subset a.d \subset a. \supset :c \subset d. \vee .d \subset e$

which in virtue of T3 and T9 is inferentially equivalent to T35.

The following thesis:

T55. $[a\,b]::a \subset b\,. \equiv :\cdot:[c\,d]:\cdot:\sim(c \subset d).c \subset a\,. \supset ::[\exists\, e\,f]::\sim(e \subset f).$
$e \subset c.e \subset b::[g\,h\,i]\therefore g \subset e.h \subset e\,. \supset :g \subset h\,. \vee .h \subset i$

could serve as a single axiom. We can easily grasp the meaning of this axiom if we compare it with

T56. $[a\,b]\therefore a \subset b\,. \equiv :[c]:c < a\,. \supset .[\exists d].d \,\varepsilon\, c.d \,\varepsilon\, b$

which is inferentially equivalent to it in virtue of T3, T28, and T54.

In connection with this result I exchanged a number of letters with Dr. Sobociński, from whom I learned that Leśniewski had had axioms for the functors of strong inclusion, weak inclusion, and partial inclusion. Sobociński could not recollect what these axioms were. He remembered, however, that T54 was the definition used by Leśniewski. Owing to his own research Sobociński knew that T49 could be used as a single axiom for the functor of partial inclusion. It is more likely than not that this was also known to Leśniewski.

While analysing T55 Sobociński noticed that a system of Ontology with the functor of weak inclusion as a single undefined term could be based on a shorter axiom, which had the form of the following expression:

T57. $[a\,b]::a \subset b\,. \equiv :\cdot:[c\,d]:\cdot:\sim(c \subset d).c \subset a::[d\,e\,f]\therefore d \subset c.e \subset c.$
$\supset :d \subset e\,. \vee .e \subset f:: \supset .e \subset b$

In virtue of T54 this axiom is inferentially equivalent to

T58. $[a\,b]\therefore a \subset b\,. \equiv :[c]:c \,\varepsilon\, a\,. \supset .c \subset b$

A further simplification of the axiom for weak inclusion took place when Sobociński proved T10. This enabled him to shorten T54 as a definition of singular inclusion in terms of weak inclusion, and to replace T57 by the following thesis:

T59. $[a\,b]::a \subset b\,. \equiv :\cdot:[c\,d]:\cdot:\sim(c \subset d).c \subset a::[d\,e]\therefore d \subset c\,. \supset :$
$c \subset d\,. \vee .d \subset e:: \supset .c \subset b$

which again is inferentially equivalent to T58 in virtue of T9, T10 and T54.

As a sequel to these results Sobociński proved T8 and

T60. $[a\,b]::a \,\varepsilon\, b\,. \equiv \therefore a < b \therefore [c]:c < a\,. \supset .a < c$

which defines singular inclusion in terms of strong inclusion. He also established that the following thesis

T61. $[a\,b]::a < b\,. \equiv :\cdot:[\exists c].c < a:\cdot:[c]::c < a\therefore [d]:d < c\,. \supset .$
$c < d\therefore \supset .c < b$

sufficed as a single axiom for the system of Ontology based on strong inclusion.

It is evident from T2 and T60 that T61 is inferentially equivalent to

T62. $[ab]::a < b. \equiv \therefore \mathrm{ex}(a) \therefore [c]:c \varepsilon a. \supset .c < b$

At this stage Sobociński remarked that curiously enough T36, T49 and T61 — which are axioms for singular inclusion, partial inclusion, and strong inclusion respectively — consisted of six ontological units each[13].

In 1956 I obtained further results concerning the axiomatic foundations of Ontology. I found that T59 could be replaced by the following shorter thesis:

T63. $[a\,b]::{\cdot}{:}a \subset b. \equiv :{\cdot}:[c]:{\cdot}:c \subset a::[d\,e]\therefore d \subset c. \supset :c \subset d. \vee .d \subset e::$
$\supset .c \subset b$

This axiom consists of six ontological units and is inferentially equivalent to

T64. $a\,b \therefore a \subset b. \equiv :[c]:c \subset a.\mathrm{sol}(c). \supset .c \subset b$

as can be seen from T10.

I also found that systems of Ontology could be constructed on the basis of each of the remaining functors of exclusion. The theses given below can serve as the respective definitions of singular inclusion and as axioms:

T65. $[a\,b]:{\cdot}:a \varepsilon b. \equiv ::[\exists c].a \not< c: \sim(a \not< b)::[c\,d]\therefore c \not< d. \supset :$
$a \not< c. \vee .a \not< d$

T66. $[a\,b]::{\cdot}{:}a \not< b. \equiv ::{:}[\exists c\,d].c \not< d. \sim(c \not< a)::{:}[c\,d]:{\cdot}:c \not< d::$
$[e\,f]\therefore e \not< f. \supset :c \not< e. \vee .c \not< f:: \supset :c \not< a. \vee .c \not< b$

T67. $[a\,b]:{\cdot}:a \varepsilon b. \equiv :: \sim(a \not\subset b)::[c\,d]\therefore c \not\subset d. \supset :c \not\subset a. \vee .d \not\subset a$

T68. $[a\,b]:{\cdot}:a \not\subset b. \equiv ::[c]::[d\,e]\therefore d \not\subset e. \supset :d \not\subset c. \vee .e \not\subset c \therefore \supset :$
$c \not\subset a. \vee .c \not\subset b$

T69. $[a\,b]::a \varepsilon b. \equiv \therefore [\exists c].a \overline{\triangle} c. \sim(a \overline{\triangle} b) \therefore [c\,d]:c \overline{\triangle} d.a \overline{\triangle} c.$
$\supset .c \overline{\triangle} a$

T70. $[a\,b]:{\cdot}:a \overline{\triangle} b. \equiv ::[\exists c]:: \sim(c \overline{\triangle} a).c \overline{\triangle} b \therefore [d\,e]:d \overline{\triangle} e.c \overline{\triangle} d.$
$\supset .d \overline{\triangle} c$

It is interesting to note that T68, *i.e.*, the axiom for weak exclusion, and T70, *i.e.*, the axiom for partial exclusion, consist of six ontological units each. I have not, however, succedded in finding six-unit axioms for singular exclusion or for strong exclusion.

The functor of singular exclusion deserves our attention because with its aid we can formulate a fairly simple frame for writing so-called ontological definitions. On adapting the rule of ontological definitions and the

---

[13] An ontological unit is a simple propositional function formed with the aid of an ontological functor.

rule of extensionality to certain properties of singular exclusion I was able to prove that following thesis:

T71. $[a\, b]{:}\, a \,\bar{\varepsilon}\, b. \equiv .[c\, d].\, a \,\bar{\varepsilon}\, d.\, c \,\bar{\varepsilon}\, b. \sim (c \,\bar{\varepsilon}\, a)$

could be used as a single axiom of a system of Ontology. This result dates from 1954[14].

Proof that a System of Ontology Can Be Based on Partial Inclusion

My summary of the results obtained in connection with the study of the axiomatic foundations of Ontology is far from being exhaustive. In particular it gives no insight into the character and amount of deductive considerations which have to be completed before a thesis can be declared to rank as a single axiom of a system of Ontology based on a single undefined functor other than the functor of singular inclusion. In what follows I propose to show how it can be established that T49, for instance, can be used as a single axiom of the system of Ontology with the functor of partial inclusion as the only undefined term. The problem is solved by proving that such a system is inferentially equivalent to the original system of Ontology based on singular inclusion. Since the functor of singular inclusion and the functor of partial inclusion are two different functors we have to deduce T49 and T48, which is a definition of singular inclusion in terms of partial inclusion, from T36 and T20, *i.e.*, from a possible axiom of the original system of Ontology and from a definition of partial inclusion in terms of singular inclusion. Then we have to show that T36 and T20 follow from T49 and T48. To put it in self-explanatory symbols we have to establish that

$$\{T36, T20\} \rightleftarrows \{T49, T48\}$$

Our deductions proceed as follows:

Axiom:

A. = T36. $[a\, b]{::}\, a \,\varepsilon\, b. \equiv {\therefore}[\exists c].\, c \,\varepsilon\, a.\, c \,\varepsilon\, b {\therefore} [c d]{:}\, c \,\varepsilon\, a.\, d \,\varepsilon\, a. \supset .\, c \,\varepsilon\, d$

Definition:

D. = T20. $[a\, b]{:}\, a \triangle b. \equiv .[\exists\, c].\, c \,\varepsilon\, a.\, c \,\varepsilon\, b$

[14] For the purpose of establishing the adequacy of the axioms for '$\subset$', '$\bar{\varepsilon}$', '$\not\prec$', and '$\triangle$' use is made of ontological definitions, whose nature is discussed in the last section of this paper.

TT1. $[a\, b\, c] : a \varepsilon b . c \varepsilon a . \supset . c \varepsilon b$

Proof: $[a\, b\, c] ::$

(1) $a \varepsilon b .$

(2) $c \varepsilon a . \supset \therefore$

(3) $[d\, e] : d \varepsilon a . e \varepsilon a . \supset . d \varepsilon e \therefore$ (follows from A and 1)

(4) $a \varepsilon a .$ (from A, 2, 3)

(5) $a \varepsilon c \therefore$ (3, 4, 2)

(6) $[d\, e] : d \varepsilon c . e \varepsilon c . \supset . d \varepsilon e \therefore$ (A, 2)

$c \varepsilon b$ (A, 5, 1, 6)

TT2. $[a\, b] : a \triangle b . \supset . b \triangle a$ (D)

TT3. $[a\, b] : a \varepsilon b . \supset . a \triangle b$

$[a\, b] \therefore$

(1) $a \varepsilon b . \supset :$

(2) $[\exists\, c] . c \varepsilon a . c \varepsilon b :$ (A, 1)

$a \triangle b$ (D, 2)

TT4. $[a\, b\, c\, d] : a \varepsilon b . c \triangle a . d \triangle a . \supset . c \triangle d$

$[a\, b\, c\, d] ::$

(1) $a \varepsilon b .$

(2) $c \triangle a .$

(3) $d \triangle a . \supset \therefore$

$[\exists\, e] :$

(4) $e \varepsilon c .$ (D, 2)

(5) $e \varepsilon a :$ (D, 2)

$[\exists\, f] .$

(6) $f \varepsilon d .$ (D, 3)

(7) $f \varepsilon a :$ (D, 3)

(8) $e \varepsilon f :$ (A, 1, 5, 7)

(9) $e \varepsilon d \therefore$ (TT1, 6, 8)

$c \triangle d$ (D, 4, 9)

TT5. $[a\, b\, c] : a \varepsilon b . b \varepsilon c . \supset . b \varepsilon a$

$[a\, b\, c] ::$

(1) $a \varepsilon b .$

(2) $b \varepsilon c . \supset \therefore$

(3) $[d\, e] : d \varepsilon b . e \varepsilon b . \supset . d \varepsilon e \therefore$ (A, 2)

(4) $b \varepsilon b \therefore$ (A, 1, 3)

$b \varepsilon a$ (3, 4, 1)

TT6. $[a\, b\, c] :: [d\, e] : d \triangle a . e \triangle a . \supset . d \triangle e \therefore b \varepsilon a . c \varepsilon a \therefore \supset . b \varepsilon c$

$[a\, b\, c] ::$

(1) $[d\, e] : d \triangle a . e \triangle a . \supset . d \triangle e \therefore$

(2) $b \varepsilon a.$
(3) $c \varepsilon a .\therefore \supset :$
(4) $b \triangle a.$ (TT3, 2)
(5) $c \triangle a.$ (TT3, 3)
(6) $b \triangle c:$ (1, 4, 5)
$[\exists d].$
(7) $d \varepsilon b.$ } (D, 6)
(8) $d \varepsilon c.$ } (D, 6)
(9) $b \varepsilon d:$ (TT5, 7, 2)
$b \varepsilon c$ (TT1, 8, 9)

TT7. $[a\, b]::a \triangle b .\therefore [c\, d]:c \triangle a . d \triangle a . \supset . c \triangle d .\therefore \supset . a \varepsilon b$
$[a\, b]::$
(1) $a \triangle b .\therefore$
(2) $[c\, d]:c \triangle a . d \triangle a . \supset . c \triangle d .\therefore \supset \therefore$
(3) $[c\, d]:c \varepsilon a . d \varepsilon a . \supset . c \varepsilon d .\therefore$ (TT6, 2)
(4) $[\exists c] . c \varepsilon a . c \varepsilon b .\therefore$ (D, 1)
$a \varepsilon b$ (A, 4, 3)

TT8. = T48. $[a\, b]::a \varepsilon b . \equiv \therefore a \triangle b .\therefore [c\, d]:c \triangle a . d \triangle a . \supset . c \triangle d$[15] (TT3, TT4, TT7)

TT9. $[a\, b]::a \triangle b . \supset \therefore [\exists c] .\therefore c \triangle a . c \triangle b .\therefore [d\, e]:d \triangle c . e \triangle c . \supset . d \triangle e$
$[a\, b]::$
(1) $a \triangle b . \supset ::$
$[\exists c] .\therefore$
(2) $c \varepsilon a.$ } (D, 1)
(3) $c \varepsilon b.$ } (D, 1)
(4) $c \triangle a.$ (TT3, 2)
(5) $c \triangle b .\therefore$ (TT3, 3)
(6) $[d\, e]:d \triangle c . e \triangle c . \supset . d \triangle e ::$ (TT4, 2)
$[\exists c] .\therefore c \triangle a . c \triangle b .\therefore [d\, e]:d \triangle c . e \triangle c . \supset . d \triangle e$ (4, 5, 6)

TT10. $[a\, b\, c]::c \triangle a . c \triangle b .\therefore [d\, e]:d \triangle c . e \triangle c . \supset . d \triangle e .\therefore \supset . a \triangle b$
$[a\, b\, c]::$
(1) $c \triangle a.$
(2) $c \triangle b .\therefore$
(3) $[d\, e]:d \triangle c . e \triangle c . \supset . d \triangle e .\therefore \supset .$
(4) $a \triangle c.$ (TT2, 1)
(5) $b \triangle c.$ (TT2, 2)
$a \triangle b$ (3, 4, 5)

---

[15] In Słupecki [1955] (*Editorial Note*: Also in the present volume) this thesis is deduced from T34.

TT11. = T49. $[a\,b]::a \triangle b. \equiv \therefore [\exists\, c] \therefore c \triangle a . c \triangle b \therefore [d\,e]:d \triangle c . e \triangle c . \supset . d \triangle e$ (TT9, TT10)

It is evident from TT11 and TT8 that T49 and T48 are deducible as theses within the framework of the original system of Ontology. It remains to show that T36 and T20 can be derived from T49 and T48, if the latter theses are regarded as the axiom and a definition respectively.

Axiom:

TTA. = TT11 $[a\,b]::a \triangle b. \equiv \therefore [\exists\, c] \therefore c \triangle a . c \triangle b \therefore [d\,e]:d \triangle c . e \triangle c . \supset . d \triangle e$

Definition:

TTD. = TT8 $[a\,b]::a\, \varepsilon\, b. \equiv \therefore a \triangle b \therefore [c\,d]:c \triangle a . d \triangle a . \supset . c \triangle d$

TT11.1 $[a\,b]:a \triangle b. \supset . [\exists\, c] . c\, \varepsilon\, a . c\, \varepsilon\, b$

$[a\,b]::$

(1) $a \triangle b . \supset ::$

$[\exists\, c] \therefore$

(2) $c \triangle a .$ (TTA, 1)

(3) $c \triangle b \therefore$ (TTA, 1)

(4) $[d\,e]:d \triangle c . e \triangle c . \supset . d \triangle e \therefore$ (TTA, 1)

(5) $c\, \varepsilon\, a .$ (TTD, 2, 4)

(6) $c\, \varepsilon\, b ::$ (TTD, 3, 4)

$[\exists\, c] . c\, \varepsilon\, a . c\, \varepsilon\, b$ (5,6)

TT11.2 $[a\,b\,c]:c\, \varepsilon\, a . c\, \varepsilon\, b . \supset . a \triangle b$

$[a\,b\,c]::$

(1) $c\, \varepsilon\, a .$

(2) $c\, \varepsilon\, b . \supset \therefore$

(3) $c \triangle a .$ (TTD, 1)

(4) $c \triangle b \therefore$ (TTD, 2)

(5) $[d\,e]:d \triangle c . e \triangle c . \supset . d \triangle e \therefore$ (TTD, 2)

$a \triangle b$ (TTA, 3, 4, 5)

TT11.3 = T20 $[a\,b]:a \triangle b . \equiv . [\exists\, c] . c\, \varepsilon\, a . c\, \varepsilon\, b$ (TT11.1, TT11.2)

TT11.4 $[a\,b]:a\, \varepsilon\, b . \supset . [\exists\, c] . c\, \varepsilon\, a . c\, \varepsilon\, b$

$[a\,b] \therefore$

(1) $a\, \varepsilon\, b . \supset :$

(2) $a \triangle b :$ (TTD, 1)

$[\exists\, c] . c\, \varepsilon\, a . c\, \varepsilon\, b$ (TT11.1, 2)

TT11.5 $[a\,b\,c\,d]:a\, \varepsilon\, b . c\, \varepsilon\, a . d\, \varepsilon\, a . \supset . c\, \varepsilon\, d$

$[a\,b\,c\,d]::$

(1) $a\, \varepsilon\, b .$

(2) $c\, \varepsilon\, a .$

(3) $d\, \varepsilon\, a . \supset \therefore$

(4) $[e f]: e \triangle a . f \triangle a . \supset . e \triangle f \therefore$ (TTD, 1)
(5) $c \triangle a \therefore$
(6) $[e f]: e \triangle c . f \triangle c . \supset . e \triangle f \therefore$ (TTD, 2)
(7) $d \triangle a .$ (TTD, 3)
(8) $c \triangle d \therefore$ (4, 5, 7)
$c \varepsilon d$ (TTD, 8, 6)

TT11.6 $[a b c] :: [d e] : d \varepsilon a . e \varepsilon a . \supset . d \varepsilon e \therefore b \triangle a . c \triangle a \therefore \supset . b \triangle c$
$[a b c] ::$
(1) $[d e] : d \varepsilon a . e \varepsilon a . \supset . d \varepsilon e \therefore$
(2) $b \triangle a .$
(3) $c \triangle a \therefore \supset ::$
$[\exists d] ::$
(4) $d \triangle b .$
(5) $d \triangle a \therefore$
(6) $[e f] : e \triangle d . f \triangle d . \supset . e \triangle f \therefore$ (TTA, 2)
(7) $d \varepsilon a ::$ (TTD, 5, 6)
$[\exists e] \therefore$
(8) $e \triangle c .$
(9) $e \triangle a \therefore$
(10) $[f g] : f \triangle e . g \triangle e . \supset . f \triangle g \therefore$ (TTA, 3)
(11) $e \varepsilon a .$ (TTD, 9, 10)
(12) $e \varepsilon d .$ (1, 11, 7)
(13) $e \triangle d ::$ (TTD, 12)
(14) $d \triangle c ::$ (TTA, 13, 8, 10)
$b \triangle c$ (TTA, 4, 14, 6)

TT11.7 $[a b c] :: c \varepsilon a . c \varepsilon b \therefore [d e] : d \varepsilon a . e \varepsilon a . \supset . d \varepsilon e \therefore \supset . a \varepsilon b$
$[a b c] ::$
(1) $c \varepsilon a .$
(2) $c \varepsilon b \therefore$
(3) $[d e] : d \varepsilon a . e \varepsilon a . \supset . d \varepsilon e \therefore \supset \therefore$
(4) $[d e] : d \triangle a . e \triangle a . \supset . d \triangle e \therefore$ (TT11.6, 3)
(5) $c \triangle a .$ (TTD, 1)
(6) $c \triangle b \therefore$
(7) $[d e] : d \triangle c . e \triangle c . \supset . d \triangle e \therefore$ (TTD, 2)
(8) $a \triangle b \therefore$ (TTA, 5, 6, 7)
$a \varepsilon b$ (TTD, 8, 4)

TT11.8 = T36 $[a b] :: a \varepsilon b . \equiv \therefore [\exists c] . c \varepsilon a . c \varepsilon b \therefore [d e] : d \varepsilon a . e \varepsilon a . \supset$
$. d \varepsilon e$ (TT11.4, TT11.5, TT11.7)

From TT11.8 and TT11.3 we see that T49 and T48 imply T36 and T20. This completes the proof that a system of Ontology inferentially equivalent to the original system can be based on T49 as a single axiom with the functor of partial inclusion as a single undefined form.

Definitions in Systems of Ontology

Now that we have a suitable axiom for the functor of partial inclusion it remains to adapt the rule for writing ontological definitions to the properties of this functor.

There are two types of definition in Ontology: protothetical or propositional definitions on the one hand and ontological or nominal definitions on the other. Propositional definitions are called so because the most characteristic functors introduced by them are proposition-forming functors. The most characteristic functors introduced by means of nominal definitions are name-forming functors. Without any claim to precision the corresponding two rules can be briefly characterized as follows[16]:

(*a*) *The rule of protothetical definitions.* On the assumption that at the given stage in the development of a system of Ontology a thesis T is the last thesis, the rule of protothetical definitions entitles us to add to the system as a new thesis an expression of the type:

(i) $$[\ldots].\ \alpha \equiv \beta$$

provided the following conditions are fulfilled: (1) 'α' stands for a propositional function. The functor of this function is a constant term which does not occur in T or in any of the theses preceding T. The arguments of the function are all variables each of which is of the same semantical category as an expression occurring in T or in a thesis preceding T. Variables which are of the same semantical category as certain expressions obtainable exclusively within the logic of propositions are not allowed in 'α'. All the variables in 'α' are free in 'α' and no variable in 'α' occurs more than once; (2) 'β' stands for a propositional expression. Every constant in 'β' belongs to the logic of propositions or occurs in T or in at least one thesis preceding T. Every variable in 'β' is of the same semantical category as an expression occurring in T or in a thesis preceding T. Every variable occurring in 'α' occurs in 'β' as a free variable in 'β'. Every free variable in 'β' occurs in 'α'. No variable is free in (i).

It is easy to see that in accordance with this rule any of the expressions: T1, T5, T6, T16, T–18T27, could be added as a definition to the system of Ontology with T34, T36, T42 or T43 as the axiom. It would suffice to assume that the axiom was the last thesis of the system. But we would have to assume that T1 and T5 were already in the system if we wanted

---

[16] In the present formulation many-link functions are not taken into consideration because no definitions involving such functions are required for the purpose of establishing the adequacy of the axioms discussed in this paper, except in the case of T42, T43, T71 and T77. For the authoritative statement of the rules of definition for the system based on T43 see Leśniewski [1930].

to add T17 as a definition. The system of Ontology based on T49 as the axiom could be immediately extended by adding T11, T30, and T48 as definitions.

(*b*) *The rule of ontological definitions.* On the assumption that at the given stage in the development of the original system of Ontology a thesis T is the last thesis, the rule of ontological definitions enables us to add to the system as a new thesis an expression of the type:

(ii) $$[a \ldots]:a \varepsilon \Psi. \equiv .a \varepsilon a.\varphi(a)$$

provided the following conditions are fulfilled: (1) 'Ψ' stands for a constant name, which does not occur in T or in a thesis preceding T, or it stands for a nominal function. If the latter is the case then the functor of this function is a constant term which does not occur in T or in any of the theses preceding T, while the arguments of the function are all variables each of which is of the same semantical category as an expression occurring in T or in a thesis preceding T. Variables which are of the same semantical category as certain expressions obtainable exclusively within the logic of propositions are not allowed in 'Ψ'. All the variables in '$a \varepsilon \Psi$' are free in '$a \varepsilon \Psi$' and none of them occurs in '$a \varepsilon \Psi$' more than once. (2) '$\varphi(a)$' stands for a propositional expression. Every constant in '$\varphi(a)$' belongs to the logic of propositions or occurs in T or in at least one thesis preceding T. Every variable in '$\varphi(a)$' is of the same semantical category as an expression occurring in T or in a thesis preceding T. Every variable occurring in '$a \varepsilon \Psi$' occurs in '$a \varepsilon a.\varphi(a)$' as a free variable. Every free variable occurring in '$a \varepsilon a.\varphi(a)$' occurs in '$a \varepsilon \Psi$'. No variable is free in (ii).

The following theses are examples of expressions which can be added as ontological definitions to the system based on T36 as the axiom:

T72. $[a]:a \varepsilon \mathsf{V}. \equiv .a \varepsilon a$
i.e., for all a, — a is an object if and only if a is a,

T73. $[a]:a \varepsilon \Lambda. \equiv .a \varepsilon a.\sim(a \varepsilon a)$
i.e., for all a, — a is an object which does not exist if and only if (a is a and it is not true that a is a),

T74. $[a\, b]:a \varepsilon \mathsf{N}(b). \equiv .a \varepsilon a.\sim(a \varepsilon b)$
i.e., for all a and b, — a is non-b if and only if (a is a and it is not true that a is b).

The rule of protothetical definitions remains the same for any system of Ontology irrespective of which functor is used as the undefined ontological term. This, however, does not hold for the rule of ontological definitions. Schema (ii) above represents what might be called the definitional frame for a system with the functor of singular inclusion as the primitive term. A different definitional frame is required for a system based on a functor other than the functor of singular inclusion.

A trivial way of obtaining such a frame would be to 'translate' schema (ii) with the aid of the definition of 'ε' in terms of the functor involved. But the frame secured in this manner proves to be long and clumsy. Thus the problem arises of finding a shortest definitional frame for a given functor.

Since the following thesis:

T75. $[b\,\varphi]::[a]:a\,\varepsilon\,b.\equiv.a\,\varepsilon\,a.\varphi(a)\therefore\equiv::[a]::a\triangle b.\equiv\therefore[\exists\,c]\therefore c\triangle a\therefore[d\,e]:d\triangle c.e\triangle c.\supset.d\triangle e\therefore\varphi(c)$

can be proved in Ontology, we conclude that, in the case of the system based on T49 as the axiom, the definitional frame for ontological definitions can take the form of the following schema:

(iii) $[a\ldots]::a\triangle\Psi.\equiv\therefore[\exists\,c]\therefore c\triangle a\therefore[d\,e]:d\triangle c.e\triangle c.\supset.d\triangle e\therefore\varphi(c)$

On the assumption that at the given stage in the development of the system a thesis T is the last thesis, an expression of type (iii) can be added to the system as a definition provided 'Ψ' and '$\varphi(c)$' satisfy the conditions on the lines of those discussed under (b) above. In particular it is required that every variable occurring in 'Ψ' should occur in '$\varphi(c)$' as a free variable, and that every free variable in '$\varphi(c)$' should occur in 'Ψ' or be equiform with 'c'.

Thus in the system under consideration the functor 'N' could be defined as follows:

T76. $[a\,b]::a\triangle N(b).\equiv\therefore[\exists\,c]\therefore c\triangle a\therefore[d\,e]:d\triangle c.e\triangle c.\supset.d\triangle e\therefore\sim(c\triangle b)$

It is not unlikely that by making full use of the rule of ontological definitions in the system based on partial inclusion, and by exploiting the rule of extensionality, whose adaptation to the system could not be discussed in this paper, we may obtain a simpler axiom or a simpler frame for ontological definitions.

In connection with the rule of ontological definitions for systems with the functor of singular inclusion as the primitive term I was anxious to make clear that in the formulation given the rule presupposed T34 or T36 as the axiom. Interestingly enough this formulation proves to be too weak if we want to use T43 as the axiom[17]. T43 has many attractions as a single axiom. This is why Leśniewski eventually decided in favour of strengthening the rule of ontological definitions, which he formulated with the aid of the following frame:

(iv) $[a\ldots].a\,\varepsilon\,\Psi.\equiv.\varphi(a)$

---

[17] Dr. Sobociński drew my attention to this fact.

In this schema '$\varphi(a)$' is of the from '$a \varepsilon \Omega$' or it is a conjunction which contains at least one conjunct of the form '$a \varepsilon \Omega$'. It is not difficult to see that an expression which is a definition in accordance with schema (ii) is also a definition in accordance with schema (iv). But the converse does not hold.

If we wanted to retain the weaker rule of ontological definitions then T43 would have to be replaced by

T77. $[a\, b] \therefore a \varepsilon b . \equiv : a \varepsilon a : [\exists c] . a \varepsilon c . c \varepsilon b$

This was established by Sobociński in 1936.

John Thomas Canty

Ontology: Leśniewski's Logical Language

Leśniewski's system of ontology developed out of his investigations of the logical paradoxes which were of concern to logicians at the beginning of this century. In discussing Russell's paradox, he distinguished collective and distributive uses of nouns in a manner not unlike that of the medieval theory of supposition. Roughly speaking, he attributed the existence of the syntactic paradoxes, at least in part, to the failure to treat separately the uses he distinguished. His objection can be paraphrased by maintaining that the existence of paradoxes in Frege, for instance, can be traced to his axioms about classes which assert some of the properties of classes where 'class' is taken by Frege in a distributive sense and some of the properties of classes where 'class' is taken in a collective sense.[1]

After his investigation of the paradoxes, Leśniewski developed a theory of collective classes which he eventually called mereology. In his expositions of this system, certain nouns and nominal phrases were clearly indicated as collective, but their logic was couched in a language which employed nouns in a distributive sense. That is, in formally presenting the logic of collective classes, Leśniewski relied on a theory of distributive predication which for a time lacked any formal exposition of its own.

In the interest of making his investigations precise, Leśniewski then developed his theory of distributive predication which he called ontology. If, following Leśniewski's analysis of propositions about individuals, one maintains that such propositions are composed of a subject, copula and predicate, then his theory is captured by asserting that any proposition about an individual subject is true only if the subject of the proposition is unique and unempty, while the copula of the proposition is transitive.

Such a position would make the following true: *Si Socrates est coniunx Xantippae et coniunx Xantippae est homo, Socrates est homo et coniunx Xantippae est Socrates.* Here '*est*' is the copula, '*Socrates*' is the subject and '*coniunx Xantippae*' the predicate of the first simple proposition that occurs in the conditional. Moreover, '*Socrates*' and '*coniunx Xantippae*'

[1] Sobociński [1949-1950].

occur sometimes as subjects and sometimes as predicates, while '*homo*' only occurs as a predicate. Symbolically, the form of the proposition is given by:

$$s \varepsilon x . x \varepsilon h . \supset . s \varepsilon h . x \varepsilon s$$

One of the first things to note about the above analysis is that predicates are construed in the traditional manner, that is, as one of the *termini* of propositions, rather than in accord with more recent usage which would render the form of the above as:

$$\varepsilon(s, x) . \varepsilon(x, h) . \supset . \varepsilon(s, h) . \varepsilon(x, s)$$

and where the only predicate recognized would be the binary one '*est*'.

Now if, following traditional terminology, one calls the epsilon which is used by nearly all logicians, a copula, then Leśniewski's copula differs from most in that its predicate place can be occupied by a term for an individual, for example '*Socrates*', as well as a term for many individuals, for example '*homo*', while for other logicians the predicate of their copula is grammatically restricted to terms for classes of objects. In particular, for Leśniewski's copula it is quite meaningful to have propositions of the form:

$$a \varepsilon a$$

Moreover, Leśniewski's copula is transitive while this is invariably not the case for other logicians. It is just not the case in set theory that members of members of a given class are *ipso facto* members of that class; while one obtains as a thesis of ontology:

$$[a, b, c] : a \varepsilon b . b \varepsilon c . \supset . a \varepsilon c$$

And this is just as it should be since it is the case that *Si Socrates est coniunx Xantippae et coniunx Xantippae est homo, Socrates est homo.* In short, Leśniewski's copula more closely approximates the linguistic uses of verbs which connote being (whence 'ontology') than do the copulas of his contemporaries.

If on the other hand, one follows modern terminology and considers Leśniewski's epsilon as a predicate, that is, as a proposition forming functor for two nominal arguments, another interesting difference among Leśniewski's analysis and others appears. In constructing predicate calculi, logicians distinguish at least two basic types of expressions: names of individuals and proposition forming functors which take these as arguments. But from what has been said about the epsilon of ontology, it is clear that if it is considered as a predicate for names, then Leśniewski's category is broader than others: constants belonging to his category are not necessarily names of individuals, they may denote many objects or even no objects at all. Thus, although for most logicians a particular quantification in respect to a nominal argument carries with it 'existential import', this is not the case for Leśniewski since his nouns are not necessarily unique, nor are they necessarily unempty. For Leśniewski, the only propositions which assert the existence of individuals are those of the form:

$$a \varepsilon b$$

or ones which imply this form, or at least:

$$[\exists x] . x \varepsilon b$$

It is not the particular quantifier as such that carries existential import, but rather the particular quantification of a certain type of expression. Now all such propositions as these are independent of ontology. In short, in as much as ontology fails to be committed to the existence of individuals (is, in current parlance, a free logic), Leśniewski's analysis of propositions is free from commitment to the existence of individuals.

Leśniewski's final step in the process of formulating his logical insights was to present formally a logical system which dealt with propositions, proposition forming functors whose arguments are propositions, and functors definable in terms of these. Given this system, which he called protothetic, the formalization was complete. Protothetic forms a basis for ontology which, in turn, is used to expound mereology (or other theories, for example, Peano's arithmetic). And from this brief sketch of Leśniewski's development, it is clear that he was not a formalist — at least not in the sense of 'first formulation, then interpretation'. Rather, his formulations grew out of his need to express clearly and consistently the theory of collective classes and his systems always remained interpreted, that is, precise languages (whence, for his theory of types: 'semantical' categories, rather than 'grammatical' or 'syntactical' categories). In short, Leśniewski, in formulating systems, brought precision to his language, rather than language to his precision.

With the completion of the program of formalization, Leśniewski was able to present, in a quite formal manner, the terminological explanations for each of his systems.[2] Naturally enough, since the terminological explanations should be as accurate as possible, Leśniewski chose to present them by making use of protothetic, ontology and mereology. Thus, though there exists a formal description of Leśniewskian systems, the description presupposes familiarity with the systems. Actually, the terminological explanations are not given axiomatically, but if one wished to give an axiomatic presentation of this meta-logic, it could be achieved by subjoining the necessary axioms, incorporating the requisite primitive meta-logical terms, to some given development of mereology.

---

[2] Leśniewski [1929a] together with [1930].

This is so, since the terminological explanations as given are understood, first of all, as relativized to a given extant development of the system to be explained. And when reference is made to, say, an axiom of the system to be explained, it is understood that this is a given inscription, not a form or type of an inscription, and certainly not a potential inscription or form of potential inscription, which if inscribed would be an axiom. Indeed, all references that are made are made to concrete individuals, and the relationships among the individuals that are discussed are basically spatial.

Leśniewski's terminological explanations, like the formation and transformation rules of others, culminate in a set of directives which guide the expansion of the system being explained. However, the directives are incorporated into a single rule which is given only relative to the last thesis of the system and one which is self-adjusting to the expansions of the system which it justifies. At any given stage the systems are always finite, but like natural languages they can develop indefinitely and any development that occurs is controlled by the directives of the system. And, as is the case in natural languages, because the directives are relativized to an extant development of the language and self-adjusting to expansions of the language, these expansions are contextually determinate.

However, unlike a natural language in which meaning generally depends on context and so the development of the language is directed, Leśniewski's directives are thorough. For instance, in Leśniewskian languages, as in natural languages, expressions of the same form can enjoy a variety of significance. But these directives do not merely reduce the ambiguity latent in using analogous forms of expression within the same language, they eliminate ambiguity.[3] Thus, one has an example of a language in which contextual determinacy is achieved without making the significance of an expression depend on the form of the expression as is so often the case with formalized languages. And in as much as linguistics is concerned with the contextual determinacy of expressions which have various significances though the same form, Leśniewski's terminological explanations provide an excellent model.[4]

The directives of the systems need not be considered in detail. They incorporate transformation rules which are generally well known. The single rule for protothetic has five directives: one for detachment, one for handling quantifiers, and a directive each for substitution, definition, and extensionality. The single rule for ontology incorporates all the directives for protothetic, understood as adjusted to the semantical categories of

---

[3] Machover [1966].

[4] But unfortunately a model which is little known even up to: Bar-Hillel [1950].

ontology, together with a directive for definitions of nouns and functors which form nouns and a directive for the extensionality of nouns and noun forming functors. It is worth noting that Leśniewski gives an adequate statement of substitution and a thorough set of criteria for good definitions as these are hard to come by in the literature.[5] The directives for extensionality are designed to guarantee the extensionality of any extant semantical category of a given system. And since definitions, and for that matter axioms, can introduce new semantical categories into a system, extensionality must be given as a directive.[6]

The key to the contextual determinacy of Leśniewskian languages is the theory of semantical categories. The directives of the terminological explanations cannot be understood without reference to the semantical categories, and the approximation to natural languages achieved by Leśniewskian languages is due to the richness of the categories and their independence from any predetermined or selected forms of expressions.

Like the language itself, the extant semantical categories of any given development are finite. But these categories are, like the language, potentially infinite. New categories are introduced into the system by defining a constant which does not belong to any extant category; primitive categories are introduced axiomatically. Thus, the criteria of good definitions direct and control the semantical development of the system.

Leśniewski's theory of semantical categories is in some ways similar to the simple theory of types, but it has much in common with the categories of a natural language and is perhaps best considered as a theory of 'parts of speech'. Historically, as Leśniewski acknowledged, it stems most directly from Husserl's *Bedeutungskategorien*, that is, a linguistic classification to be used to prevent nonsense from occurring in a language, rather than contradiction.[7] Like many other logicians of the twentieth century, Leśniewski turned to a theory of types when paradoxes approached. But unlike many who incorporate some recognition of types into their work only as a necessary evil which is to be avoided as much as possible (witness the typical ambiguity of *Principia Mathematica*[8] and Quine's stratification), Leśniewski set out to construct the most natural and the most complete type theory imaginable, and once satisfied with the construction, he unfailingly adhered to it in all its complexity. Indeed, he claimed that even if he had been unaware of the paradoxes he would not be any less committed to his theory of semantical categories, since it incorporates the means to preserve a consistent determinacy as to what shall count as senseful.

---

[5] Luschei [1962, pp. 36–38].

[6] For a full discussion of the directives of the systems see Luschei [1962] or Canty [1967].

[7] Bar-Hillel [1957].

[8] Whitehead and Russell [1963] (hereafter *PM*).

Once the terminological explanations are given, a meausre of the semantical categories allowable in ontology can be attained. It is clear from the explanations that Leśniewski takes two categories as basic in ontology: propositions and nouns. In protothetic only propositions are taken as basic — one of the significant features of ontology is that it expands the basic categories to include nouns. The axiom of mereology, although it introduces primitive terms not definable in ontology, does not expand the basic semantical categories, and for Leśniewski, no axiomatic extension of ontology would expand the number of basic categories.

All categories other than the two basic ones are understood ultimately in terms of these. In order to determine the semantical category of any term it is only necessary to specify the number, the order, and the categories of its arguments, together with the category of the functor produced by it. For example, 'it is not the case that ...' is completely determined by indicating that it is a propositional functor formed from one propositional argument, while '... is unempty' is determined by indicating that it is a propositional functor formed from one nominal argument.

Now given the above two categories as basic, the directives for definitions yield the information needed to determine the range of the general theory: the directives determine the constants definable in the system, and so, the categories allowable. In order to avoid ambiguity, Leśniewski's directives for definitions require that different styles of brackets be used in association with different semantical categories: if a new semantical category is introduced by a definition, then a new style of brackets must be employed in association with the constants being defined. Thus, the determination of the categories can be achieved by determining what kinds of brackets can be employed in ontology.

Let all brackets be expressed as subscripted parentheses:

$$(_\alpha,)_\alpha, (_\beta,)_\beta, (_\gamma,)_\gamma, \ldots$$

where the subscripts are natural numbers. It is then a question of establishing a one-to-one correspondence between the allowable semantical categories and some subset of the natural numbers.

Propositions and nouns are assigned the numbers one and two respectively. If n is the number of arguments used to form a function, $n+2$ is associated with that number of arguments. In this way prime factorizations of numbers can be used to code the semantical categories and the number of the category used as a subscript to determine unique parentheses for each semantical category. For example, if one writes 'a is unempty' as:

$$!(_\alpha a)_\alpha$$

where $\alpha = 2^3 \times 3^2 \times 5^1$, then the sign of unemptiness is determined to be a propositional functor formed from one nominal argument: the first exponent of α indicating the number of arguments of the functor, the second the category of the single argument, and the last the category of the functor which is formed by the exclamation point. Again, if one writes '*a* is *b*' as:

$$\varepsilon\,({}_{\beta}a,\ b)_{\beta}$$

where $\beta = 2^4 \times 3^2 \times 5^2 \times 7^1$, then the primitive epsilon of ontology is determined to be a proposition forming functor for two nominal arguments.

In general, a type-*t* functor forming functor of *n*-arguments is represented by a prime factorization of the form:

$$2^{n+2} \times C_1 \times C_2 \times \cdots \times C_n \times \tau$$

where each C_i (and τ) is a prime factorization beginning with the next prime in order of magnitude not yet appearing in the representation and indicating the category of the argument (of *t*, that is, the category of the functor formed by the term in question). Thus, if one writes 'object that ψ' as:

$$1\,({}_{\gamma}\psi)_{\gamma}$$

where $\gamma = 2^3 \times 3^3 \times 5^2 \times 7^1 \times 11^2$, then the constant is a noun forming functor of one argument whose category has the exponents 3, 2, 1 in that order, that is, the constant is a description operator which forms a noun from an α-functor.

As a final example, consider the following definition where the definiendum occurs on the right:

$$[a,\ b]{:}\varepsilon\,({}_{\beta}a,\ b)_{\beta}\,.\equiv.\,\varepsilon\,({}_{\delta}b)_{\delta}\,({}_{\alpha}a)_{\alpha}$$

where $\delta = 2^3 \times 3^2 \times 5^3 \times 7^2 \times 11^1$. Here the constant being defined forms an α-functor from one nominal argument and the semantical category of the constant being defined is δ.

This example, by the way, also illustrates how constants of the same form can be used with different significances without fear of ambiguity. The semantical category of the constant being defined, namely δ, is completely determined in accord with the directive for definitions by previous semantical categories, namely α, β and the category of the biconditional sign. And since this constant will always carry with it the mark of its category, one can use as the form to express the constant, a form which is also used to express a constant of a different semantical category: in this case the epsilon of the β-category. And this points up the fact that what is needed to achieve determinacy of meaning is not necessarily various forms of expression for various significances, but a theory of semantical categories and a means for determining which expressions, of whatever form, belong to which categories. In a Leśniewskian language this means is given by the context in which the expression occurs: in this example by the definition as a whole together with any other propositions needed to fix the categories of all terms in the definition which are not in question.[9]

Here prime factorizations have been used to represent the semantical categories. The representation is akin to that of Curry[10] who discusses 'grammatical' categories. And Ajdukiewicz[11] takes up this question which he considers to be a linguistic problem of 'syntactical' coherence, and in doing so uses a numerical representation developed later by Bar-Hillel[12] but different from the above.

As Ajdukiewicz pointed out, in order to achieve a general theory of coherence one "must make no decision as to the number and kind of basic semantic categories and functorial categories, since they may be different in different languages". It is not the purpose of this essay to attempt a general theory of coherence, but only to provide an appreciation for the coherence that Leśniewski achieved by means of his semantical categories. Even though he had only two basic categories, the richness of the resulting categories and the uses to which they can be put may well point the way to a general theory. And the richness of these categories is perhaps most graphically illustrated by seeing the uses to which they can be put. Therefore, an informal comparison of a Leśniewskian treatment and the exposition of *PM* will be given. The points of comparison will be familiar logical concepts.

Consider for the moment:

$$*22.34 \vdash .\therefore x \in \alpha \cup \beta . \equiv : x \in \alpha . \vee . x \in \beta$$

of *PM*. This theorem follows from the definition of what the authors of *PM* call the 'logical sum' of two classes α and β. But as Quine points out,[13] Russell's contextual definition of classes "is a technical construction enabling us to speak ostensibly of identical classes by way of shorthand for discourse about coincident but perhaps non-identical attributes". That is to say, that although the theorem treats the logical sum as nominal (that is, as a class), since for the authors of *PM* the use of their epsilon and class abstractor is only a notational device for speaking extensionally about predicates, see for instance *20 of *PM*, classes are no more than 'fictitious objects'. Now ontology incorporates the means of speaking about predicates (that is, proposition forming functors for a nominal argument) extensionally in its rule. And since there are two directives for extension-

---

[9] Machover [1966] supplies a general method for analysing given contexts and establishing whether they are coherent.

[10] As given in the section on grammatics in Curry [1963].

[11] Ajdukiewicz [1935].

[12] Bar-Hillel [1953].

[13] Quine [1951].

ality, one protothetical, the other ontological, there are also two analogues for logical sum:

$$[x, \alpha, \beta] \therefore \alpha(x) . \vee . \beta(x) : \equiv . \cup[\alpha, \beta](x)$$
$$[x, \alpha, \beta] \therefore x \varepsilon \alpha . \vee . x \varepsilon \beta : \equiv . x \varepsilon \cup\langle\alpha, \beta\rangle$$

where the functor defined in the latter thesis forms a noun from two nominal arguments, but where the functor defined in the former thesis is a closer analogue to the logical sum of classes. This is so since the former functor forms a predicate from two predicative arguments, and, as has been pointed out, the classes of *PM* are really predicative and only conventionally nominative.

Now there are other refinements that would have to be given in order to make the comparison of the expositions exact, for instance, the difference of arguments of predicates in the two systems: ontology allowing nouns in general, *PM* restricting its arguments to names for individuals; or the difference of considerations of type theory in the two systems: ontology using analogous symbols for different constants, *PM* using typically ambiguous theorems. These points will be touched upon. But for now this much is apparent: any comparison of the two systems will issue in ontology having two analogues for each concept of *PM*: one nominal or ontological, the other predicative or protothetical; and the protothetical analogue will more closely approximate the concepts of *PM*.

One can define a special set of predicates which are proposition forming functors for one nominal argument, namely, those which hold only for names of individuals:

$$[\psi] \therefore [a] : \psi(a) . \supset . a \varepsilon a : \equiv . \mathrm{R}\{\psi\}$$

With these predicates distinguished, one now defines a new logical sum:

$$[x, \alpha, \beta] \therefore \mathrm{R}\{\alpha\} . \mathrm{R}\{\beta\} : \alpha(x) . \vee . \beta(x) : \equiv . \cup_{\mathrm{Ind'v}}[\alpha, \beta](x)$$

where the subscript is added to the sign for logical sum to point up the restriction of the predicate to names for individuals.

This last definition gives a closer approximation to the logical sum of *PM* than does the first candidate put forth. However, it could be argued that this functor is only one of the enumerably many envisaged by the assertion of *22.34. This is so since the logical sum of *22.34 is typically ambiguous. That is, the authors see the theorem as representing not only the logical sum of classes of individuals, for which an analogue has just been given, but also of classes of classes of individuals, etc. But even this disparity can be accounted for by a further refinement in comparison.

What is here at issue is Whitehead and Russell's willingness to exhibit theorems which are typically ambiguous and Leśniewski's unwillingness to do so. But by relying on Leśniewski's use of analogous symbols for different constants, the effect of employing typical ambiguity can be obtained in ontology. At this point it is clear that (given the reproducibility in ontology of the effects of typical ambiguity) even the protothetical treatment of functors in ontology is more general than the treatment of functors in *PM*. Not only are there two possible ways of treating functors in ontology: ontologically or protothetically, there is even a variety of protothetical treatments available: one can define constant predicates for nouns in general or constant predicates which hold only for names of individuals.

From what has been shown about the relationship of the expositions of ontology and *PM*, anyone interested in developing the calculus of classes or the calculus of relations within ontology is in a position to do so. The means of translating the relevant theorems of *PM* into their ontological counterparts is not a recent discovery; it was well known to Leśniewski and his students. And within ontology one can follow closely the work of Frege and Whitehead and Russell, going on to the concepts of equinumerosity, cardinality, and the addition of cardinal numbers.

However, if one wished to obtain Peano's axioms for arithmetic in a Leśniewskian language, certain provisos must be considered. First of all, if one has an eye towards doing things in a purely Leśniewskian manner, one would simply add Peano's axioms to ontology. The effect of recursive definitions of numerical operations could be achieved through the directives for definition which are already available by using Frege's method of 'impredicative' definitions in order to obtain the counterparts of recursive definitions as theses. And in adding Peano's axioms to ontology, the primitive notions that are needed, namely, zero, successor and number, would be introduced as members of the basic category of nouns. For the first time one would have in the language a thesis of the form:

$$a \varepsilon b$$

namely, the axiom which states that zero is a number. The development would not follow the spirit of logicism. Rather it displays from the start its 'ontic commitment', presenting numbers as individuals which are subject to the consequences of the arithmetical axioms — axioms containing primitive notions unanalysed in the language.

Of course, this is not the only way that Peano's arithmetic could be produced in a Leśniewskian language. One could add an axiom of infinity to the effect that infinitely many individuals exist and follow the lines of *PM* in order to 'construct' the natural numbers.[14] However, if this course is taken an interesting contrast between the expositions of *PM* and ontology arises.

As with any arithmetical operation, the result of adding numbers must be guaranteed to exist. This guarantee is established in *PM* by de-

[14] See Chapter II of Canty [1967] where this construction is carried out in detail.

fining numerical addition as a special case of logical sum and relying on the hierarchy of types present in the system of *PM* to supply the disjoint classes requisite to construct a class representing the numerical addition of two classes. This practice is common to many expositions of arithmetic, for instance, set theory or Quine's work. Now in *PM*, the successor of a class is not required to be of the same type as the class (*110). Since this is so, it can be shown in *PM*, without the aid of an axiom of infinity, that the successor of a class of some given type is existent if it is a higher type than the given class. This is formulated in *PM* as:

$$\alpha \in \mathrm{NC} . \supset . \alpha +_c 1 \in \mathrm{NC}$$

and the proposition is actually used as a lemma in the proof of:

$$*120.14 \vdash . \mathrm{NCinduct} \subset \mathrm{NC}$$

However, in a Leśniewskian development of successor, the successor of a proposition forming functor for one name argument, is itself a proposition forming functor for one name argument. But if the above lemma is to be maintained without the axiom of infinity, the cardinality mentioned in the consequent must in general be of a different type than the cardinality mentioned in the antecedent. That is, the successor of a cardinal number is of a different type than the cardinal number.

Now in *PM* *120.14 is in fact established without the axiom of infinity. Of course the authors note immediately after the statement of the theorem that "this proposition does not show that every inductive cardinal is an *existent* cardinal; to obtain this, we require the axiom of infinity". So, the most that can be claimed for *120.14 is that it implies for every inductive cardinal that there is a type in which it exists. On the other hand, the Leśniewskian analogue of this theorem has as a corollary that every inductive cardinal is existent. But this should come as no surprise, since the constants of ontology cannot be ambiguous as to type (the successor of a cardinal must be of the same type as the cardinal), and so the proof of the analogue of *120.14 relies on the axiom of infinity.

In effect, the above lemma implies that for a cardinal number of a given type there is a successor, albeit of a higher type, which is a cardinal number. And the ambiguity of the lemma is carried into *120.14 leaving us with the statement that for any inductive cardinal of a given type, there is a cardinal number of a higher type. That is, simply and ambiguously, the inductive cardinals are included in the cardinal numbers.

Now it is precisely this ambiguity of expression which allows the authors of *PM* to avoid the use of an axiom of infinity in establishing *120.14. By relying on the unending hierarchy of types, rather than the axiom of infinity, the authors claim in their summary of the section on Finite and Infinite that they "can show of every inductive cardinal that there is a type in which it exists". But as soon as the ambiguous term 'inductive cardinal' is made definite, the claim, if meaningful, requires the axiom of infinity. A typically definite inductive cardinal, if it exists at all, exists in its own type. Thus, to show that a typically definable inductive cardinal is existent, the proper number of objects of the appropriate type must be available. And to show that any typically definite inductive cardinal, however large, is existent, an infinity of objects must be available.

Having seen how the use of ambiguous terminology has allowed *120.14 to be established without an axiom of infinity, even though an unambiguous statement that inductive cardinals are included in cardinal number requires such an axiom, it is perhaps best to conclude that the argument is simply fallacious.[15] But there are other motives for using ambiguous symbols.

When presenting an analysis of number in a language in which the types of expressions are carefully noted, there is the haunting fear that the analysis never really achieves its purpose. Constant numbers can be defined, but if they are typically definite a duplication of constants occurs. There will be, for instance, the number two for classes, the number two for classes of classes, etc. One has then, instead of the number two, only instances of it — the cardinal remains forever inaccessible. Thus, in order to achieve the standpoint of ordinary arithmetic which ignores differences of type in numbers, one must attain a liberality in symbolism. The liberality enjoyed by *PM* in reaching the standpoint of ordinary arithmetic carries with it undesirable consequences — witness the above discussion. However, the meta-theory of ontology contains theses which indicate that the Leśniewskian development of arithmetic is as fully general as that of *PM* and moreover, that the Leśniewskian development in its generality avoids the ambiguities latent in *PM*.

As has been mentioned, ontology is based on protothetic by adding to a given development of that system a single axiom which introduces the semantical category of nouns and two directives: one countenancing ontological extensionality, the other ontological definition. The axiom employs an undefined proposition forming functor for two noun arguments: the primitive epsilon. Thus, ontological definition and extensionality can be considered as principles for defining nouns and reasoning extensionally in respect to them. Now although the semantical categories of nouns and propositions are distinguished in ontology there is no commitment as to what the nouns refer to. Thus, the possibility arises of defining a binary functor whose arguments are not nouns, but which, by playing a role in ontology very similar to the primitive epsilon of the system, can be construed as a proposition forming functor for 'nouns' — such a functor is called a higher epsilon since its arguments belong to non-primitive semantical categories.

---

[15] See Luschei [1962, p. 63] where this question is raised.

In order to make explicit the similarity between the primitive epsilon and any such higher epsilon it is required that there be derivable for the higher epsilon: (a) a thesis identical in shape to the axiom of ontology except for the parentheses associated with the defined epsilon; (b) theses identical in shape, except for the associated parentheses, to the theses of extensionality for nouns and noun forming functors; and (c) theses identical in shape, except for the parentheses, to the ontological definitions for nouns and noun forming functors. By completing this task, it is evident that the similarity between the primitive epsilon and the higher epsilon is a complete analogy, for all of ontology is derivable for the higher epsilon in the sense that one is guaranteed an internal ontological model for the defined constant.

Indeed, a higher epsilon exists for each and every semantical category definable in ontology.

It is outside the scope of this work to prove the above assertion or to discuss at any length all the various higher epsilons definable in ontology. The proof of the existence of higher epsilons was originally given by Leśniewski and it will be taken as established in order to bring to a close this comparison of *PM* and ontology.

Since there is a higher epsilon for each and every semantical category definable in ontology, the exposition for the cardinality of nouns, or the calculus of nouns, or any theory developed in respect to the basic category of ontology, serves as an exposition for that same theory in respect to any semantical category definable in ontology. For instance, the cardinality of any semantical category definable in ontology is to be developed in exactly the manner in which the cardinality of nouns is given. This is merely a question of considering the theses which contain the primitive epsilon as theses containing some higher epsilon and making the appropriate adjustments in parentheses for all the terms occurring in the theses.

It is important in this respect to note that an analogue for the axiom of infinity for each higher epsilon also exists. After all, the exposition for the cardinality of nouns relies on an axiom of infinity for nouns, and thus, expositions for the cardinality of other semantical categories will also require axioms of infinity for those categories. But this presents no insurmountable problem, since it can be shown that the axiom of infinity is reproducible at higher levels, and thus it is certain that the expositions for the cardinality of any category are available.

At this point a precise exposition of arithmetic for each semantical category definable in ontology has been attained. No doubt this is a far cry from the standpoint of ordinary arithmetic which ignores differences of type in numbers, but on the basis of these different types of arithmetic one can safely reach the standpoint of ordinary arithmetic.

The only remaining requirement in the quest for arithmetic is the ability to handle arithmetical operations whose arguments are non-homogeneous. The possibility already exists of producing, for instance, the sum of homogeneous cardinals: the sum of the cardinality of a predicate and the cardinality of some other predicate is once again the cardinality of a predicate. Now consider for the moment the problem of adding the cardinality of a noun to the cardinality of a predicate. Since the operation of addition has homogeneous arguments, it is necessary to secure elements of the same semantical category in order to perform the operation. But this can always be done in a satisfactory manner by producing an element of the higher semantical category equinumerous to the element of the lower category: in this case by producing a predicate equinumerous to the noun in question.

Of course, this requires expanding the notion of equinumerosity so that it countenances non-homogeneous arguments. But the higher epsilons afford a precise basis for such an extension of the notion. Once again, it can be seen that it is a simple matter, in the presence of the higher epsilons, to perform arithmetical operations, now, for arguments from any semantical category whatsoever. One simply finds a representative from the higher semantical category for the element of the lower category and performs the operation in the arithmetic of the higher level.

The concern with types that has so complicated the sense in which arithmetic is developed in this or that system, has not been without its commentators. For instance, Gödel[16] states that:

> if ... one considers concepts as real objects, the theory of simple types is not very plausible, since what one would expect to be such a concept (such as, e.g., "transitivity" or the number two) would seem to be something behind all its various "realizations" on the different levels and therefore does not exist according to the theory of types. Nevertheless, there seems to be some truth behind this idea of realizations of the same concept on various levels, and one might, therefore, expect the theory of simple types to prove useful or necessary at least as a stepping stone for a more satisfactory system, a way in which it has already been used by Quine. Also Russell's "typical ambiguity" is a step in this direction. Since, however, it only adds certain simplifying symbolic conventions to the theory of types, it does not *de facto* go beyond this theory.

But it could now be argued that the meta-logical consideration given above forms a precise basis from which to go beyond the limitations imposed by adherence to a theory of types in ontology.

At any rate, it is now clear in what sense the standpoint of ordinary arithmetic which ignores the difference in type among numbers can be attained in ontology. A sufficiently high semantical category is selected and all work is carried on in the arithmetic of that level. It is not that numbers are ambiguous as to type, but that any given arithmetical problem can be handled in a single sufficiently high semantical category. And at this point, the sense in which arithmetic is derivable in ontology

---

[16] Gödel [1944, pp. 125-153].

is clear. Indeed, the discussion of the higher epsilons, and particularly the results obtaining from that discussion, closely parallels the discussion in *PM* on 'typically indefinite inductive cardinals', while the advantage of the Leśniewskian exposition lies in the fact that the standpoint of ordinary arithmetic has been achieved without any of the ambiguities latent in the development of *PM*.

Leśniewski's ontology remains an early source of a language whose terminology is thoroughly explained; whose coherence is contextually determinate and unambiguous; whose type theory adheres closely to categories which must be recognized in ordinary language; and whose directives for development mirror the contextually determinate development that is to be expected of a vehicle for communication.

Bogusław Iwanuś

On Leśniewski's Elementary Ontology

S. Leśniewski's calculus of names, often referred to as ontology, originated in 1920. This system like his other systems, mereology and protothetics, was constructed with the aim on the one hand of bringing logic closer to the intuitions of natural language and on the other of searching for foundations for mathematics. Leśniewski's ontology, in spite of numerous intuitive and formal advantages and in spite of its considerable expressive potential, has been underrated and little known for a long time; although half a century has passed since the construction of the system no precise elaboration of it has yet appeared which takes account of its methodological aspect. This situation seems fundamentally to be due to the fact that Leśniewski published no paper presenting his system in more or less final form. The manuscripts that Leśniewski left and which covered the results of the years of his investigations into ontology were destroyed during world war II. Nor should one ignore the fact that most of the published papers, in which Leśniewski presented his system at the stage of formalisation, were written in a difficult and not easily intelligible style. Leśniewski's complicated symbols, although abounding in interesting ideas, differ from the familiar logical and set-theoretic symbols and thus create an obstacle to the appreciation of his ideas. This explains the scarcity of extensive discussions of the ontology. Those that exist are as a rule based on Leśniewski's lecture notes and are primarily concerned with providing intelligible presentation of the problem of language and terminology (cf. Luschei [1962] and Słupecki [1955]*). Most of the discussions of the ontology has been concerned with the attempt to simplify the only axiom of the system and replace its only primitive term with another collection of primitive terms (Lejewski [1958]**, Sobociński [1934] and Słupecki [1955]*). These problems were of considerable importance to Leśniewski himself but have currently lost much of their import.[1]

* *Editorial Note*: Also in the present volume.

** *Editorial Note*: Also in the present volume.

[1] They have in my opinion no relevance for ontology. None of the equivalents, that I know, of Leśniewski's axiom suffices even to axiomatize the elementary ontology in view of creativity of definitions acceptable in the system.

In recent years there has been more interest in this system. Apart from methodological elaborations devoted to ontology alone (see Canty [1969]*, Canty [1967], Canty [1969a]) the attempts to use the bases of this system for constructing other logical calculi deserve attention (see Lebiediewa [1969] and Kubiński [1958]). Nowadays too, more place is given to this system in handbooks of formal logic (e.g. Borkowski [1970]).

The present paper is concerned with a part of the system of ontology, viz. with the elementary ontology. This is a particularly interesting aspect of the system in view of its closeness to the intuitions of natural language. The problems discussed here are closely related to that property of the definitions accepted in ontology which makes the sysetm differ from the majority of standardly constructed systems. I shall briefly characterize this property.

It is a familiar fact that in this system definitions are creative, i.e. with their help one can prove certain sentences containing no defined terms, which cannot be proved without the definitions.[2] The effect of this is that ontology results in a sequence of richer and richer systems where each subsequent system contains one axiom and one specific constant more than the previous one. This property makes it difficult to carry on methodological investigations into this calculus, i.e., it is an obstacle to comparing this system with other logical calculi. The latter difficulty seems most important for the following reason. In spite of its originality, ontology has many features in common with the systems of other authors. Comparing ontology with other theories which are better known and have been analysed methodologically may throw more light on some features of ontology.

The present article consists of three parts:

Part I which is most extensive is devoted to the study of definitions in Leśniewski's elementary ontology. I prove in this part that after adding two new axioms to the only axiom of this system, another system is obtained which is equivalent to elementary ontology with definitions satisfying the non-creativity condition.

In Part II I discuss the relationships between elementary

---

* *Editorial Note*: Also in the present volume.

[2] For instance the definition of product of names is creative:

$$X \varepsilon Y \cdot Z \equiv X \varepsilon Y \wedge X \varepsilon Z. \tag{a_1}$$

As an immediate consequence of this definition we arrive at the sentence:

$$\prod_{X,Y} \sum_{Z} \prod_{U} \{U \varepsilon Z \equiv U \varepsilon X \wedge U \varepsilon Y\}. \tag{a_2}$$

The sentence (a_2) does not follow from the axiom of ontology which may be observed when we interpret Leśniewski's ontology in his system of protothetics: nominal variables of ontology are understood to be sentential variables and the constant 'ε' is interpreted as equivalence of sentences ($\equiv$). In this interpretation both the axiom of ontology and the negation of the sentence (a_2) are theses of protothetics.

ontology and other deductive systems. I give there for instance the proof of the fact that the elementary ontology is equivalent to a version of atomic Boolean algebra.

In part III I present the decidability proof for elementary ontology based on rejection rules and its semantical completeness proof.

In this paper I make use of the following symbols:

a) functors of the sentential calculus: negation ($\sim$), conjunction($\wedge$), alternation ($\vee$), implication ($\rightarrow$) and equivalence ($\equiv$);

b) quantifiers: universal quantifier ($\prod$) and existential quantifier ($\sum$);

c) nominal variables occurring in the expressions of the calculi: X, Y, Z, ..., X_1, X_2, ...;

d) variables standing for the sentential expressions of the calculi: $\alpha, \beta, \gamma, \ldots, \alpha_1, \alpha_2, \ldots$;

e) variables standing for the nominal expressions of the calculi: ω, $\delta, \ldots, \delta_1, \delta_2, \ldots$;

f) variables ranging over the set of natural numbers: i, j, k, l, m, n, $n_1, n_2, \ldots$;

g) constants: $a, b, c, \ldots, a_1, a_2, \ldots$, occurring exclusively in the proof lines when the rule of omitting the existential quantifier is applied.

I also utilize the following abbreviations:

$X \neq Y$ for $\sim X = Y$,

$X_1 \neq \ldots \neq X_n$ for $X_1 \neq X_2 \wedge \ldots \wedge X_1 \neq X_n \wedge X_2 \neq X_3 \wedge \ldots \wedge X_2 \neq X_n \wedge \ldots \wedge X_{n-1} \neq X_n$,

$X \not\subset Y$ for $\sim X \subset Y$,

$X \bar{\varepsilon} Y$ for $\sim X \varepsilon Y$.

The proofs of theses are given here in the form of proofs from assumptions. The expressions appearing in brackets at the right side of proof lines are, respectively, the abbreviations of the following:

'a.' — 'assumption(s)',

'a.i.p.' — 'assumption of indirect proof',

'i.a.' — 'inductive assumption',

'ad.a.' — 'additional assumption of a proof',

'contr.' — 'contradiction',

'(1.1), (1.2)→contr.' — 'contradiction results from proof lines (1.1) and (1.2)'.

While reading the expressions of ontology I mostly utilize the terminology of the algebra of sets, e.g. instead of speaking about the product of names I speak about the product of sets. When putting down the theses and definitions I omit except in some rare cases, the universal quantifier signs occurring at the beginning of formulae.

In the present paper I make essential use of the results contained in the following publications: Skolem [1919], Słupecki [1955]*, Tarski [1933]**.

---

* *Editorial Note*: Also in the present volume.

** *Editorial Note*: English translation in Tarski [1956].

I. Definitions in the Elementary Ontology

1. I give briefly the characteristics of elementary ontology following with slight modifications that version of the system presented in Słupecki [1955][3].

Elementary ontology is a calculus of names constructed over the narrower predicate calculus without identity, treated metalogically (i.e., without the variables standing for predicates). The only primitive term of ontology is the constant 'ε', its meaning is determined by the following axiom:

A1. $$\prod_{X,Y}\{X \varepsilon Y \equiv \sum_{Z}\{Z \varepsilon X\} \wedge \prod_{Z,U}\{Z \varepsilon X \wedge U \varepsilon X \rightarrow Z \varepsilon U\} \wedge \prod_{Z}\{Z \varepsilon X \rightarrow Z \varepsilon Y\}\}.$$

The symbol "ε" which occurs in the above axiom is a sentence-forming functor of two nominal arguments where both arguments are of the same syntactic category.

In ontology the rules are valid for joining to the system definitions of the following entities: (I) name-forming functors of nominal arguments, (II) nominal constants and (III) sentence-forming functors of nominal arguments. The respective schemes of these definitions are as follows:

(I) $$X \varepsilon f|X_1, \ldots, X_n| \equiv X \varepsilon X \wedge \alpha(X, X_1, \ldots, X_n),$$

(II) $$X \varepsilon A \equiv X \varepsilon X \wedge \alpha(X),$$

(III) $$\varphi /X_1, \ldots, X_n/ \equiv \alpha(X_1, \ldots, X_n).$$

The defined terms are: 'f', 'A' and 'φ' in (I), (II) and (III), respectively. $X, X_1, \ldots, X_n$ are here the nominal variables while α is a sentential expression of the elementary ontology. Equivalences (I), (II) and (III) satisfy the familiar conditions usually imposed on definitions with the exception for the definitions (I) and (II) on which the uniqueness and existence conditions are not set (cf. p. 191). After Leśniewski, I want to

---

[3] In this article (included in the present volume — *Editorial Note*) there is an extensive discussion of the system together with its intuitive foundations (see also Leśniewski [1930], Leśniewski [1931], Luschei [1962], Kotarbiński [1929] (and Kotarbiński [1966] — *Editorial Note*)). I make use of the symbols taken from Słupecki [1955] with some minor changes introduced. In particular, the expressions appearing in the scope of the quantifier are put here into brackets {...}, arguments of name-forming functors are put in between vertical strokes |...|, and I employ oblique strokes /.../ in the case of the arguments of the sentence-forming functors. Binary functors are the exception since they appear between their arguments, and so is the sign of negation for names which is put after the argument.

stipulate that no defined terms may occur in the definiens of any definition above.

The expression $f|X_1, \ldots, X_n|$ as well as the constant A occurring in the equivalences (I) and (II) is of the same syntactic category as the variable X. In the elementary ontology all variables, the expressions formed of name-forming functors and their arguments, the arguments of functors different from the connectives of the sentential calculus and nominal variables too, belong to the same syntactic category, viz. the category of names. In this calculus there occur neither functor variables nor functors depending on parameters.

The set of meaningful sentential expressions of this calculus is defined in a way analogous to other deductive systems but this set will also include the expressions in which the defined terms occur.

After this brief presentation of elementary ontology I now move to stating and proving the main theorem of this paper, which refers to the rules of definition valid in the system. The definitions of sentence-forming functors in the elementary ontology are non-creative and they do not differ from the definitions of predicates accepted in the narrower functional calculus. Only the definitions of name-forming functors and the definitions of nominal constants are creative (cf. note 2). I shall attempt instead to prove that the following theorem is true:

Theorem 1.I. *If we base elementary ontology on the following set of axioms*:

A1. $$\prod_{X,Y} \{X \varepsilon Y \equiv \sum_{Z} \{Z \varepsilon X\} \wedge \prod_{Z,U} \{Z \varepsilon X \wedge U \varepsilon X \rightarrow Z \varepsilon U\} \wedge \prod_{Z} \{Z \varepsilon X \rightarrow Z \varepsilon Y\}\},$$

A2. $$\prod_{X} \sum_{Y} \prod_{Z} \{Z \varepsilon Y \equiv Z \varepsilon Z \wedge Z \bar{\varepsilon} X\},$$

A3. $$\prod_{X,Y} \sum_{Z} \prod_{U} \{U \varepsilon Z \equiv U \varepsilon X \wedge U \varepsilon Y\},$$

then the definitions of name-forming functors (I) *and the definitions of nominal constants* (II) *are non-creative too.*

Sentences A1–A3 form an independent axiom set. It is also worth-while to observe that axioms A2 and A3, where the first of them assumes that there exists a complement for any set and where the second assumes that for any two sets there exists their product, do not lead outside the system of elementary ontology with definitions. They are the immediate consequences of the following definitions accepted in ontology:

(a) $$X \varepsilon Y' \equiv X \varepsilon X \wedge X \bar{\varepsilon} Y,$$

(b) $$X \varepsilon Y \cdot Z \equiv X \varepsilon Y \wedge X \varepsilon Z.$$

In the proof of theorem 1.I I make use of a non-creative extension of ontology based on axioms A1–A3, which will be defined in the sequel.

The following two lemmas stating, respectively, the existence of the empty set and the existence of the union set, are consequences of axioms A1–A3.

L1.1. $$\sum_{Y}\prod_{X}\{X \varepsilon Y \equiv X \varepsilon X \wedge X \bar{\varepsilon} X\}$$

L1.2. $$\prod_{X,Y}\sum_{Z}\prod_{U}\{U \varepsilon Z \equiv U \varepsilon X \vee U \varepsilon Y\}$$

I add to the system under discussion the following definitions[4]:

Df.1.1. $$X \subset Y \equiv \prod_{Z}\{Z \varepsilon X \rightarrow Z \varepsilon Y\}$$

Df.1.2. $$X = Y \equiv \prod_{Z}\{Z \varepsilon X \equiv Z \varepsilon Y\}$$

Df.1.3. $$X \varepsilon \emptyset \equiv X \varepsilon X \wedge X \bar{\varepsilon} X$$

Df.1.4. $$X \varepsilon Y' \equiv X \varepsilon X \wedge X \bar{\varepsilon} Y$$

Df.1.5. $$X \varepsilon Y \cdot Z \equiv X \varepsilon Y \wedge X \varepsilon Z$$

Df.1.6. $$X \varepsilon Y + Z \equiv X \varepsilon Y \vee X \varepsilon Z$$

The system determined by the axiom set A1–A3 and by definitions 1.1–1.6 will be symbolized by EO.

Definitions 1.1–1.6 define respectively, the relation of inclusion and of extensional equality of sets, the empty set, the complement of a set, the intersection and the union of sets. While introducing these definitions to the system of elementary ontology we extend at the same time the set of nominal expressions (terms) and the set of (meaningful) sentential expressions of the system. By the set of terms we mean the smallest set containing nominal variables and the constant Ø, which is closed under the operations ′, +, and · . The set of atomic expressions of the calculus is composed of formulae of the form: $\delta_1 \varepsilon \delta_2$, $\delta_1 \subset \delta_2$, and $\delta_1 = \delta_2$, where δ_1 and δ_2 are arbitrary terms and where the set of the sentential expressions coincides with the smallest set containing the atomic formulae and closed under the operations of forming the sentential expressions with the help of the sentential connectives and quantifiers. The mea-

[4] The right side of the equivalence Df.1.1 appears in Leśniewski's ontology as the definiens of the categorical universal affirmative sentence *XaY* (cf. Słupecki [1948, p. 28]). Instead of the symbol '*a*' I utilized the symbol '⊂' since the relation defined by definition 1.1 has, in my opinion, more in common with the set theoretic inclusion than with the Aristotelian *a*.

On the right sides of definitions 1.5 and 1.6 I have omitted the factor $X \varepsilon X$. This factor is superfluous by virtue of the following equivalences resulting from A1:

$$X \varepsilon X \wedge (X \varepsilon Y \wedge X \varepsilon Z) \equiv X \varepsilon Y \wedge X \varepsilon Z,$$
$$X \varepsilon X \wedge (X \varepsilon Y \vee X \varepsilon Z) \equiv X \varepsilon Y \vee X \varepsilon Z.$$

ningful sentential expressions in which free variables occur are called sentential functions while the sentential expressions without free variables are called sentences.

When definitions 1.1–1.6, and particularly definitions 1.3–1.6 are added to axioms A1–A3, we obtain an extension of the calculus of names. This extension is a non-creative one in view of A2, A3, L1.1 and L1.2, in the sense that every thesis of the system EO which is put down without using the defined terms is a consequence of axioms A1–A3 alone (cf. Grzegorczyk [1961], Chapter III, § 6). The proof of the theorem 1.I may as well be carried without the help of definitions 1.1–1.6 although these allow for a considerable shortening of the proof. In particular, the relation $=$, the set $\emptyset$, the operations $'$, $+$ and $\cdot$, defined by means of these definitions satisfy in the system EO the postulates of Boolean algebra (cf. Słupecki [1955], pp. 41–46).* Moreover, there holds the equivalence $X \subset Y \equiv X + Y = Y$, which is often considered in Boolean algebra as a definition of inclusion. Therefore I omit below the proofs of these theses of the system EO, produced with the help of the constants: $\subset$, $=$, $\emptyset$, $'$, $+$ and $\cdot$, which are the equivalents of familiar theorems of Boolean algebra.

The proof of theorem 1.I still needs one essential remark.

It is familiar fact that all definitions of elementary ontology satisfy the conditions of non-creativity if we add to the system an axiom equivalent to the definition axiom of Russell's simple theory of types (see Słupecki [1955], p. 65). **This axiom allows for adding to the system any sentence of the form:

$$(*) \qquad \prod_{X_1} \dots \prod_{X_n} \sum_{Y} \prod_{Z} \{Z \varepsilon Y \equiv Z \varepsilon Z \wedge \alpha(Z, X_1, \dots, X_n)\},$$

where α is a meaningful expression of the elementary ontology containing no defined terms, and variables $Z, X_1, \dots, X_n$ are all the free variables of the expression α and the free variable Y does not occur in it. In particular cases expression α may contain no occurrence of the variable Z and likewise it may contain no free variable whatsoever.

To prove theorem 1.I it is enough to demonstrate:

Theorem 1.Ia. *Every sentential expression of the system* EO *which falls under the scheme* (*) *is a thesis of the system.*

2. The present section is entirely devoted to the proof of theorem 1.Ia. In the proof I shall make use of the following metasystem definitions and abbreviations:

Let $X_1, \dots, X_n$ be an arbitrary n-element sequence of variables which are not equiform.

Assume that $X_i^0 = X_i$ and $X_i^1 = X_i'$ for $1 \leqslant i \leqslant n$.

* *Editorial Note*: Also the present volume, pp. 93–98.

** *Editorial Note*: Also the present volume, p. 117.

Definition 1.I. Every expression

$$X_1^{i_1} \cdot X_2^{i_2} \cdot \ldots \cdot X_n^{i_n},$$

where $i_k = 1$ or $i_k = 0$, for $1 \leqslant k \leqslant n$, will be called *constituent in the variables* $X_1, \ldots, X_n$.

Two constituents δ_1, δ_2 in the variables $X_1, \ldots, X_n$ are equal (but not necessarily equiform) provided that they are products of the same factors, i.e. each of the variables $X_1, \ldots, X_n$ occurs in the constituent δ_1 (not) preceded by negation sign if and only if it occurs (not) preceded by negation sign in the constituent δ_2.

Formulating the following definition and some lemmas as well, I make use of the following abbreviation. Instead of:

$$\sum_{Y_1} \cdots \sum_{Y_n} \{Y_1 \neq Y_2 \wedge \ldots \wedge Y_1 \neq Y_n \wedge \ldots \wedge Y_{n-1} \neq Y_n \wedge Y_1 \varepsilon \delta \wedge \ldots \wedge Y_n \varepsilon \delta\}$$

I write:

$$\sum_{Y_1 \neq Y_n} \cdots \sum \{Y_1, \ldots, Y_n \varepsilon \delta\}.$$

Definition 1.II.

$$\delta \succ n = \begin{cases} \ulcorner \sum_{Y} \{Y \varepsilon \delta\} \urcorner, & \text{for } n = 1, \\ \ulcorner \sum_{Y_1 \neq Y_n} \cdots \sum \{Y_1, \ldots, Y_n \varepsilon \delta\} \urcorner, & \text{for } n > 1. \end{cases}$$

The variable n occurring in definition 1.II, which is a metasystem abbreviation, represents an arbitrary natural number other than zero while δ represents an arbitrary term of the calculus EO. The expressions of the form: $\delta \succ n$ and $\sim\delta \succ n$ are called quantitative formulae. They state respectively, that the set δ contains at least n or at most n-1 elements.

The following expressions composed of the quantitative formulas and of the connectives of the sentential calculus are, respectively, the abbreviations of the following expressions:

$\sim\delta \succ n - \ulcorner \sim \sum_{Y_1 \neq Y_n} \cdots \sum \{Y_1, \ldots, Y_n \varepsilon \delta\} \urcorner$,

$\delta_1 \succ n \wedge \delta_2 \succ m - \ulcorner \sum_{Y_1 \neq Y_n} \cdots \sum \{Y_1, \ldots, Y_n \varepsilon \delta_1\} \wedge \sum_{Y_1 \neq Y_m} \cdots \sum \{Y_1, \ldots, Y_m \varepsilon \delta_2\} \urcorner$,

$\delta_1 \succ n \vee \delta_2 \succ m -$ " $\vee$ "

$\delta_1 \succ n \rightarrow \delta_2 \succ m -$ " $\rightarrow$ "

$\delta_1 \succ n \equiv \delta_2 \succ m -$ " $\equiv$ "

While formulating some lemmas of this section I use the symbol "$\Rightarrow$" as the sign of a metasystem implication. A formula of the form:

$$n \geqslant m \Rightarrow (X \succ n \rightarrow X \succ m)$$

is read as follows: if $n \geqslant m$, then the expression: $X \succ n \rightarrow X \succ m$ is a thesis of the calculus EO.

Let $X_1, \ldots, X_n$ be all the free variables and let $\delta_1, \ldots, \delta_l$ be all the terms occurring in the sentential function α, each of them being a constituent in the variables $X_1, \ldots, X_n$.

Definition 1.III. Expression a has a *normal alternative form of the quantitative formulae made up of the constituents in the variables* $X_1, \ldots, X_n$ if and only if α is an alternation of a certain number of the conjunctions $\alpha_1, \alpha_2, \ldots, \alpha_k$, the conjuncts being formulae of the form: $\delta \succ m$ or $\sim \delta \succ m$, where δ is one of the constituents $\delta_1, \ldots, \delta_l$, and m is an arbitrary number other than zero.

It is to be easily seen that variables $X_1, \ldots, X_n$, which are all the free variables of the sentential function α, occur as well in every conjunct of the conjunction α_i $(i = 1, 2, \ldots, k)$. In case $k = 1$, the expression α is called the elementary conjunction of quantitative formulas.

In the proof of theorem 1.Ia an important role is played by Lemma 1.V according to which every sentential function α of the calculus EO with free variables $X_1, \ldots, X_n$, may be transformed into an equivalent function β that has its normal alternative form of quantitative formulae made up of the constituents in the variables $X_1, \ldots, X_n$. The concept of a quantitative formula, which is going to be utilized in the proof of the above lemma as well as in the proofs of some theorems in III, is taken from Skolem [1919]. I also make use of the method of eliminating quantifiers which was used in that paper.

The following theses of the calculus EO which will be applied in the proof of lemma 1.V are now given without proof.[5]

L1.3a. $X \varepsilon Y \rightarrow X \varepsilon X$ b. $X \varepsilon Y \wedge Y \varepsilon Z \rightarrow X \varepsilon Z$

c. $X \varepsilon Y \wedge Y \varepsilon Z \rightarrow Y \varepsilon X$ d. $X \varepsilon Y \wedge Y \varepsilon Z \rightarrow X = Y$

e. $X \varepsilon Y \rightarrow X \varepsilon \emptyset'$

L1.4a. $X \varepsilon Y \equiv X \varepsilon X \wedge \sum_{Z} \{Z \varepsilon X \wedge Z \varepsilon Y\}$

b. $X \varepsilon X \equiv \sum_{Y} \{Y \varepsilon X\} \wedge \sim \sum_{Y \neq Z}\sum \{Y, Z \varepsilon X\}$

L1.5a. $X \varepsilon Y \rightarrow X \subset Y$ b. $X \varepsilon Y \rightarrow X \cdot Y' = \emptyset$

L1.6a. $X \varepsilon X \wedge X \cdot Y' = \emptyset \rightarrow X \varepsilon Y$ b. $X \varepsilon Y \rightarrow X \cdot Y \varepsilon X \cdot Y$

c. $X \cdot Y \varepsilon X \cdot Y \wedge X \cdot Y' = \emptyset \rightarrow X \varepsilon Y$

L7. $X \varepsilon X \rightarrow (X \bar{\varepsilon} Y \equiv X \varepsilon Y')$

L8. $X \subset \emptyset \equiv X = \emptyset \equiv \sim \sum_{Y} \{Y \varepsilon X\}$

[5] Proofs of some of these theses are contained in Słupecki [1955]. (The paper is included in the present volume – *Editorial Note.*) The remaining proofs may be easily deduced from axioms A1–A3 and definitions 1.1–1.6.

In view of the reasons given earlier (cf. p. 171), I provide the following algebraic lemmas omitting the proofs.

L1.9a. $X \cdot \emptyset = \emptyset$ b. $X \cdot \emptyset' = X$
c. $X + \emptyset = X$ d. $X + \emptyset' = \emptyset'$
e. $X + X' = \emptyset'$ f. $X \cdot X' = \emptyset$
L1.10a. $\emptyset \subset X$ b. $X \subset \emptyset'$
L1.11. $X'' = X$
L1.12a. $X + Y = Y + X$ b. $X \cdot Y = Y \cdot X$
L1.13a. $X + X = X$ b. $X \cdot X = X$
L1.14a. $(X + Y) + Z = Z + (Y + Z)$ b. $(X \cdot Y) \cdot Z = X \cdot (Y \cdot Z)$
L1.15a. $(X + Y) \cdot Z = X \cdot Z + Y \cdot Z$ b. $X + Y \cdot Z = (X + Y) \cdot (X + Z)$
L1.16a. $(X + Y)' = X' \cdot Y'$ b. $(X \cdot Y)' = X' + Y'$
L1.17a. $X \subset Y \equiv X + Y = Y$ b. $X \subset Y \equiv X \cdot Y = X$
c. $X \subset Y \equiv X \cdot Y' = \emptyset$
L1.18a. $X \subset X + Y$ b. $X \cdot Y \subset X$
L1.19a. $X = X$ b. $X = Y \rightarrow Y = X$
c. $X = Y \wedge Y = Z \rightarrow X = Z$

According to L1.19a–c the extensional equality defined by definition 1.2 is an equivalence relation. It may be also demonstrated that it fulfills the extensionality condition (cf. Słupecki [1955, p.14])* which states that the theses of the system EO include all expressions of the form:

L1.19d. $X = Y \wedge \alpha(X) \rightarrow \alpha(Y)$.

$\alpha(X)$ represents in the above scheme any sentential function of the calculus EO in which the variable X does not occur in the scope of the quantifier binding the variable Y, while $\alpha(Y)$ results from $\alpha(X)$ by substituting the variable Y for the variable X at all or some places.[6]

L1.20. $Y_1, \ldots, Y_n \, \varepsilon \, X \rightarrow (Y_1 + \ldots + Y_n) \subset X$
L1.21. $(Y_1, \ldots, Y_n \, \varepsilon \, X) \wedge X \subset Y \rightarrow Y_1, \ldots, Y_n \, \varepsilon \, Y$
L1.22a. $X \, \varepsilon \, (Y_1 + \ldots + Y_n) \equiv X \, \varepsilon \, Y_1 \vee X \, \varepsilon \, Y_2 \vee \ldots \vee X \, \varepsilon \, Y_n$
b. $X \, \varepsilon \, (Y_1 \cdot \ldots \cdot Y_n) \equiv X \, \varepsilon \, Y_1 \wedge X \, \varepsilon \, Y_2 \wedge \ldots \wedge X \, \varepsilon \, Y_n$
L1.23a. $Y_1, \ldots, Y_n \, \varepsilon \, X \cdot Y \equiv Y_1, \ldots, Y_n \, \varepsilon \, X \wedge Y_1, \ldots, Y_n \, \varepsilon \, Y$
b. $Y_1, \ldots, Y_n \, \varepsilon \, X + Y \equiv (Y_1, \ldots, Y_n \, \varepsilon \, X) \vee (Y_1, \ldots, Y_n \, \varepsilon \, Y) \vee (Y_{i_1}, \ldots, Y_{i_k} \, \varepsilon \, X \wedge Y_{i_{k+1}}, \ldots, Y_{i_n} \, \varepsilon \, Y)$, where $i_1, \ldots, i_k, \ldots, i_n$ is a permutation of the numbers $1, 2, \ldots, n$.

---

* *Editorial Note*: Also in the present volume, p. 105.

[6] Lemmas 1.19a–d and the condition of extensionality in particular, meet numerous applications in proving the theses of the calculus EO. To make the proofs of theses shorter I omit these proof lines in which I make use of lemmas 1.19a–d.

Lemma 1.19d is obviously a metathesis of the system EO. The further lemmas of this paragraph which I make use of in the proof of theorem 1.Ia are also the metatheses stating that the expressions of the respective form are theses of the system EO. Most of them however, are numbered like the previously noted theses of the system.

L1.24. $(Y_1, \ldots, Y_n \varepsilon X) \wedge Y \varepsilon X' \to Y_1, \ldots, Y_n \varepsilon Y'$

L1.25. $(Y_1 \varepsilon Y_1 \wedge \ldots \wedge Y_n \varepsilon Y_n) \wedge (Y_1, \ldots, Y_n \bar{\varepsilon} X) \to Y_1, \ldots, Y_n \varepsilon X'$[7]

L1.26. $X \varepsilon X \wedge (Y_1 \varepsilon Y_1 \wedge \ldots \wedge Y_n \varepsilon Y_n) \wedge (X \neq Y_1 \wedge \ldots \wedge X \neq Y_n)$
$\to Y_1, \ldots, Y_n \varepsilon X'$

L1.27. $Y_1, \ldots, Y_n \varepsilon X' \to X \neq Y_1, \wedge \ldots \wedge X \neq Y_n$

L1.28. $(Y_1, \ldots, Y_n \varepsilon X) \wedge (Z_1, \ldots, Z_m \varepsilon Y) \wedge X \cdot Y = \emptyset$
$\to Y_1 \neq Z_1 \wedge \ldots \wedge Y_1 \neq Z_m \wedge Y_2 \neq Z_1 \wedge \ldots \wedge Y_n \neq Z_m$

L1.29. $(X = Y_1 + \ldots + Y_n) \wedge (Y_1 \varepsilon Y_1 \wedge \ldots \wedge Y_n \varepsilon Y_n) \to Y_1, \ldots, Y_n$
εX

L1.30. $(X = Y_1 + \ldots + Y_n) \wedge (Y_1, \ldots, Y_n \varepsilon Y) \to X \cdot Y = X$

L1.31. $(Y_1, \ldots, Y_n \varepsilon X) \wedge (Y_1 \neq \ldots \neq Y_n) \to (X = Y_1 + \ldots + Y_n$
$\equiv \sim \sum\limits_{Z_1 \neq Z_{n+1}} \ldots \sum \{Z_1, \ldots, Z_{n+1} \varepsilon X\})$

L1.32. $(X = Y_1 + \ldots + Y_m + \ldots + Y_{m+n}) \wedge (Y_1 \varepsilon Y_1 \wedge \ldots \wedge Y_{m+n}$
$\varepsilon Y_{m+n}) \wedge (Y_1 \neq \ldots \neq Y_{m+n}) \to (X \cdot Y = Y_1 + \ldots + Y_m$
$\equiv X \cdot Y' = Y_{m+1} + \ldots + Y_{m+n})$

L1.33. $(Y_1, \ldots, Y_m, \ldots, Y_{m+n} \varepsilon X) \wedge (Y_1 \neq \ldots \neq Y_{m+n})$
$\wedge \sim \sum\limits_{Z_1 \neq Z_m} \ldots \sum \{Z_1, \ldots, Z_m \varepsilon X \cdot Y\}$
$\to \sum\limits_{Z_1 \neq Z_{n+1}} \ldots \sum \{Z_1, \ldots, Z_{n+1} \varepsilon X \cdot Y'\}$

The proofs by induction of lemmas 1.20–1.33 are quite easy on the grounds of the previously proved lemmas.

Let $\delta_1, \ldots, \delta_{2n}$ be all unequal constituents in the variables $X_1, \ldots, X_n$.

L1.34a. $i \neq j \Rightarrow \delta_i \cdot \delta_j = \emptyset$

b. $X = \delta_1 \cdot X + \delta_2 X + \ldots + \delta_{2n} \cdot X$

c. $\delta_j \cdot \sum\limits_{i=1}^{m} (\delta_i \cdot X_i^{k_i}) = \delta_j \cdot X_j^{k_j}$, where $m \leqslant 2^n$, $X^{k_j} = X$
or $X^{k_j} = X'$, for j $= 1, 2, \ldots, m$.

Lemmas 1.34a–b of which the first states that two unequal constituents are disjoint while the other states that the sum of all products of the form: $\delta_i \cdot X$, where δ_i are constituents, is equal to X, are the counterparts of the familiar theorems of Boolean algebra or of the algebra of sets and may be proved on the grounds of the previously stated algebraic lemmas (Kuratowski and Mostowski [1967, pp. 20–21]). Each particular case of lemma (c) is a thesis of the calculus EO which may be easily seen in view of the following remark.

---

[7] The expression: $Y_1, \ldots, Y_n \bar{\varepsilon} X$ is an abbreviation of the conjunction: $Y_1 \bar{\varepsilon} X \wedge$ $\wedge Y_2 \bar{\varepsilon} X \wedge \ldots \wedge Y_n \bar{\varepsilon} X$.

From the definition of the sum $\sum_{i=1}^{m}(\delta_i \cdot X_i^{k_i})$, by the respective lemmas (L1.14a–b, L1.15a) we obtain:

(1) $$\delta_j \cdot \sum_{i=1}^{m}(\delta_i \cdot X_i^{k_i}) = (\delta_j \cdot \delta_1)\cdot X_1^{k_1} + \ldots + (\delta_j \cdot \delta_m)\cdot X_m^{k_m}.$$

According to L1.34a, L1.9a and L1.13b

(2) $(\delta_j \cdot \delta_i)\cdot X_i^{k_i} = \emptyset$, for $i \neq j$ and $(\delta_j \cdot \delta_i)\cdot X_i^{k_i} = \delta_j \cdot X_i^{k_i}$, for $i = j$.

Hence by L1.9c we obtain L1.34c:

$$\delta_j \cdot \sum_{i=1}^{m}(\delta_i \cdot X_i^{k_i}) = \delta_j \cdot X_j^{k_j}.$$

Making use of the fact that the relation $=$, the set $\emptyset$, operations $'$, $+$ and $\cdot$ satisfy in ontology the postulates of Boolean algebra, we may prove similarly to L1.34, the following lemma which is of considerable importance for our further considerations:

Lemma 1.I. *Every nominal expression (term) $\omega(X_1, \ldots, X_n)$ of the calculus* EO *which contains exclusively the free variables $X_1, \ldots, X$ satisfies one of the following equalities*:

(a) $\omega = \emptyset$,

(b) $\omega = \delta_1 + \delta_2 + \ldots + \delta_m$,

where $\delta_1, \ldots, \delta_m$ ($m \leqslant 2^n$) are different constituents in the variables $X_1, .., X_n$ (cf. Kuratowski [1967, pp. 20–23]).

The following lemmas characterize quantitative formulae:

L1.35a. $X \varepsilon X \equiv X \succ 1 \wedge \sim X \succ 2$

Proof.	(1.1)	$X \varepsilon X$	{ad. a.}
	(1.2)	$\sum_{Y}\{Y \varepsilon X\} \wedge \sim \sum_{Y \neq Z}\sum\{Y, Z \varepsilon X\}$	{L1.4b, 1.1}
	(1.3)	$X \succ 1 \wedge \sim X \succ 2$	{Df.1.II, 1.2}
	(2.1)	$X \succ 1 \wedge \sim X \succ 2$	{ad. a.}
	(2.2)	$\sum_{Y}\{Y \varepsilon X\} \wedge \sim \sum_{Y \neq Z}\sum\{Y, Z \varepsilon X\}$	{Df.1.II, 2.1}
	(2.3)	$X \varepsilon X$	{L1.4b, 2.2}
		$X \varepsilon X \equiv X \succ 1 \wedge \sim X \succ 2$	{1.1→1.3, 2.1→2.3}

L1.35b. $X \varepsilon Y \equiv X \cdot Y \succ 1 \wedge \sim X \cdot Y \succ 2 \wedge \sim X \cdot Y' \succ 1$

Proof.	(1.1)	$X \varepsilon Y$	{ad. a.}
	(1.2)	$X \cdot Y \varepsilon X \cdot Y$	{L1.6b, 1.1}
	(1.3)	$X \cdot Y' = \emptyset$	{L1.5b, 1.1}

(1.4) $X \cdot Y \succ 1 \wedge \sim X \cdot Y \succ 2 \wedge \sim X \cdot Y' \succ 1$ {L1.35a. 1.2, L1.8, Df.1.II, 1.3}

(2.1) $X \cdot Y \succ 1 \wedge \sim X \cdot Y \succ 2 \wedge \sim X \cdot Y' \succ 1$ {ad. a.}

(2.2) $X \cdot Y \varepsilon X \cdot Y$ {L1.35a, 2.1}

(2.3) $X \cdot Y' = \emptyset$ {Df.1.II, L1.8, 2.1}

(2.4) $X \varepsilon Y$ {L1.6c. 2.2, 2.1}

$X \varepsilon Y \equiv X \cdot Y \succ 1 \wedge \sim X \cdot Y \succ 2 \wedge \sim X \cdot Y' \succ 1$ {1.1 → 1.4, 2.1 → 2.4}

L1.35c. $X \subset Y \equiv \sim X \cdot Y' \succ 1$

Proof. (1.1) $X \subset Y$ {ad. a.}

(1.2) $X \cdot Y' = \emptyset$ {L1.17c, 1.1}

(1.3) $\sim X \cdot Y' \succ 1$ {Df.1.II, L1.8, 1.2}

(2.1) $\sim X \cdot Y' \succ 1$ {ad. a.}

(2.2) $X \cdot Y' = \emptyset$ {Df.1.II, L1.8, 2.1}

(2.3) $X \subset Y$ {L1.17c, 2.1}

$X \subset Y \equiv \sim X \cdot Y' \succ 1$ {1.1→1.3, 2.1→2.3}

L1.35d. $X = Y \equiv \sim X \cdot Y' \succ 1 \wedge \sim X' \cdot Y \succ 1$ {L1.35c, Df.1.1–1.2}

It may easily be seen that lemmas 1.35a–d assign to each atomic formula of the calculus EO an equivalent quantitative formula or an equivalent conjunction of quantitative formulae.

L1.36a. $\sim \emptyset \succ n$ {Df.1.II, L1.8}

b. $\emptyset \succ n \equiv X \succ 1 \wedge \sim X \succ 1$ {L1.36a}

L1.37a. $X \cdot Y \succ n \rightarrow X \succ n$

Proof. (1) $X \cdot Y \succ n$ {a.}

(2) $a_1 \neq \ldots \neq a_n$[8]

(3) $a_1, \ldots, a_n \varepsilon X \cdot Y$ {Df.1.II, 1}

(4) $a_1, \ldots, a_n \varepsilon X$ {L1.23a, 3}

$X \succ n$ {Df.1.II, 2, 4}

L1.37b. $X \succ n \rightarrow X + Y \succ n$

The proof of lemma (b) is analogous to that of (a).

L1.38a. $n \geqslant m \Rightarrow (X \succ n \rightarrow X \succ m)$

b. $n < m \Rightarrow (X \succ m \rightarrow X \succ n)$ {Df.1.II}

---

[8] Proof lines (2) and (3) have been obtained from (1) by virtue of 1.II and by the rule for omitting the existential quantifier which was applied to the expression:

$$\sum_{X_1 \neq X_n} \ldots \sum \{Y_1, \ldots, Y_n \varepsilon X \cdot Y\}.$$

Writing the constants $a_1, \ldots, a_n$ I have, however, omitted the factors X and Y (cf. Słupecki and Borkowski [1967, p. 99]). In the majority of the subsequent proofs I also do that.

L1.39a. $n \geqslant m \Rightarrow (X \succ n \equiv X \succ n \wedge X \succ m)$
b. $n < m \Rightarrow (X \succ m \equiv X \succ n \wedge X \succ m)$
c. $n \geqslant m \Rightarrow (\sim X \succ m \equiv \sim X \succ n \wedge \sim X \succ m)$
d. $n < m \Rightarrow (\sim X \succ n \equiv \sim X \succ n \wedge \sim X \succ m)$ {L1.38a–b}

L1.40a. $X \succ n+1 \rightarrow (Y \varepsilon X \rightarrow X \cdot Y' \succ n)$

Proof.
(1) $X \succ n+1$
(2) $Y \varepsilon X$ {a.}
(3) $a_1 \neq \ldots \neq a_{n+1}$
(4) $a_1, \ldots, a_{n+1} \varepsilon X$ {Df.1.II, 1}
(1.1) for some $1 \leqslant i \leqslant n+1 : a_i \varepsilon Y$ {ad. a.}
(1.2) $Y = a_i$ {L1.3d, 1.1, 2}
(1.3) $Y \neq a_1 \wedge \ldots \wedge Y \neq a_{i-1} \wedge Y \neq a_{i+1} \wedge \ldots \wedge Y \neq a_{n+1}$ {3, 1.2}
(1.4) $a_1, \ldots, a_{i-1}, a_{i+1}, \ldots, a_{n+1} \varepsilon Y'$ {L1.3a, L1.26, 2, 4, 1.3}
(1.5) $a_1, \ldots, a_{i-1}, a_{i+1}, \ldots, a_{n+1} \varepsilon X \cdot Y'$ {L1.23a, 4, 1.4}
(1.6) $X \cdot Y' \succ n$ {Df.1.II, 3, 1.5}
(2.1) $a_1, \ldots, a_{n+1} \bar{\varepsilon} Y$ {ad. a.}
(2.2) $a_1, \ldots, a_{n+1} \varepsilon Y'$ {L1.25, L1.3a, 4, 2.1}
(2.3) $a_1, \ldots, a_{n+1} \varepsilon X \cdot Y'$ {L1.23a, 4, 2.2}
(2.4) $X \cdot Y' \succ n+1$ {Df.1.II, 2.3, 4}
(2.5) $X \cdot Y' \succ n$ {L1.38b, 2.4}
$X \cdot Y' \succ n$ {1.1→1.6, 2.1→2.5}

L1.40b. $X \succ n \rightarrow (Y \varepsilon X' \rightarrow X \cdot Y' \succ n)$

The proof of lemma (b) is analogous to that of (a).

L1.41a. $X \cdot Y \succ n \wedge X \cdot Y' \succ m \rightarrow X \succ n+m$

Proof.
(1) $X \cdot Y \succ n \wedge X \cdot Y' \succ m$ {a.}
(2) $a_1 \neq \ldots \neq a_n$
(3) $b_1 \neq \ldots \neq b_m$
(4) $a_1, \ldots, a_n \varepsilon X \cdot Y$
(5) $b_1, \ldots, b_m \varepsilon X \cdot Y'$ {Df.1.II, 1}
(6) $a_1 \neq b_1 \wedge \ldots \wedge a_1 \neq b_m \wedge \ldots \wedge a_n \neq b_m$ {L1.23a. L1.9f, L1.28, 4, 5}
(7) $a_1, \ldots, a_n, b_1, \ldots, b_m \varepsilon X$ {L1.23a. 4, 5}
$X \succ n+m$ {Df.1.II, 2, 3, 6, 7}

L1.41b. $\sim X \cdot Y \succ n \wedge \sim X \cdot Y' \succ m \rightarrow \sim X \succ n+m-1$

The proof of lemma (b) does not differ in any essential way from that of lemma (a).

L1.42. $n > m \Rightarrow$ *there exists a natural number i such that the expression:*

$$X \succ m \wedge \sim X \succ n \rightarrow X \succ m+i \wedge \sim X \succ m+i+1$$

is a thesis of the system EO.

I omit an easy proof by induction of the lemma 1.42.

L1.43. $X \cdot Y = \emptyset \rightarrow (X \succ n \wedge Y \succ m \rightarrow X + Y \succ n + m)$

Proof.
(1) $X \cdot Y = \emptyset$
(2) $X \succ n \wedge Y \succ m$ {a.}
(3) $a_1 \neq \ldots \neq a_n$
(4) $b_1 \neq \ldots \neq b_m$
(5) $a_1, \ldots, a_n \varepsilon X$
(6) $b_1, \ldots, b_m \varepsilon Y$ {Df.1.II, 2}
(7) $a_1 \neq b_1 \wedge \ldots \wedge a_1 \neq b_m \wedge a_2 \neq b_1 \wedge \ldots \wedge a_n \neq b_m$ {L1.28, 5, 6, 1}
(8) $a_1, \ldots, a_n, b_1, \ldots, b_m \varepsilon X + Y$ {L1.23b, 5, 6}
$X + Y \succ n + m$ {Df.1.II, 7, 8, 3, 4}

L1.44a. $n = 1 \Rightarrow (X + Y \succ n \equiv X \succ n \vee Y \succ n)$ {Df.1.II, Df.1.6}

b. $n > 1 \Rightarrow (X \cdot Y = \emptyset \rightarrow (X + Y \succ n \equiv (X \succ n) \vee (X \succ n-1 \wedge Y \succ 1) \vee \ldots \vee (X \succ 1 \wedge Y \succ n-1) \vee (Y \succ n)))$

Proof.
(1) $n > 1$
(2) $X \cdot Y = \emptyset$ {a.}
(1.1) $X + Y \succ n$ {ad. a.}
(1.2) $a_1 \neq \ldots \neq a_n$
(1.3) $a_1, \ldots, a_n \varepsilon X + Y$ {Df.1.II, 1.1}
(1.4) $(a_1, \ldots, a_n \varepsilon X) \vee (a_1 \ldots, a_n \varepsilon Y) \vee (a_{i_1}, \ldots, a_{i_k} \varepsilon X \wedge a_{i_{k+1}}, \ldots, a_{i_n} \varepsilon Y)$ {L1.23b, 1.3}
(1.5) $(X \succ n) \vee (X \succ n-1 \wedge Y \succ 1) \vee \ldots \vee (X \succ 1 \wedge Y \succ n-1) \vee (Y \succ n)$ {Df.1.II, 1.4, 1.2}
(2.1) $(X \succ n) \vee (X \succ n-1 \wedge Y \succ 1) \vee \ldots \vee (X \succ 1 \wedge Y \succ n-1) \vee (Y \succ n)$ {ad. a.}
(3.1) $X \succ n \vee Y \succ n$ {ad. a.}
(3.2) $X + Y \succ n$ {L1.37b, 3.1}
(4.1) $X \succ n - i \wedge Y \succ i$, for $1 \leqslant i \leqslant n-1$ {ad. a.}
(4.2) $X + Y \succ n$ {L1.43, 2, 4.1}
(2.2) $X + Y \succ n$ {2.1, 3.1→3.2, 4.1 →4.2}
$X + Y \succ n \equiv (X \succ n) \vee (X \succ n-1 \wedge Y \succ 1) \vee \ldots \vee (X \succ 1 \wedge Y \succ n-1) \vee (Y \succ n)$ {1.1→1.5, 2.1→2.2}

I shall now give the lemmas which allow for eliminating the existential quantifier whose scope contains a conjunction of quantitative formulae (see also Skolem [1919, pp. 32–33]).

L1.45a. $n_1 < n_3 \wedge n_2 < n_4 \Rightarrow (\sum_Y \{X \cdot Y \succ n_1 \wedge X \cdot Y' \succ n_2 \wedge \sim X \cdot Y \succ n_3 \wedge \sim X \cdot Y' \succ n_4\} \equiv (X \succ n_1 + n_2 \wedge \sim X \succ n_3 + n_4 - 1))$

Proof.

(I)

(1) $n_1 < n_3 \wedge n_2 < n_4$

(2) $\sum_{Y}\{X \cdot Y \succ n_1 \wedge X \cdot Y' \succ n_2 \wedge \sim X \cdot Y \succ n_3 \wedge \sim X \cdot Y' \succ n_4\}$ {a.}

(3) $X \cdot a \succ n_1 \wedge X \cdot a' \succ n_2$

(4) $\sim X \cdot a \succ n_3 \wedge \sim X \cdot a' \succ n_4$ {2}

(5) $X \succ n_1 + n_2 \wedge \sim X \succ n_3 + n_4 - 1$ {L1.41a–b, 3, 4}

(II)

(1) $n_1 < n_3 \wedge n_2 < n_4$

(2) $(X \succ n_1 + n_2) \wedge (\sim X \succ n_3 + n_4 - 1)$ {a.}

(3) $(X \succ n_1 + n_2 + i_1) \wedge (\sim X \succ n_1 + n_2 + i_1 + 1)$ {L1.42. 1, 2}

(1.1) $i_1 = 0$ {ad. a.}

(1.2) $\sim X \succ n_1 + n_2 + 1$ {3, 1.1}

(1.3) $a_1 \neq \ldots \neq a_{n_1} \neq \ldots \neq a_{n_1+n_2}$

(1.4) $a_1, \ldots, a_{n_1}, \ldots, a_{n_1+n_2} \varepsilon X$ {Df.1.II, 2}

(1.5) $a_1 \varepsilon a_1 \wedge \ldots \wedge a_{n_1+n_2} \varepsilon a_{n_1+n_2}$ {L1.3a, 1.4}

(1.6) $X = (a_1 + \ldots + a_{n_1} + \ldots + a_{n_1+n_2})$ {L1.31, Df.1.II, 1.2, 1.3 1.4}

(1.7) $Y_1 = (a_1 + \ldots + a_{n_1})$ {definition}

(1.8) $X \cdot Y_1 = (a_1 + \ldots + a_{n_1})$ {L1.30, 1.7, 1.4}

(1.9) $X \cdot Y_1' = (a_{n_1+1} + \ldots + a_{n_1+n_2})$ {L1.32, 1.6, 1.5, 1.4, 1.8}

(1.10) $a_1, \ldots, a_{n_1} \varepsilon X \cdot Y_1$

(1.11) $a_{n_1+1}, \ldots, a_{n_1+n_2} \varepsilon X \cdot Y_1'$ {L1.29, 1.8, 1.9, 1.5}

(1.12) $X \cdot Y_1 \succ n_1 \wedge X \cdot Y_1' \succ n_2$ {Df.1.II, 1.3, 1.10, 1.11}

(1.13) $\sim X \cdot Y_1 \succ n_1 + 1 \wedge \sim X \cdot Y_1' \succ n_2 + 1$ {L1.31, Df.1.II, 1.5, 1.3, 1.8, 1.9}

(1.14) $\sim X \cdot Y_1 \succ n_3 \wedge \sim X \cdot Y_1' \succ n_4$ {L1.38a, 1, 1.13}

(1.15) $\sum_{Y}\{X \cdot Y \succ n_1 \wedge X \cdot Y' \succ n_2 \wedge \sim X \cdot Y \succ n_3 \wedge \sim X \cdot Y' \rightarrow n_4\}$ {1.12, 1.14}

(2.1) $i_1 \neq 0 \wedge n_1 + i_1 < n_3$ {ad. a.}

(2.2) $a_1 \neq \ldots \neq a_{n_1+n_2+i_1}$

(2.3) $a_1, \ldots, a_{n_1+i_1}, \ldots, a_{n_1+n_2+i_1} \varepsilon X$ {Df.1.II, 3}

(2.4) $a_1 \varepsilon a_1 \wedge \ldots \wedge a_{n_1+n_2+i_1} \varepsilon a_{n_1+n_2+i_1}$ {L1.3a, 2.3}

(2.5) $X = (a_1 + \ldots + a_{n_1+n_2+i_1})$ {L1.31, Df1.II, 2.3,. 2.2, 3}

(2.6) $Y_2 = (a_1 + \ldots + a_{n_1+i_1})$ {definition}

(2.7) $X \cdot Y_2 = (a_1 + \ldots +_{n_1+i_1})$ {L1.30, 2.6, 2.3}

(2.8) $X \cdot Y_2' = (a_{n_1+1+i_1} + \ldots + a_{n_1+n_2+i_1})$ {L1.32, 2.5, 2.4, 2.7, 2.2}

(2.9) $a_1, \ldots, a_{n_1+i_1} \varepsilon X \cdot Y_2$

(2.10) $a_{n_1+1+i_1}, \ldots, a_{n_1+n_2+i_1} \varepsilon X \cdot Y_2'$ {L1.29, 2.7, 2.8, 2.4}

(2.11) $X \cdot Y_2 \succ n_1 + i_1 \wedge X \cdot Y_2' \succ n_2$ {Df.1.II, 2.2, 2.9, 2.10}

(2.12) $(\sim X \cdot Y_2 \succ n_1 + i_1 + 1) \wedge (\sim X \cdot Y_2' \succ n_2 + 1)$ {L1.31, Df.1.II, 2.4, 2.2, 2.7, 2.8}

(2.13) $\sim X \cdot Y_2 \succ n_3 \wedge \sim X \cdot Y_2' \succ n_4$ {L1.38a, 2.1, 2.12}

(2.14) $\sum_{Y} \{X \cdot Y \succ n_1 \wedge X \cdot Y' \succ n_2 \wedge \sim X \cdot Y \succ n_3 \wedge \sim X \cdot Y' \succ n_4\}$ {L1.38a, 2.11, 2.13}

(3.1) $i_1 \neq 0 \wedge n_1 + i_1 \geqslant n_3$ {ad. a.}

(3.2) $a_1 \neq \ldots \neq a_{n_1+n_2+i_1}$

(3.3) $a_1, \ldots, a_{n_3}, \ldots, a_{n_1+i_1}, \ldots, a_{n_1+n_2+i_1} \varepsilon X$ {Df.1.II, 3} (for 3.2–3.3)

(3.4) $a_1 \varepsilon a_1 \wedge \ldots \wedge a_{n_1+n_2+i_1} \varepsilon a_{n_1+n_2+i_1}$ {L1.3a, 3.3}

(3.5) $X = (a_1 + \ldots + a_{n_3} + \ldots + a_{n_1+n_3+i_1})$ {L1.31, Df1.II, 3.3, 3.4, 3}

(3.6) $Y_3 = (a_1 + \ldots + a_{n_3-1})$ {definition}

(3.7) $X \cdot Y_3 = (a_1 + \ldots + a_{n_3-1})$ {L1.30, 3.6, 3.3}

(3.8) $X \cdot Y_3' = (a_{n_3} + a_{n_3-1} + \ldots + a_{n_1+n_2+i_1})$ {L1.32, 3.5, 3.4, 3.2, 3.7}

(3.9) $a_1, \ldots, a_{n_3-1} \varepsilon X \cdot Y_3$

(3.10) $a_{n_3}, \ldots, a_{n_1+n_2+i_1} \varepsilon X \cdot Y_3'$ {L1.29, 3.7, 3.8, 3.4} (for 3.9–3.10)

(3.11) $(X \cdot Y_3 \succ n_3 - 1) \wedge (X \cdot Y_3' \succ n_1 + n_2 + i_1 - (n_3 - 1))$ {Df.1.II, 3.2, 3.9, 3.10}

(3.12) $(n_3 - 1 \geqslant n_1) \wedge (n_1 + n_2 + i_1 - (n_3 - 1) \geqslant n_2)$ {1, 3.1}

(3.13) $X \cdot Y_3 \succ n_1 \wedge X \cdot Y_3' \succ n_2$ {L1.38a, 3.12, 3.11}

(3.14) $(\sim X \cdot Y_3 \succ n_3) \wedge (\sim X \cdot Y_3' \succ n_1 + n_2 + i_1 - (n_3 - 1) + 1)$ {L1.31, Df.1.II, 3.4, 3.2, 3.7, 3.8}

(3.15) $n_1 + n_2 + i_1 < n_3 + n_4 - 1$ {L1.38a, 3, 2}

(3.16) $(n_1 + n_2 + i_1 - (n_3 - 1) + 1) \leqslant n_4$ {3.15}

(3.17) $\sim X \cdot Y_3' \succ n_4$ {L1.38a, 3.14, 3.16}

(3.18) $\sum_{Y} \{X \cdot Y \succ n_1 \wedge X \cdot Y' \succ n_2 \wedge \sim X \cdot Y \succ n_3 \wedge \sim X \cdot Y' \succ n_4\}$ {3.13, 3.14, 3.17}

(4) $\sum_{Y} \{X \cdot Y \succ n_1 \wedge X \cdot Y' \succ n_2 \wedge \sim X \cdot Y \succ n_3 \wedge \sim X \cdot Y' \succ n_4\}$ {1.1→1.15, 2.1→2.14, 3.1→3.18}

From (I) and (II) there follows lemma 1.45a.

L1.45b.

$$n_1 \geqslant n_3 \vee n_2 \geqslant n_4 \Rightarrow \left(\sum_{Y} \{X \cdot Y \succ n_1 \wedge X \cdot Y' \succ n_2 \wedge \sim X \cdot Y \succ n_3 \wedge \sim X \cdot Y' \succ n_4\} \equiv (X \succ n_1 \wedge \sim X \succ n_1)\right)$$

Proof. According to L1.38a: if $n_1 \geqslant n_3$ or $n_2 \geqslant n_4$, for n_1, n_2, n_3, n_4 different from zero, then the negation of the sentential function:

$$\sum_{Y}\{X\cdot Y \succ n_1 \wedge X\cdot Y' \succ n_2 \wedge \sim X\cdot Y \succ n_3 \wedge \sim X\cdot Y' \succ n_4\}$$

is a thesis of the calculus EO. It is convenient for our further considerations to assign to the above function a contradictory expression which is equivalent to it and has the form:

$$(X \succ n_1 \wedge \sim X \succ n_1).$$

L1.46a. $n_1 < n_3 \Rightarrow \left(\sum_{Y}\{X\cdot Y \succ n_1 \wedge X\cdot Y' \succ n_2 \wedge \sim X\cdot Y \succ n_3\} \equiv (X \succ n_1 + n_2)\right)$

Proof. (1) $n_1 < n_3$ {a.}

(1.1) $\sum_{Y}\{X\cdot Y \succ n_1 \wedge X\cdot Y' \succ n_2 \wedge \sim X\cdot Y \succ n_3\}$ {ad. a.}

(1.2) $X\cdot a \succ n_1 \wedge X\cdot a' \succ n_2$ {1.1}

(1.3) $X \succ n_1 + n_2$ {L1.41a, 1.2}

(2.1) $X \succ n_1 + n_2$ {ad. a.}

(2.2) $a_1 \neq \ldots \neq a_{n_1+n_2}$

(2.3) $a_1, \ldots, a_{n_1}, \ldots, a_{n_1+n_2} \varepsilon X$ {Df.1.II, 2.1}

(2.4) $a_1 \varepsilon a_1 \wedge \ldots \wedge a_{n_1} \varepsilon a_{n_1} \wedge \ldots \wedge a_{n_1+n_2} \varepsilon a_{n_1+n_2}$ {L1.3a, 2.3}

(2.5) $Y_1 = (a_1 + \ldots + a_{n_1})$ {definition}

(2.6) $X\cdot Y_1 = Y_1$ {L1.30, 2.5, 2.3}

(2.7) $X\cdot Y_1 = (a_1 + \ldots + a_{n_1})$ {2.5, 2.6}

(2.8) $a_1, \ldots, a_{n_1} \varepsilon X\cdot Y_1$ {L1.29, 2.7, 2.4}

(2.9) $X\cdot Y_1 \succ n_1 \wedge \sim X\cdot Y_1 \succ n_1 + 1$ {Df.1.II, L1.31, 2.8, 2.2, 2.7}

(2.10) $\sim X\cdot Y_1 \succ n_3$ {L1.38a, 2.9, 1}

(2.11) $X\cdot Y_1' \succ n_2$ {L1.33, Df.1.II, 2.3, 2.9}

(2.12) $\sum_{Y}\{X\cdot Y \succ n_1 \wedge X\cdot Y' \succ n_2 \wedge \sim X\cdot Y \succ n_3\}$ {2.9, 2.11, 2.10}

$\sum_{Y}\{X\cdot Y \succ n_1 \wedge X\cdot Y' \succ n_2 \wedge \sim X\cdot Y \succ n_3\} \equiv (X \succ n_1 + n_2)$ {1.1→1.3, 2.1→2.12}

L1.46b. $n_1 \geqslant n_3 \Rightarrow \left(\sum_{Y}\{X\cdot Y \succ n_1 \wedge X\cdot Y' \succ n_2 \wedge \sim X\cdot Y \succ n_3\} \equiv (X \succ n_1 \wedge \sim X \succ n_1)\right)$

L1.47a. $n_2 < n_3 \Rightarrow \left(\sum_{Y}\{X\cdot Y \succ n_1 \wedge X\cdot Y' \succ n_2 \wedge \sim X\cdot Y' \succ n_3\} \equiv (X \succ n_1 + n_2)\right)$

b. $n_2 \geqslant n_3 \Rightarrow \left(\sum_{Y}\{X\cdot Y \succ n_1 \wedge X\cdot Y' \succ n_2 \wedge \sim X\cdot Y' \succ n_3\} \equiv (X \succ n_1 \wedge \sim X \succ n_1)\right)$

L1.48a. $n_1 < n_2 \Rightarrow \big(\sum_Y \{X \cdot Y \succ n_1 \wedge \sim X \cdot Y \succ n_2 \wedge \sim X \cdot Y' \succ n_3\}$

$\equiv (X \succ n_1 \wedge \sim X \succ n_2 + n_3 - 1)\big)$

b. $n_1 \geqslant n_2 \Rightarrow \big(\sum_Y \{X \cdot Y \succ n_1 \wedge \sim X \cdot Y \succ n_2 \wedge \sim X \cdot Y' \succ n_3\}$

$\equiv (X \succ n_1 \wedge \sim X \succ n_1)\big)$

L1.49a. $n_1 < n_3 \Rightarrow \big(\sum_Y \{X \cdot Y' \succ n_1 \wedge \sim X \cdot Y \succ n_2 \wedge \sim X \cdot Y' \succ n_3\}$

$\equiv (X \succ n_1 \wedge \sim X \succ n_2 + n_3 - 1)\big)$

b. $n_1 \geqslant n_3 \Rightarrow \big(\sum_Y \{X \cdot Y' \succ n_1 \wedge \sim X \cdot Y \succ n_2 \wedge \sim X \cdot Y' \succ n_3\}$

$\equiv (X \succ n_1 \wedge \sim X \succ n_1)\big)$

The proofs of lemmas 1.46b–1.49b are similar to those of lemmas 1.45a–b and 1.46a.

L1.50. $\sum_Y \{X \cdot Y \succ n_1 \wedge X \cdot Y' \succ n_2\} \equiv (X \succ n_1 + n_2)$

Proof. (1.1) $\sum_Y \{X \cdot Y \succ n_1 \wedge X \cdot Y' \succ n_2\}$ {a.}

(1.2) $X \succ n_1 + n_2$ {L1.41a, 1.1}

(2.1) $X \succ n_1 + n_2$ {a.}

(2.2) $a_1 \neq \ldots \neq a_{n_1+n_2}$ {Df.1.II, 2.1}

(2.3) $a_1, \ldots, a_{n_1}, \ldots, a_{n_1+n_2} \,\varepsilon\, X$ {Df.1.II, 2.1}

(2.4) $a_1 \,\varepsilon\, a_1 \wedge \ldots \wedge a_{n_1+n_2} \,\varepsilon\, a_{n_1+n_2}$ {L1.3a, 2.3}

(2.5) $Y_1 = (a_1 + \ldots + a_{n_1})$ {definition}

(2.6) $X \cdot Y_1 = Y_1 = (a_1 + \ldots + a_{n_1})$ {L1.30, 2.5, 2.3}

(2.7) $a_1, \ldots, a_{n_1} \,\varepsilon\, X \cdot Y_1$ {L1.29, 2.6, 2.4}

(2.8) $X \cdot Y_1 \succ n_1 \wedge \sim X \cdot Y_1 \succ n_1 + 1$ {L1.31, 2.7, 2.2, 2.6, Df.1.II}

(2.9) $X \cdot Y_1' \succ n_2$ {L1.33, Df.1.II, 2.2, 2.3, 2.8}

(2.10) $\sum_Y \{X \cdot Y \succ n_1 \wedge X \cdot Y' \succ n_2\}$ {2.8, 2.9}

$\sum_Y \{X \cdot Y \succ n_1 \wedge X \cdot Y' \succ n_2\} \equiv (X \succ n_1 + n_2)$

{1.1→1.2, 2.1→2.10}

L1.51a. $n_1 < n_2 \Rightarrow \big(\sum_Y \{X \cdot Y \succ n_1 \wedge \sim X \cdot Y \succ n_2\} \equiv (X \succ n_1)\big)$

b. $n_1 \geqslant n_2 \Rightarrow \big(\sum_Y \{X \cdot Y \succ n_1 \wedge \sim X \cdot Y \succ n_2\} \equiv (X \succ n_1$

$\wedge \sim X \succ n_1)\big)$

L1.52. $\sum_Y \{X \cdot Y \succ n_1 \wedge \sim X \cdot Y' \succ n_2\} \equiv (X \succ n_1)$

L1.53. $\sum_Y \{X \cdot Y' \succ n_1 \wedge \sim X \cdot Y \succ n_2\} \equiv (X \succ n_1)$

L1.54a. $n_1 < n_2 \Rightarrow \left(\sum_Y \{X\cdot Y' \succ n_1 \wedge \sim X\cdot Y' \succ n_2\} \equiv (X \succ n_1)\right)$

b. $n_1 \geqslant n_2 \Rightarrow \left(\sum_Y \{X\cdot Y' \succ n_1 \wedge \sim X\cdot Y' \succ n_2\} \equiv (X \succ n_1 \wedge \right.$

$\left. \wedge \sim X \succ n_1)\right)$

L1.55. $\sum_Y \{\sim X\cdot Y \succ n_1 \wedge \sim X\cdot Y' \succ n_2\} \equiv (\sim X \succ n_1 + n_2 - 1)$

L1.56. $\sum_Y \{X\cdot Y \succ n\} \equiv (X \succ n)$

L1.57. $\sum_Y \{X\cdot Y' \succ n\} \equiv (X \succ n)$

L1.58. $\sum_Y \{\sim X\cdot Y \succ n\} \equiv (X \succ n \vee \sim X \succ n)$

L1.59. $\sum_Y \{\sim X\cdot Y' \succ n\} \equiv (X \succ n \vee \sim X \succ n)$

The proofs of lemmas 1.51a–1.59 which are omitted do not differ essentially from the proofs of the previous lemmas.

Lemma 1.II. *Every sentential function $\alpha(X_1, \ldots, X_n)$ with the variables $X_1, \ldots, X_n$, which does not contain any quantifiers, may be transformed into an equivalent function $\beta(X_1, \ldots, X_n,)$ that has a normal alternative form of the quantitative formulae composed of the constituents in the variables $X_1, \ldots, X_n$.*

Proof. The atomic expressions that occur in the sentential function $\alpha(X_1, \ldots, X_n)$ are formulae of the following form:

(1) $\omega_1 \varepsilon \omega_2, \omega_1 \subset \omega_2, \omega_1 = \omega_2.$

According to lemmas 1.35a–d and 1.I, each of such formulae may be transformed into an equivalent conjunction or alternation of the quantitative formulae of the form:

(2) $\delta \succ n, \sim\delta \succ n,$

where δ is a sum of a certain number of different constituents in the variables $X_1, \ldots, X_n$ or $\delta = \emptyset$.

Let us symbolize by δ_1 that constituent in which no variable occurs with the negation sign:

(3) $\delta_1 = X_1\cdot X_2\cdot \ldots \cdot X_n.$

It may be deduced from lemma 1.36b that if $\delta = \emptyset$, then the equivalence:

(4) $\delta \succ n \equiv (\delta_1 \succ 1 \wedge \sim\delta_1 \succ 1)$

is a thesis of the system. By virtue of (4), (3) and by lemmas 1.44a–b and 1.34a we may transform each of the formulae (2) into an equivalent expression which has its normal alternative form of the quantitative formulae that are composed of the constituents in the variables $X_1, \ldots, X_n$. Hence and from the theorem stating that every sentential expression of the sentential calculus is equivalent to some expression in the normal alternative form, we immediately obtain lemma 1.II.

Let $Y, X_1, \ldots, X_m$ be all the free variables which occur in the expression α, and let α be an arbitrary conjunction of the quantitative formulae of the form:

$$\delta \cdot Y \succ n, \ \delta \cdot Y' \succ n, \ \delta \cdot Y \succ n \text{ or } \sim \delta \cdot Y' \succ n,$$

each of these conjunctions has the same factor δ which is a certain constituent in the variables $X_1, \ldots, X_m$. Then the following lemma holds:

Lemma 1.III. *There exist natural numbers n_1, n_2, n_3 and n_4 different from zero such that the sentential function α is equivalent to one of the following fifteen elementary conjunctions of quantitative formulae:*

(1) $\delta \cdot Y \succ n_1 \wedge \delta \cdot Y' \succ n_2 \wedge \sim \delta \cdot Y \succ n_3 \wedge \sim \delta \cdot Y' \succ n_4$
(2) $\delta \cdot Y \succ n_1 \wedge \delta \cdot Y' \succ n_2 \wedge \sim \delta \cdot Y \succ n_3$
(3) $\delta \cdot Y \succ n_1 \wedge \delta \cdot Y' \succ n_2 \wedge \sim \delta \cdot Y' \succ n_3$
(4) $\delta \cdot Y \succ n_1 \wedge \sim \delta \cdot Y \succ n_2 \wedge \sim \delta \cdot Y' \succ n_3$
(5) $\delta \cdot Y' \succ n_1 \wedge \sim \delta \cdot Y \succ n_2 \wedge \sim \delta \cdot Y' \succ n_3$
(6) $\delta \cdot Y \succ n_1 \wedge \delta \cdot Y' n_2$
(7) $\delta \cdot Y \succ n_1 \wedge \sim \delta \cdot Y \succ n_2$
(8) $\delta \cdot Y \succ n_1 \wedge \sim \delta = Y' \succ n_2$
(9) $\delta \cdot Y' \succ n_1 \wedge \sim \delta \cdot Y \succ n_2$
(10) $\delta \cdot Y' \succ n_1 \wedge \sim \delta \cdot Y' \succ n_2$
(11) $\sim \delta \cdot Y \succ n_1 \wedge \sim \delta \cdot Y' \succ n_2$
(12) $\delta \cdot Y \succ n_1$
(13) $\delta \cdot Y' \succ n_1$
(14) $\sim \delta \cdot Y \succ n_1$
(15) $\sim \delta \cdot Y' \succ n_1$

Lemma 1. III follows from lemmas 1.39a–d.

It is worth observing that in the case where α is an elementary conjunction of the quantitative formulae that are composed exclusively of the constituents in the variable Y, then α is equivalent to one of the formulae which may be obtained from formulae (1)–(15) by omitting the constituent δ.

Lemma 1.IV. *Every sentential function $\sum_{Y} \{\alpha(Y, X_1, \ldots, X_n)\}$ where $\alpha(Y, X_1, \ldots, X_n)$ is an elementary conjunction of the quantitative formulae that are composed of the constituents in the variables $Y, X_1, \ldots, X_n$ is equivalent to a certain expression $\beta(X_1, \ldots, X_n)$ in its normal alternative form of the quantitative formulae that are composed of the constituents in the variables $X_1, \ldots, X_n$.*

Proof. In view of the associative and commutative laws for the sets (L1. 14b, L1.12b) we may assume that every constituent occurring in the expression α has the form; $\delta\cdot Y$ or $\delta\cdot Y'$; where δ is a certain constituent in the variables $X_1, \ldots, X_n$. Assume that $\delta_1, \ldots, \delta_l$ are all the different constituents in the variables $X_1, \ldots, X_n$ that occur in α, and furthermore, that

$$(1)\qquad \ulcorner\alpha(Y, X_1, \ldots, X_n)\urcorner = \ulcorner\alpha_1(\delta_1, Y)\wedge \ldots \wedge\alpha_l(\delta_l, Y)\urcorner,$$

at the same time each factor $\alpha_i(\delta_i, Y)$, $1\leqslant i\leqslant l$, being a conjunction of the quantitative formulae having the same constituent δ_i.

$$(1.1)\qquad \sum_Y\{\alpha_1, (\delta_1, Y)\}\wedge \ldots \wedge\sum_Y\{\alpha_l(\delta_l, Y) \qquad \{\text{a.}\}$$

$$(1.2)\qquad \alpha_1(\delta_1, a_1)\wedge \ldots \wedge\alpha_l(\delta_l, a_l) \qquad \{1.1\}$$

Assume:

$$(1.3)\qquad a = \sum_{i=1}^{l}(\delta_i\cdot a_i) \qquad \{\text{definition}\}$$

$$(1.4)\qquad \delta_j\cdot a = \delta_i\cdot a_j,\quad 1\leqslant j\leqslant l \qquad \{\text{L1.34c, 1.3}\}$$

$$(1.5)\qquad \delta_j\cdot a' = \delta_i\cdot a_j' \quad ,, \qquad \{1.4\}$$

Replacing in the conjunction (1.2) every term of the form $\delta_j\cdot a_j$ by the term of the form $\delta_j\cdot a$, and every term of the form $\delta_j\cdot a_j'$ by the term $\delta_j\cdot a'$ we obtain

$$(1.6)\qquad \alpha_1(\delta_1, a)\wedge \ldots \wedge\alpha_l(\delta_l, a) \qquad \{1.2, 1.4, 1.5\}$$

$$(1.7)\qquad \sum_Y\{\alpha_1(\delta_1, Y)\wedge \ldots \wedge\alpha_l(\delta_l, Y)\} \qquad \{1.6\}$$

$$(2)\qquad \sum_Y\{\alpha_1(\delta_1, Y)\}\wedge \ldots \wedge\sum_Y\{\alpha_l(\delta_l, Y)\}$$
$$\rightarrow\sum_Y\{\alpha_1, (\delta_1, Y)\wedge \ldots \wedge\alpha_l(\delta_l, Y)\} \qquad \{1.1\rightarrow 1.7\}$$

From (2) and from the fact that the implication inverse to (2) is a substitution of a tautology of the predicate calculus, we have:

$$(3)\qquad \sum_Y\{\alpha_1(\delta_1, Y)\}\wedge \ldots \wedge\sum_Y\{\alpha_l(\delta_l, Y)\}$$
$$\equiv\sum_Y\{\alpha_1(\delta_1, Y)\wedge \ldots \wedge\alpha_l(\delta_l, Y)\}.$$

By virtue of lemmas 1.III and 1.45a–1.59, for every expression $\sum_Y\{\alpha_i(\delta_i, Y)\}$, $1\leqslant i\leqslant l$, there exists an expression in the normal alternative form of the quantitative formulae composed of the constituent δ_i, which is equivalent to the former expression. Hence, from (1), (3) and from L1.II it follows that there exists such a sequence $\beta_1(X_1, \ldots, X_n), \ldots, \beta_m(X_1, \ldots, X_n)$ of the conjunctions of the elementary quantitative formulae which are composed of the constituents in the variables $X_1, \ldots, X_n$, that the equivalence

$$(4)\qquad \sum_Y\{\alpha(Y, X_1, \ldots, X_n)\} \equiv \beta_1(X_1, \ldots, X_n)\wedge \ldots \wedge\beta_m(X_1, \ldots, X_n)$$

is a thesis of ontology.

Equivalence (4) completes the proof of lemma 1.IV.

Lemma 1.V. *Every sentential function* $\alpha(X_1, \ldots, X_n)$ *of the calculus* EO *with the free variables* $X_1, \ldots, X_n$ *is equivalent to a certain expression* $\beta(X_1, \ldots, X_n)$ *in the normal alternative form of the quantitative formulae composed of the constituents in the variables* $X_1, \ldots, X_n$ (cf. Skolem [1919, p. 34]).

Lemma 1.V is a simple consequence of lemmas 1.II–1.IV and definition 1.III.

Theorem 1.Ia whose proof will be presented now, will be stated in the following form:

Theorem 1.Ia. *If* $\alpha(Z, X_1, \ldots, X_n)$ *is a sentential function of* EO *and it contains neither any defined terms nor the variable* Y, *and at most the free variables* $Z, X_1, \ldots, X_n$ *occur in this function, then the sentence*

$$(*) \qquad \prod_{X_1} \cdots \prod_{X_n} \sum_{Y} \prod_{Z} \{Z \varepsilon Y \equiv Z \varepsilon Z \wedge \alpha(Z, X_1, \ldots, X_n)\}$$

is a thesis of the system.

Proof. We assume that the variable Z is distinct from all the variables $X_1, \ldots, X_n$ and we distinguish two cases.

(I) Z does not occur in α. Then the proof of (*) may be constructed as follows:

(1.1) $\alpha(X_1, \ldots, X_n)$ {ad. a.}

(1.2) $Z \varepsilon \varnothing' \equiv Z \varepsilon Z$ {Df.1.3, Df.1.4}

(1.3) $Z \varepsilon \varnothing' \equiv Z \varepsilon Z \wedge \alpha(K_1, \ldots, X_n)$ {1.2, 1.1}

(1.4) $\sum_{Y} \prod_{Z} \{Z \varepsilon Y \equiv Z \varepsilon Z \wedge \alpha(X_1, \ldots, X_n)\}$ {1.3}

(2.1) $\sim \alpha(X_1, \ldots, X_n)$ {ad. a.}

(2.2) $Z \varepsilon \varnothing \equiv Z \varepsilon Z \wedge Z \bar{\varepsilon} Z$ {Df.1.3}

(2.3) $Z \varepsilon \varnothing \equiv Z \varepsilon Z \wedge \alpha(X_1, \ldots, X_n)$ {2.1, 2.2}

(2.4) $\sum_{Y} \prod_{Z} \{Z \varepsilon Y \equiv Z \varepsilon Z \wedge \alpha(X_1, \ldots, X_n)\}$ {2.3}

$$\prod_{X_1} \cdots \prod_{X_n} \sum_{Y} \prod_{Z} \{Z \varepsilon Y \equiv Z \varepsilon Z \wedge \alpha(X_1, \ldots, X_n)\}$$

{1.1→1.4, 2.1→2.4}

It is worth stressing that the deductive steps from (1.1) and (1.2) to (1.3), and from (2.1) and (2.2) to (2.3) were made in view of the following tautologies:

$$p \rightarrow ((q \equiv r) \equiv (q \equiv r \wedge p)),$$
$$\sim p \rightarrow ((q \equiv r \wedge \sim r) \equiv (q \equiv r \wedge p)).$$

(II) Z occurs in α. In this case our proof is inductive. By virtue of lemma 1.V we may assume that the expression $\alpha(Z, X_1, \ldots, X_n)$ has a normal alternative form of the quantitative formulas that are composed of the constituents in the variables $Z, X_1, \ldots, X_n$.

(A) In the first step of the proof by induction it is then sufficient to consider the cases when α is equiform with one of the following quantitative formulas:

(a) $\delta\cdot Z \succ m$, (b) $\delta\cdot Z' \succ m$, (c) $Z \succ m$, (d) $Z' \succ m$,

where δ is a particular constituent in the variables $X_1, \ldots, X_n$, and m is an arbitrary natural number other than zero (cf. Df.1.III, Df.1.II). (c) and (d) cover the cases when Z is the only free variable occurring in the expression α.

(a) (1) $\ulcorner\alpha(Z, Z_1, \ldots, X\ )\urcorner = \ulcorner\delta\cdot Z \succ m\urcorner$ {a.}

(1.1) $m = 1$ {ad. a.}

(1.2) $Z \varepsilon \delta \equiv Z \varepsilon Z \wedge \delta\cdot Z \succ m$ {L1.4a, Df.1.5, Df.1.II 1.1}

(1.3) $\sum_{Y}\prod_{Z}\{Z \varepsilon Y \equiv Z \varepsilon Z \wedge \delta\cdot Z \succ m\}$ {1.2}

(2.1) $m > 1$ {ad. a.}

(2.2) $\sim(Z \varepsilon Z \wedge \delta\cdot Z \succ m)$ {L1.4b, Df.1.II, L1.37a, 2.1}

(2.3) $Z \varepsilon \emptyset \equiv Z \varepsilon Z \wedge Z \bar{\varepsilon} Z$ {Df.1.3}

(2.4) $Z \varepsilon \emptyset \equiv Z \varepsilon Z \wedge \delta\cdot Z \succ m$ {2.3, 2.2}

(2.5) $\sum_{Y}\prod_{Z}\{Z \varepsilon Y \equiv Z \varepsilon Z \wedge \delta\cdot Z \succ m\}$ {2.4}

(2) $\prod_{X_1}\ldots\prod_{X_n}\sum_{Y}\prod_{Z}\{Z \varepsilon Y \equiv Z \varepsilon Z \wedge \alpha(Z, X_1, \ldots, X_n)\}$

{1, 1.1→1.3, 2.1→2.5}

(b) (1) $\ulcorner\alpha(Z, X_1, \ldots, X_n)\urcorner = \ulcorner\delta\cdot Z' \succ m\urcorner$ {a.}

(2) $(\delta \succ m \wedge \sim\delta \succ m+1) \vee (\sim\delta \succ m) \vee (\delta \succ m+1)$

{$(p \wedge \sim q) \vee \sim p \vee q$}

(1.1) $\delta \succ m \wedge \sim\delta \succ m+1$ {ad. a.}

(2.1) $Z \varepsilon \delta'$ {ad. a.}

(2.2) $Z \varepsilon Z \wedge \delta\cdot Z' \succ m$ {L1.40b, 1.1, L1.3a, 2.1}

(3.1) $Z \varepsilon Z \wedge \delta\cdot Z' \succ m$ {ad. a.}

(3.2) $\sim Z \varepsilon \delta'$ {ad. i. a.}

(3.3) $Z \varepsilon \delta$ {L1.7, 3.1, 3.2}

(3.4) $a_1 \neq \ldots \neq a_m$

(3.5) $a_1, \ldots, a_m \varepsilon \delta\cdot Z'$ {Df.1.II, 3.1} (for 3.4 and 3.5)

(3.6) $Z \neq a_1, \wedge \ldots \wedge Z \neq a_m$ {L1.23a, 3.5, L1.27}

(3.7) $Z, a_1, \ldots, a_m \varepsilon \delta$ {3.3, L1.23a, 3.5}

(3.8) $\delta \succ m+1$ {Df.1.II, 3.4, 3.6, 3.7}

contr. {1.1, 3.8}

(1.2) $Z \varepsilon \delta' \equiv Z \varepsilon Z \wedge \delta \cdot Z' \succ m$ {2.1→2.2, and 3.2→ →contr.}

(1.3) $\sum_{Y} \prod_{Z} \{Z \varepsilon Y \equiv Z \varepsilon Z \wedge \delta \cdot Z' \succ m\}$ {1.2}

(4.1) $\sim \delta \succ m$ {ad. a.}

(4.2) $\sim \delta \cdot Z' \succ m$ {L1.37a, 4.1}

(4.3) $Z \varepsilon \emptyset \equiv Z \varepsilon Z \wedge \delta \cdot Z' \succ m$ {Df.1.3, 4.2}

(4.4) $\sum_{Y} \prod_{Z} \{Z \varepsilon Y \equiv Z \varepsilon Z \wedge \delta \cdot Z' \succ m\}$ {4.3}

(5.1) $\delta \succ m+1$ {ad. a.}

(6.1) $Z \varepsilon \emptyset'$ {ad. a.}

(6.2) $Z \varepsilon \delta \rightarrow \delta \cdot Z' \succ m$ {L1.40a, 5.1}

(6.3) $Z \bar{\varepsilon} \delta \rightarrow \delta \cdot Z' \succ m$ {L1.40b, L1.38a, 5.1, L1.7, 6.1}

(6.4) $\delta \cdot Z' \succ m$ {6.2, 6.3}

(5.2) $Z \varepsilon \emptyset' \rightarrow Z \varepsilon Z \wedge \delta \cdot Z' \succ m$ {L1.3a, 6.1→6.4}

(5.3) $Z \varepsilon Z \wedge \delta \cdot Z' \succ m \rightarrow Z \varepsilon \emptyset'$ {Df.1.3, L1.7}

(5.4) $Z \varepsilon \emptyset' \equiv Z \varepsilon Z \wedge \delta \cdot Z' \succ m$ {5.2, 5.3}

(5.5) $\sum_{Y} \prod_{Z} \{Z \varepsilon Y \equiv Z \varepsilon Z \wedge \delta \cdot Z' \succ m\}$ {5.4}

(3) $\prod_{X_1} \cdots \prod_{X_n} \sum_{Y} \prod_{Z} \{Z \varepsilon Y \equiv Z \varepsilon Z \wedge \alpha(Z, X_1, \ldots, X_n)\}$

{1.1→1.3, 4.1→4.4, 5.1→5.5, 2, 1}

The proofs of cases (c) and (d) are omitted. They may be obtained in a mechanical manner from the proofs of (a) and (b) by replacing the constituent δ by the constant $\emptyset'$ and utilizing the following thesis of ontology: $Z \cdot \emptyset' = Z$(L.1.9b).

(B) In the inductive step of the proof of theorem 1.Ia it will suffice in view of definition 1.III and lemma 1.V, to demonstrate that if this theorem is true for two expressions α and β, then it is true for the negation, conjunction and alternation of these expressions, too.

(1) $\prod_{X_1} \cdots \prod_{X_n} \sum_{Y} \prod_{Z} \{Z \varepsilon Y \equiv Z \varepsilon Z \wedge \alpha(Z, X_1, \ldots, X_n)\}$ {i. a.}

(2) $\prod_{Y_1} \cdots \prod_{Y_m} \sum_{Y} \prod_{Z} \{Z \varepsilon Y \equiv Z \varepsilon Z \wedge \beta(Z, Y_1, \ldots, Y_m)\}$ {i. a.}

(3) $\prod_{Z} \{Z \varepsilon a_{X_1, \ldots, X_n} \equiv Z \varepsilon Z \wedge \alpha(Z, X_1, \ldots, X_n)\}$ {1}

(4) $\prod_{Z} \{Z \varepsilon b_{Y_1, \ldots, Y_m} \equiv Z \varepsilon Z \wedge \alpha(Z, Y_1, \ldots, Y_m)\}$ {2}

(5) $\sum_{Y} \prod_{Z} \{Z \varepsilon Y \equiv Z \varepsilon Z \wedge Z \bar{\varepsilon}\, a_{X_1, \ldots, X_n}\}$ {A2}

(6) $\prod_{X_1} \cdots \prod_{X_n} \sum_{Y} \prod_{Z} \{Z \varepsilon Y \equiv Z \varepsilon Z \wedge \sim\alpha(Z, X_1, \ldots, X_n)\}$

{5, 3}

(7) $\sum_{Y}\prod_{Z}\{Z \varepsilon Y \equiv Z \varepsilon a_{X_1, \ldots, X_n} \wedge Z \varepsilon b_{Y_1, \ldots, Y_m}\}$ {A3}

(8) $\sum_{Y}\prod_{Z}\{Z \varepsilon Y \equiv Z \varepsilon a_{X_1, \ldots X_n} \vee Z \varepsilon b_{Y_1, \ldots, Y_m}\}$ {L1.2}

(9) $\prod_{X_1} \ldots \prod_{X_n}\prod_{Y_1} \ldots \prod_{Y_m}\sum_{Y}\prod_{Z}\{Z \varepsilon Y \equiv Z \varepsilon Z \wedge (\alpha(Z, X_1, \ldots, X_n) \wedge \beta(Z, Y_1, \ldots, Y_m))\}$ {7, 3, 4}

(10) $\prod_{X_1} \ldots \prod_{X_n}\prod_{Y_1} \ldots \prod_{Y_m}\sum_{Y}\prod_{Z}\{Z \varepsilon Y \equiv Z \varepsilon Z \wedge (\alpha(Z, X_1, \ldots, X_n) \vee \beta(Z, Y_1, \ldots, Y_m))\}$ {8, 3, 4}

Proof lines (6), (9) and (10) complete the proof of theorem 1.Ia for these sentential functions α of the calculus EO which contain also the other free variables apart from Z. It is easy to see that the proof of this theorem for the expressions in which there occurs exclusively the free variable Z or no free variable at all, will be quite analogous. In view of non-creativity of definitions 1.1–1.6 of the system EO (cf. p. 171), theorem 1.I is thus proved.

This theorem brings a solution to the problem of creativity of definitions with regard to elementary ontology. In the proof presented above I made an essential use of the theorem on reducing the sentential functions of elementary ontology to the normal forms of quantitative formulae. This specific theory of normal forms which I also make use of in Part III of the present paper, I have presented with slight modifications after T. Skolem [1919]. It cannot however, provide a way of solving in an analogous manner, the problem of noncreativity in the whole system of Leśniewski's ontology.

3. According to theorem 1.I the definitions of name-forming functors and nominal constants are non-creative in the system of the elementary ontology based on the set of postulates A1–A3. Some supplement is needed with regard to the problem of their translatability. The proof of translatability of definitions of the whole system of ontology has been given in Słupecki [1955, pp. 61–64]*. This proof however, is incapable of being translated into the language of the elementary ontology, i.e., because of the fact that there occur expressions in this proof which contain functor variables as well as quantifiers binding these variables.

It is a familiar fact that the definitions of the system satisfy the translatability condition provided that for any meaningful expression α of this system which is written down with the help of the defined terms, there exists an expression β which contains no defined terms and such that the equivalence $\alpha \equiv \beta$ is a thesis of the system. Definitions of name-forming functors and of nominal constants satisfy the above condition in the elementary ontology which may be seen in the following outline of the proof.

* *Editorial Note*: Also in the present volume, pp. 112–116.

Assume that the definition of the form

(I) $X \varepsilon f|X_1, \ldots, X_n| \equiv X \varepsilon X \wedge \alpha(X, X_1, \ldots, X_n)$

has been added to ontology, and that the functor "f" defined in the above manner occurs in the following two atomic expressions:

(a) $\delta \varepsilon f|\delta_1, \ldots, \delta_n|$ and (b) $f|\delta_1, \ldots, \delta_n| \varepsilon \delta$,

where $\delta, \delta_1, \ldots, \delta_n$ are the terms that may contain the defined terms in particular cases.

The following three equivalences are obviously theses of the elementary ontology; the second of them is a substitution of axiom A1:

(II) $\delta \varepsilon f|\delta_1, \ldots, \delta_n| \equiv \delta \varepsilon \delta \wedge \alpha(\delta, \delta_1, \ldots, \delta_n)$ {I}

(III) $f|\delta_1, \ldots, \delta_n| \varepsilon \delta \equiv \sum_X \{X \varepsilon f|\delta_1, \ldots, \delta_n|\} \wedge \prod_{X,Y} \{X, Y \varepsilon f|\delta_1, \ldots$

$\ldots, \delta_n| \to X \varepsilon Y\} \wedge \prod_X \{X \varepsilon f|\delta_1, \ldots, \delta_n| \to X \in \delta\}$ {A1}

(IV) $f|\delta_1, \ldots, \delta_n| \varepsilon \delta \equiv (\sum_X \{X \varepsilon X \wedge \alpha(X, \delta_1, \ldots, \delta_n)\}$

$\wedge \prod_{X,Y} \{X \varepsilon X \wedge \alpha(X, \delta_1, \ldots, \delta_n) \wedge Y \varepsilon Y \wedge \alpha(Y, \delta_1, \ldots, \delta_n)$

$\to X \varepsilon Y\} \wedge \prod_X \{X \varepsilon X \wedge \alpha(X, \delta_1, \ldots, \delta_n) \to X \varepsilon \delta\})$ {II, III}

I assume that expressions (II) and (III) are correct substitutions of the scheme (I) and axiom A1, and that the variables X, Y do not occur in the expressions (a) and (b). It may easily be seen that with the help of the above three equivalences we may eliminate any functor 'f' introduced to the system by means of definition (I), from the sentential function (a) and (b). Nominal constants may be treated as 0-ary functors.

To conclude this chapter I want to discuss the relationship between the manner of defining name-forming functors that was given by Leśniewski and a particular standard way of defining this kind of functors.

Definitions of n-ary name-forming functors of nominal arguments treated as the theses of the system are usually formulated in the form of equivalences:[9]

(IV) $Y = f|X_1, \ldots, X_n| \equiv \alpha(Y, X_1, \ldots, X_n)$.

Here the term 'f' is a defined functor and α is a sentential function in which the variables $Y, X_1, \ldots, X_n$ occur as the only free variables. It is assumed at the same time, that equivalence (IV) satisfies the following conditions:

[9] Where the expression α contains no free variables $X_1, \ldots, X_n$, equivalence (IV) is a definition of the nominal constant. The symbol '$=$' stands for the identity sign here.

(a) On the right side of equivalence (IV), i.e. in the definiens, there appear neither the defined term 'f', nor any term which has been defined with the help of the term defined in this definition.

(b) On the left side of equivalence (IV), i.e. in the definiendum, every variable occurs only once.

(c) Every free variable that occurs on either side of the equivalence (IV) should occur as a free variable on the other side of it.

Definitions of the type (IV) may be added to the system provided that they also satisfy the so called existence and uniqueness condition, i.e. provided that the sentences:

(d) $$\prod_{X_1} \cdots \prod_{X_n} \sum_{Y} \{\alpha(Y, X_1, \ldots, X_n)\},$$

(e) $$\prod_{X_1} \cdots \prod_{X_n} \prod_{Y_1} \prod_{Y_2} \{\alpha(Y_1, X_1, \ldots, X_n) \wedge \alpha(Y_2, X_1, \ldots, X_n) \rightarrow Y_1 = Y_2\}.$$

are theses of the system.

As has already been noted above the definition of an analogous functor has the form of the following equivalence in ontology:

(I) $$X \varepsilon f|X_1, \ldots, X_n| \equiv X \varepsilon X \wedge \alpha(X, X_1, \ldots, X_n).$$

It is assumed at the same time that equivalence (I) also satisfies the conditions (a), (b) and (c), and that there are no defined terms occurring in the expression α. If the free variables $X_1, \ldots, X_n$ do not occur in α, then equivalence (I) is treated, similarly to equivalence (IV), as a definition of a nominal constant.

The following theorem which states that both ways of defining are equivalent is valid.

Theorem 1.II. A) *For any sentential function* $\alpha(X, X_1, \ldots, X_n)$ *of the calculus* EO *a sentential function* $\beta(Y, X_1, \ldots, X_n)$ *may be chosen such that the following conditions are satisfied:*

(i) *the sentence*[10]

(f) $$\prod_{X_1} \cdots \prod_{X_n} \sum_{Y} {}_1\{\beta(Y, X_1, \ldots, X_n)\}$$

is a thesis of the system,

(ii) *the expressions*

(I) $$\prod_{X} \{X \varepsilon f|X_1, \ldots, X_n| \equiv X \varepsilon X \wedge \alpha(X, X_1, \ldots, X_n)\},$$

(II) $$\prod_{Y} \{Y = f|X_1, \ldots, X_n| \equiv \beta(Y, X_1, \ldots, X_n)\}$$

are equivalent in the system EO.

B) *If* $\alpha(Y, X_1, \ldots, X_n)$ *is a sentential function of the calculus* EO, *and the following sentence is a thesis of the system*

(f) $$\prod_{X_1} \cdots \prod_{X_n} \sum_{Y} {}_1\{\alpha(Y, X_1, \ldots, X_n)\},$$

[10] The sentence (f) is the conjunction of the sentences (d) and (e).

then a sentential function $\beta(X, X_1, \ldots, X_n)$ *may be chosen such that the expressions*

$$\text{(I)} \quad \prod_X \{X \varepsilon f|X_1, \ldots, X_n| \equiv X \varepsilon X \wedge \beta(X, X_1, \ldots, X_n)\},$$

$$\text{(IV)} \quad \prod_Y \{Y = f|X_1, \ldots, X_n| \equiv \alpha(Y, X_1, \ldots, X_n)\}$$

are equivalent in the system EO.

I assume that the variables $X, Y, X_1, \ldots, X_n$ that occur in the theorem formulated above are not equiform and that for every variable $X_1, \ldots, X_n$ of the sentential function α which occurs in part A(B) of theorem 1.II, there exists an equiform variable of the function β that occurs, respectively, in part A(B), and vice versa. I also assume that the equivalences (I) and (IV) satisfy the conditions (a), (b) and (c) and that no defined terms occur in the expressions α or β.

Proof. (A) Let $\alpha(X, X_1, \ldots, X_n)$ be a sentential function of the system EO and let it contain no defined terms.

Assume that

$$\text{(1)} \quad \ulcorner\beta(X, X_1, \ldots, X_n)\urcorner = \ulcorner\prod_Y \{X \varepsilon Y \equiv X \varepsilon X \wedge \alpha(X, X_1, \ldots, X_n)\}\urcorner.$$

Hence from theorem 1.Ia and from definition 1.2 it follows that sentence (2) is a thesis of the system EO:

$$\text{(2)} \quad \prod_{X_1} \cdots \prod_{X_n} \sum_Y {}_1\{\beta(Y, X_1, \ldots, X_n)\}.$$

$$\text{(1.1)} \quad \prod_X \{X \varepsilon f|X_1, \ldots, X_n| \equiv X \varepsilon X \wedge \alpha(X, X_1, \ldots, X_n)\} \qquad \{a.\}$$

$$\text{(1.2)} \quad \prod_Y \{Y = f|X_1, \ldots, X_n| \rightarrow \prod_X \{X \varepsilon Y \equiv X \varepsilon X \wedge \alpha(X, X_1, \ldots, X_n)\}\} \qquad \{1.1\}$$

$$\text{(1.3)} \quad \prod_Y \{Y = f|X_1, \ldots, X_n| \rightarrow \beta(Y, X_1, \ldots, X_n)\} \qquad \{1, 1.2\}$$

$$\text{(1.4)} \quad \prod_Y \{\beta(Y, X_1, \ldots, X_n) \rightarrow Y = f|X_1, \ldots, X_n|\} \qquad \{1, 1.1, \text{Df.1.2}\}$$

$$\text{(1.5)} \quad \prod_Y \{Y = f|X_1, \ldots, X_n| \equiv \beta(Y, X_1, \ldots, X_n)\} \qquad \{1.3, 1.4\}$$

$$\text{(2.1)} \quad \prod_Y \{Y = f|X_1, \ldots, X_n| \equiv \beta(Y, X_1, \ldots, X_n)\} \qquad \{a.\}$$

$$\text{(2.2)} \quad \prod_Y \{Y = f|X_1, \ldots, X_n| \equiv \prod_X \{X \varepsilon Y \equiv X \varepsilon X \wedge \alpha(X, X_1, \ldots, X_n)\}\} \qquad \{1, 2.1\}$$

(2.3) $\prod_X \{X \varepsilon f|X_1, \ldots, X_n| \equiv X \varepsilon X \wedge \alpha(X, X_1, \ldots, X_n)\}$ {2.2}

(3) $\prod_X \{X \varepsilon f|X_1, \ldots, X_n| \equiv X \varepsilon X \wedge \alpha(X, X_1, \ldots, X_n)\}$
$\equiv \prod_Y \{Y = f|X_1, \ldots, X_n| \equiv \beta(Y, X_1, \ldots, X_n)\}$
{1.1→1.5, 2.1→2.3}

(B) Let $\alpha(Y, X_1, \ldots, X_n)$ be an arbitrary function containing no defined terms and such that sentence

(1) $\prod_{X_1} \cdots \prod_{X_n} \sum_Y {}_1 \{\alpha(Y, X_1, \ldots, X_n)\}$

is a thesis of the system EO.

Assume that

(2) $\ulcorner\beta(X, X_1, \ldots, X_n)\urcorner = \ulcorner\prod_Y \{\alpha(Y, X_1, \ldots, X_n) \to X \varepsilon Y\}\urcorner$.

(1.1) $\prod_X \{X \varepsilon f|X_1, \ldots, X_n| \equiv X \varepsilon X \wedge \beta(X, X_1, \ldots, X_n)\}$ {a.}

(1.2) $\prod_X \{X \varepsilon f|X_1, \ldots, X_n| \equiv X \varepsilon X \wedge \prod_Y \{\alpha(Y, X_1, \ldots, X_n)$
$\to X \varepsilon Y\}\}$ {2, 1.1}

(1.3) $\alpha(a_{X_1, \ldots, X_n}, X_1, \ldots, X_n)$ {1}

(2.1) $X \varepsilon f|X_1, \ldots, X_n|$ {ad. a.}

(2.2) $\prod_Y \{\alpha(Y, X_1, \ldots, X_n) \to X \varepsilon Y\}$ {1.2, 2.1}

(2.3) $X \varepsilon a_{X_1, \ldots, X_n}$ {2.2, 1.3}

(3.1) $X \varepsilon a_{X_1, \ldots, X_n}$ {ad. a.}

(3.2) $X \bar{\varepsilon} f|X_1, \ldots, X_n|$ {ad. i. a.}

(3.3) $\sum_Y \{\alpha(Y, X_1, \ldots, X_n) \wedge X \bar{\varepsilon} Y\}$ {1.2, 3.2; L1.3a, 3.1}

(3.4) $\alpha(b_{X_1, \ldots, X_n}, X_1, \ldots, X_n)$
(3.5) $X \bar{\varepsilon} b_{X_1, \ldots, X_n}$ } {3.3}

(3.6) $a_{X_1, \ldots, X_n} = b_{X_1, \ldots, X_n}$ {1. 1.3, 3.4}

(3.7) $X \varepsilon b_{X_1, \ldots, X_n}$ {3.6, 3.1, Df.1.2}

(1.4) $f|X_1, \ldots, X_n| = a_{X_1, \ldots, X_n}$ {Df.1.2, 2.1→2.3, 3.1 and 3.2→contr.}

(1.5) $\prod_Y \{Y = f|X_1, \ldots, X_n| \equiv \alpha(Y, X_1, \ldots, X_n)\}$ {1, 1.3, 1.4}

(4.1) $\prod_Y \{Y = f|X_1, \ldots, X_n| \equiv \alpha(Y, X_1, \ldots, X_n)\}$ {a.}

(5.1) $X \varepsilon f|X_1, \ldots, X_n|$
(5.2) $\alpha(Y, X_1, \ldots, X_n)$ } {ad. a.}

(5.3) $Y = f|X_1, \ldots, X_n|$ {4.1, 5.2}

(5.4) $X \varepsilon Y$ {5.3, 5.1}

(4.2) $X \varepsilon f|X_1, \ldots, X_n| \to X \varepsilon X \wedge \prod_Y \{\alpha(Y, X_1, \ldots, X_n) \to X \varepsilon Y\}$

{5.1 and 5.2→5.4}

(6.1) $X \varepsilon X \wedge \prod_Y \{\alpha(Y, X_1, \ldots, X_n) \to X \varepsilon Y\}$ {ad. a.}

(6.2) $\alpha(f|X_1, \ldots, X_n|, X_1, \ldots, X_n)$ {4.1}

(6.3) $X \varepsilon f|X_1, \ldots, X_n|$ {6.1, 6.2}

(4.3) $\prod_X \{X \varepsilon f|X_1, \ldots, X_n| \equiv X \varepsilon X \wedge \beta(X, X_1, \ldots, X_n)\}$

{4.2, 6.1→6.3}

(3) $\prod_X \{X \varepsilon f|X_1, \ldots, X_n| \equiv X \varepsilon X \wedge \beta(X, X_1, \ldots, X_n)\}$

$\equiv \prod_Y \{Y = f|X_1, \ldots, X_n| \equiv \alpha(Y, X_1, \ldots, X_n)\}$

{1.1→1.5, 4.1→4.3}

The last line completes the proof of theorem 1.II.

This theorem needs some comment. The definitions of name-forming functors of the form (IV) are usually utilized in the calculi constructed over the narrower predicate calculus with identity; at the same time the sign '=' that occurs in the definition (IV) is treated as a symbol of logical identity. Definitions of the form (IV) considered in theorem 1.II differ from the definitions of that kind which are accepted in the other systems only in the fact that instead of logical identity we take extensional equality. That equality is an equivalence relation and it satisfies the extensionality condition in ontology (cf. p. 174). From the formal point of view there is, it seems, no essential difference between the system of ontology presented in this paper and the system which we would obtain by constructing elementary ontology over the narrower predicate calculus with identity adding at the same time, the axiom stating that extensionally equal sets are identical.[11]

Another remark should be made with regard to the restriction which I have made after Leśniewski that no defined terms may occur in the definiens of a definition of ontology. This limitation facilitates to a certain degree the methodological investigations into elementary ontology, i.e., it simplifies the proofs of certain theorems. It is visible from the proofs presented that theorems 1.I and 1.II as well as the remark on translatability of the definitions of elementary ontology will be true also in the case when the previously defined symbol is allowed to occur in the definiens of a definition under consideration.

---

[11] More considerable differences might appear if elementary ontology were enriched with new primitive terms. Extensional equality might then fail to satisfy the condition of extensionality.

II. Elementary Ontology and Boolean Algebra

It is sometimes suggested, seemingly in conformity with Leśniewski's intentions, that ontology is most closely related to the Aristotelian tradition, being supposedly a continuation and extension of that tradition (Kotarbiński [1929, p. 253]*. From the formal point of view however, ontology has more features in common with modern logical systems than with Aristotelian syllogistic. This is also true of elementary ontology which, as I shall attempt to demonstrate, is equivalent to a system of atomic Boolean algebra as well as to the algebra of sets as it has been presented by Tarski in his paper [1933]** with the addition of the sentence which he symbolized by (α). I shall now discuss in more detail the first of these equivalences.

1. The atomic Boolean algebra which I now present is an elementary system constructed over the narrower predicate calculus without identity in the same way as elementary ontology. This system I shall symbolize by EB. It is defined by means of the following set of axioms and definitions:

AB1a. $X = X$ b. $X = Y \to Y = X$ c. $X = Y \wedge Y = Z \to X = Y$

AB2a. $X = Y \to X + Z = Y + Z$ b. $X = Y \to X \cdot Z = Y \cdot Z$

AB3a. $X + \emptyset = X$ b. $X \cdot 1 = X$

AB4a. $X + X' = 1$ b. $X \cdot X' = \emptyset$

AB5a. $X + Y = Y + X$ b. $X \cdot Y = Y \cdot X$

AB6a. $X + Y \cdot Z = (X + Y) \cdot \cdot (X + Y)$ b. $X \cdot (Y + Z) = X \cdot Y + X \cdot Z$

Df.B1. $X \subset Y \equiv X + Y = Y$

Df.B2. $\mathrm{At}(X) \equiv X \neq \emptyset \wedge \prod_{Y} \{Y \neq \emptyset \wedge Y \subset X \to X = Y\}$

Df.B3. $X \varepsilon Y \equiv \mathrm{At}(X) \wedge X \subset Y$

AB7. $X \neq \emptyset \to \sum_{Y} \{\mathrm{At}(Y) \wedge Y \subset X\}$

Expressions AB1a–AB6b form one of the possible sets of postulates for Boolean algebra, in which the symbol '$=$' is treated as a specific constant of the system and not as a symbol of logical identity (cf. Mostowski [1948, p. 103]).

Df.B1 defines the relation of inclusion.

---

* *Editorial Note*: English translation – Kotarbiński [1966, p. 210].

** *Editorial Note*: English translation in Tarski [1956].

Df.B2 is a definition of atom.[12] According to this definition element X of Boolean algebra is an atom provided that $X \neq 0$ and that every element $Y \neq 0$ included in X is identical with it.

Df.B3 defines the relation ε which corresponds to relation ε of Leśniewski's ontology.

AB7 assumes that every element $X \neq 0$ includes an atom. The contents of the quoted axiom may be expressed in terms of algebra of sets in the following way: every non-empty set includes a one-element set.

Theorem 2.I. *The system of atomic Boolean algebra* EB *and the system of elementary ontology* EO *which is enriched with the definitions*

Df.1.7. $X \varepsilon 1 \equiv \sum_{Y}\{X \varepsilon Y\}$,

Df.1.8. $\mathrm{At}(X) \equiv X \varepsilon X$

are equivalent.

To prove the above theorem it is enough to demonstrate that:

(A) each of the expressions AB1a–AB6b, AB7 and Df.B1–Df.B3 of the algebra EB is a thesis of the ontology EO, and

(B) each of the expressions A1–A3, Df.1.1–Df.1.8 of the ontology EO is a thesis of the algebra EB.

Proof. (A) As has been already said in Part I of the present paper, relations $\subset$, $=$, the set $\emptyset$, and operations $'$, $+$ and $\cdot$ that were defined in definitions 1.1–1.6 satisfy the postulates of Boolean algebra in ontology. This also pertains to set 1 which satisfies in ontology the following equality: $1 = \emptyset'$. Axioms AB1a–AB6b and definition B1 of algebra EB are thus the theses of ontology. Expressions Df.B2, Df.B3 and AB7 are the simple consequences of definition 1.8 and the following theses of elementary ontology:

(a_1) $X \varepsilon Y \equiv X \varepsilon X \wedge X \subset Y$,

(a_2) $X \neq \emptyset \equiv \sum_{Y}\{Y \varepsilon X\}$,

(a_3) $X \varepsilon X \equiv X \neq \emptyset \wedge \prod_{Y}\{Y \neq \emptyset \wedge Y \subset X \rightarrow X \subset Y\}$.

We hence obtain the conclusion that algebra EB is included in ontology EO.

(B) The theorem stating that system EO is included in algebra EB needs a more detailed proof.

---

[12] The concept of atom was first introduced to Boolean algebra by Schröder [1891, pp. 318–349]. It was then utilized by Tarski in formulating the axiom set for atomic Boolean algebra (Tarski [1935]; also Tarski [1956]). Tarski's system which he obtained from the so called extended (completely additive) Boolean algebra by adding an axiom that stated that every non-empty element contains an atom (AB7), is a nonelementary theory and it is essentially richer than the algebra EB presented above.

The following theses of Boolean algebra which will be utilized in subsequent proofs are given without proof.

L2.1a. $X \subset X$ b. $X \subset Y \wedge Y \subset Z \rightarrow X \subset Z$

c. $X = Y \equiv X \subset Y \wedge Y \subset X$ d. $X = Y \wedge X \subset Z \rightarrow Y \subset Z$

L2.2a. $\emptyset \subset X$ b. $X \subset 1$

c. $X \subset \emptyset \equiv X = \emptyset$

L2.3a. $X \cdot Y \subset X$ b. $X \subset X + Y$

c. $X \subset Y \cdot Z \equiv X \subset Y \wedge X \subset Z$

L2.4a. $X = Y \rightarrow Z + X = Z + Y$ b. $X = Y \rightarrow Z \cdot X = Z \cdot Y$

L2.5a. $(X + Y)' = X' \cdot Y'$ b. $(X \cdot Y)' = X' + Y'$

L2.6a. $X \subset Y \wedge X \subset Y' \rightarrow X = \emptyset$ b. $X \subset Y \wedge Y = \emptyset \rightarrow X = \emptyset$

L2.7a. $X \subset Y \equiv X \cdot Y' = \emptyset$ b. $X \subset Y' \equiv X \cdot Y = \emptyset$

L2.8a. $X + (Y + Z) = (X + Y) + Z$ b. $X \cdot (Y \cdot Z) = (X \cdot Y) \cdot Z$

I moreover, make use of the following lemmas:

L2.9a. $\prod_{At(Z)} \{Z \subset X \rightarrow Z \subset Y\} \rightarrow X \subset Y$[13]

Proof.
(1) $\prod_{At(Z)} \{Z \subset X \rightarrow Z \subset Y\}$ {a.}
(2) $X \not\subset Y$ {a. i. p.}
(3) $X \cdot Y' \neq \emptyset$ {L2.7a, 2}
(4) $\mathrm{At}(a_1) \wedge a_1 \subset X \wedge a_1 \subset Y'$ {AB7, 3, L2.3c}
(5) $a_1 \neq \emptyset$ {Df.B2, 4}
(6) $a_1 \subset Y$ {1, 4}
(7) $a_1 = \emptyset$ {L2.6a, 6, 4}
contr. {5, 7}

L2.9b. $X \subset Y \equiv \prod_{At(Z)} \{Z \subset X \rightarrow Z \subset Y\}$ {L2.9a, L2.1b}

L2.10. $\mathrm{At}(X) \rightarrow (X \subset Y \equiv X \not\subset Y')$

Proof.
(1.1) $\mathrm{At}(X) \wedge X \subset Y$ {ad. a.}
(1.2) $X \neq \emptyset$ {Df.B2, 1.1}
(1.3) $X \not\subset Y'$ {L2.6a, 1.2, 1.1}
(2.1) $\mathrm{At}(X) \wedge X \not\subset Y'$ {ad. a.}
(2.2) $X \cdot Y \neq \emptyset$ {L2.7b, 2.1}
(2.3) $\mathrm{At}(a_1) \wedge a_1 \subset X \wedge a_1 \subset Y$ {AB7, 2.2, L2.3c}
(2.4) $a_1 = X$ {Df.B2, 2.3, 2.1}
(2.5) $X \subset Y$ {L2.1d, 2.4, 2.3}
$\mathrm{At}(X) \rightarrow (X \subset Y \equiv X \not\subset Y')$ {1.1→1.3, 2.1→2.5}

---

[13] Lemma 2.9a and axiom AB7 are equivalent on the grounds of axioms AB1a–AB6b and definition B2. Axiom AB7 follows from lemma 2.9a, definition B2 and the following thesis of Boolean algebra: $X \subset \emptyset \equiv X = \emptyset$. It becomes apparent when we substitute the constant $\emptyset$ for the variable Y in the expression L2.9a.

The remark contained in Tarski [1935, p. 194], from which it results that the expression AB7 is weaker than the expression L2.9a on the grounds of the Boolean algebra, is then false.

L2.11. $\mathrm{At}(X) \wedge X \subset Y+Z \to (X \subset Y \vee X \subset Z)$

Proof.	(1)	$\mathrm{At}(X) \wedge X \subset Y+Z$	{a.}
	(2)	$X \not\subset Y$	
	(3)	$X \cdot Y' \neq \emptyset$	{L2.7a, 2}
	(4)	$X \cdot Y' \subset X$	{L2.3a}
	(5)	$\prod_{Z} \{Z \neq \emptyset \wedge Z \subset X \to X = Z\}$	{Df.B2, 1}
	(6)	$X = X \cdot Y'$	{5, 3, 4}
	(7)	$X \cdot (Y+Z)' = \emptyset$	{L2.7a, 1}
	(8)	$X \cdot (Y+Z)' = X \cdot Z'$	{6, L2.5a, L2.4b, L2.8b}
	(9)	$X \cdot Z' = \emptyset$	{AB1c, 8, 7}
		$X \subset Z$	{L2.7a, 9}

L2.12. $X \varepsilon Y \to X \varepsilon X$ {Df.B3, L2.1a}

L2.13. $X \varepsilon Y \to \sum_{Z} \{Z \varepsilon Y\}$ {L2.12}

L2.14. $X \varepsilon Y \to \prod_{Z,U} \{Z \varepsilon X \wedge U \varepsilon X \to Z \varepsilon U\}$

Proof.	(1)	$X \varepsilon Y$	{a.}
	(2)	$Z \varepsilon X \wedge U \varepsilon X$	
	(3)	$\prod_{Z} \{Z \neq \emptyset \wedge Z \subset X \to Z = X\}$	{Df.B3, Df.B2, 1, 2}
	(4)	$Z = X \wedge U = X$	
	(3)	$\mathrm{At}(Z) \wedge Z \subset U$	{Df.B3, AB1c, AB1b, L2.1c, 2, 4}
		$Z \varepsilon U$	{Df.B3, 5}

L2.15. $X \varepsilon Y \to \prod_{Z} \{Z \varepsilon X \to Z \varepsilon Y\}$ {Df.B3, L2.1b}

L2.16. $\prod_{Y,Z} \{Y \varepsilon X \wedge Z \varepsilon X \to Y \varepsilon Z\} \equiv \prod_{Y} \{Y \neq \emptyset \wedge Y \subset X \to X = Y\}$

Proof. (a)	(1)	$\prod_{Y,Z} \{Y \varepsilon X \wedge Z \varepsilon X \to Y \varepsilon Z\}$	{a.}
	(2)	$Y \neq \emptyset \wedge Y \subset X$	
	(3)	$X \neq Y$	{a. i. p.}
	(4)	$X \cdot Y' \neq \emptyset$	{L2.1c, 2, 3, L2.7a}
	(5)	$\mathrm{At}(a_1) \wedge a_1 \subset X \wedge a_1 \subset Y'$	{AB7, 4, L2.3c}
	(6)	$\mathrm{At}(a_2) \wedge a_2 \subset Y$	{AB7, 2}
	(7)	$a_1 \varepsilon X \wedge a_2 \varepsilon X$	{Df.B3, L2.1b, 5, 6, 2}
	(8)	$a_1 \varepsilon a_2$	{1, 7}
	(9)	$a_1 \subset a_2$	{Df.B3, 8}
	(10)	$a_1 \subset Y$	{L2.1b, 9, 6}
	(11)	$a_1 = \emptyset$	{L2.6a, 10, 5}
	(12)	$a_1 \neq \emptyset$	{Df.B2, 5}
		contr. {11, 12}	

(b) (1) $\prod_{Y}\{Y \neq \emptyset \wedge Y \subset X \rightarrow X = Y\}$ {a.}

(2) $Y \varepsilon X \wedge Z \varepsilon X$

(3) $Y \neq \emptyset \wedge Z \neq \emptyset \wedge Y \subset X \wedge Z \subset X$ {Df.B3, Df.B2, 2}

(4) $Y \subset Z$ {1, 3, AB1b, AB1c, L2.1c}

(5) $Y \varepsilon Z$ {Df.B3, 2, 4}

Each of the eleven lemmas that follow is equiform with one of the expressions A1–A3 and Df.1.1–Df.1.8. of the system EO.

L2.17. $X \subset Y \equiv \prod_{Z}\{Z \varepsilon X \rightarrow Z \varepsilon Y\}$ {L2.9b, Df.B3}

L2.18. $X = Y \equiv \prod_{Z}\{Z \varepsilon X \equiv Z \varepsilon Y\}$ {L2.17, L2.1c}

L2.19. $X \varepsilon Y \equiv \sum_{Z}\{Z \varepsilon X\} \wedge \prod_{Z,U}\{Z \varepsilon X \wedge U \varepsilon X \rightarrow Z \varepsilon U\}$
$\wedge \prod_{Z}\{Z \varepsilon X \rightarrow Z \varepsilon Y\}$

Proof. (1) $X \varepsilon Y \rightarrow \sum_{Z}\{Z \varepsilon X\} \wedge \prod_{Z,U}\{Z \varepsilon X \wedge U \varepsilon X \rightarrow Z \varepsilon U\}$
$\wedge \prod_{Z}\{Z \varepsilon X \rightarrow Z \varepsilon Y\}$ {L2.13–L2.15}

(1.1) $\sum_{Z}\{Z \varepsilon X\}$

(1.2) $\prod_{Z,U}\{Z \varepsilon X \wedge U \varepsilon X \rightarrow Z \varepsilon U\}$ {ad. a.}

(1.3) $\prod_{Z}\{Z \varepsilon X \rightarrow Z \varepsilon Y\}$

(1.4) $X \neq \emptyset$ {Df.B3, Df.B2, 1.1}

(1.5) $\prod_{Z}\{Z \neq \emptyset \wedge Z \subset X \rightarrow X = Z\}$ {L2.16, 1.2}

(1.6) $X \subset Y$ {L2.17, 1.2}

(1.7) $X \varepsilon Y$ {Df.B3, Df.B2, 1.4, 1.5, 1.6}

$X \varepsilon Y \equiv \sum_{Z}\{Z \varepsilon X\} \prod_{Z,U}\{Z \varepsilon X \wedge U \varepsilon X \rightarrow Z \varepsilon U\}$
$\wedge \prod_{Z}\{Z \varepsilon X \rightarrow Z \varepsilon Y\}$ {1, 1.1→1.7}

L2.20. $X \varepsilon \emptyset \equiv X \varepsilon X \wedge X \bar{\varepsilon} X$ {Df.B2, Df.B3, L2.2c}

L2.21. $X \varepsilon 1 \equiv \sum_{Y}\{X \varepsilon Y\}$ {Df.B3, L2.1a, L2.2b}

L2.22. $X \varepsilon Y' \equiv X \varepsilon X \wedge X \bar{\varepsilon} Y$

Proof. (1.1) $X \varepsilon Y'$ {ad. a.}

(1.2) $\mathrm{At}(X) \wedge X \not\subset Y$ {Df.B3, 1.1, L2.10}

(1.3) $X \varepsilon X \wedge X \bar{\varepsilon} Y$ {Df.B3, L2.1a, 1.2}

(2.1) $X \varepsilon X \wedge X \bar{\varepsilon} Y$ {ad. a.}

(2.2) $\mathrm{At}(X) \wedge X \not\subset Y$ {Df.B3, 2.1}

(2.3) $X \subset Y'$ {L2.10, 2.2}

	(2.4)	$X \varepsilon Y'$	{Df.B3, 2.2, 2.3}
		$X \varepsilon Y' \equiv X \varepsilon X \wedge X \bar{\varepsilon} Y$	{1.1→1.3, 2.1→2.4}
L2.23.		$\prod_{X} \sum_{Y} \prod_{Z} \{Z \varepsilon Y \equiv Z \varepsilon Z \wedge Z \bar{\varepsilon} X\}$	{L2.22}
L2.24.		$X \varepsilon Y \cdot Z \equiv X \varepsilon Y \wedge X \varepsilon Z$	
Proof.	(1.1)	$X \varepsilon Y \cdot Z$	{ad. a.}
	(1.2)	$\mathrm{At}(X) \wedge X \subset Y \wedge X \subset Z$	{Df.B3, L2.3c, 1.1}
	(1.3)	$X \varepsilon Y \wedge X \varepsilon Z$	{Df.B3, 1.2}
	(2.1)	$X \varepsilon Y \wedge X \varepsilon Z$	{ad. a.}
	(2.2)	$\mathrm{At}(X) \wedge X \subset Y \wedge X \subset Z$	{Df.B3, 2.1}
	(2.3)	$X \varepsilon Y \cdot Z$	{Df.B3, L2.3c, 2.2}
		$X \varepsilon Y \cdot Z \equiv X \varepsilon Y \wedge X \varepsilon Z$	{1.1→1.3, 2.1→2.3}
L2.25.		$\prod_{X,Y} \sum_{Z} \prod_{U} \{U \varepsilon Z \equiv U \varepsilon X \wedge U \varepsilon Y\}$	{L2.24}
L2.26.		$X \varepsilon Y + Z \equiv X \varepsilon Y \vee X \varepsilon Z$	
Proof.	(1.1)	$X \varepsilon Y + Z$	{ad. a.}
	(1.2)	$\mathrm{At}(X) \wedge X \subset Y + Z$	{Df.B3, 1.1}
	(1.3)	$X \subset Y \vee X \subset Z$	{L2.11, 1.2}
	(1.4)	$X \varepsilon Y \vee X \varepsilon Z$	{Df.B3, 1.2, 1.3}
	(2.1)	$X \varepsilon Y \vee X \varepsilon Z$	{ad. a.}
	(2.2)	$\mathrm{At}(X) \wedge X \subset Y + Z$	{Df.B3, L2.3b, L2.1b, 2.1}
	(2.3)	$X \varepsilon Y + Z$	{Df.B3, 2.2}
		$X \varepsilon Y + Z \equiv X \varepsilon Y \vee X \varepsilon Z$	{1.1→1.4, 2.1→2.3}
L2.27.		$\mathrm{At}(X) \equiv X \varepsilon X$	{Df.B3, L2.1a}

Lemma 2.27 completes the proof of the fact that elementary ontology is included in algebra EB which concludes the proof of theorem 2.I.

The proof of equivalence of these systems needs some supplement. Both systems are constructed over the narrower predicate calculus without identity and thus they do not differ with respect to the rules of inference valid in the system. According to theorem 1.II whose proof I have presented in the previous chapter, every n-ary functor (or nominal constant) introduced to the system EO by means of the definition:

(I) $$X \varepsilon f|X_1, \ldots, X_n| \equiv X \varepsilon X \wedge \alpha(X, X_1, \ldots, X_n)$$

may be defined in an equivalent manner with the help of the definition:

(IV) $$Y = f|X_1, \ldots, X_n| \equiv \beta(Y, X_1, \ldots, X_n),$$

where $\beta(Y, X_1, \ldots, X_n)$ is a sentential function that satisfies the existence and uniqueness condition with regard to the variable Y. I assume that in atomic Boolean algebra there is the valid rule of defining name-forming functors (and nominal constants) of the form (IV).[14] One may thus claim

[14] All the definitions which have the form of equality in Boolean algebra (e.g. $X - Y = X \cdot Y'$), may be transformed into equivalences (e.g. $Z = X - Y \equiv Z = X \cdot Y'$).

that these calculi do not differ essentially, with respect to their rules of definition. However as far as I know, unlike ontology for which Leśniewski has precisely formulated the rules of definition, the systems of Boolean algebras have not yet been provided with precise rules of definition.

In connection with the last theorem (Th.2.I) a controversial philosophical remark might be made. Attention is often drawn to the analogies between ontology and certain systems of Boolean algebras.[15] These analogies are primarily concerned with elementary ontology, which is merely a fragment of the whole system of ontology. It seems that even in that case it is difficult to identify the system of elementary ontology with the system of atomic Boolean algebra. The following points may be made in support of this claim.

The majority of deductive systems may be classed, according to the aims that their founders had in mind, into two groups. The first group comprises the system each of which was constructed with the intention of describing one of the possible mains which are models of a given system. It seems that this group should include Leśniewski's systems since he treated them as true theories describing what actually exists. This is true particularly of ontology which may be called, following Kotarbiński [1929, p.254]* 'the general theory of objects'. The second group comprises the systems constructed with the aim of distinguishing the features common to many domains. The value of these systems consists principally in their numerous and differing interpretations. This group comprises i.e. various system of Boolean algebras (Boolean algebra, completely additive Boolean algebra, atomic Boolean algebra etc.).

The intuitions which Leśniewski followed while constructing the system of ontology and which he took to form the analysis of natural language, partially overlap with the intuitions that we combine with the atomic Boolean algebra. The model of reality which the elementary ontology describes and which underlies the whole system of ontology is one of the models of this algebra. A similar thing happens to the other theories which are distinguished from the system of Boolean algebras although these theories or their fragments have their interpretations in these algebras.

In view of the above it seems worth while to stress that the system of atomic (completely additive) Boolean algebra was formed later than ontology. Leśniewski's research in the field of mereology played a role in for-

---

[15] The connections of ontology and the systems of Boolean algebra (without the condition of atomicity) are discussed in Lejewski [1960]. These problems are also discussed in Grzegorczyk [1955] where he claims that from the formal point of view Leśniewski's ontology is a system of the atomic completely additive Boolean algebra which is enriched with the simple theory of types (Grzegorczyk [1955, pp. 89–90]). It is however, difficult to give unqualified agreement to that statement.

* *Editorial Note*: English translation — Kotarbiński [1966, p. 211].

mulating the axioms of this algebra (Tarski [1935, p. 190]). It is hard to resist the impression that some positive influence on the formation of this system was exerted by Leśniewski's ontology as well.

2. The algebra of sets which has been mentioned at the beginning of this chapter will be symbolized by AT. The sole primitive term of this algebra is the symbol of inclusion of sets '$\subset$'; its axioms are the following expressions:

AT1. $X \subset X$

AT2. $X \subset Y \wedge Y \subset Z \rightarrow X \subset Z$

AT3. $\sum_{Z}\{X \subset Z \wedge Y \subset Z \wedge \prod_{U}\{X \subset U \wedge Y \subset U \rightarrow Z \subset U\}\}$

AT4. $\sum_{Z}\{Z \subset X \wedge Z \subset Y \wedge \prod_{U}\{U \subset X \wedge U \subset Y \rightarrow U \subset Z\}\}$

AT5. $\sum_{Y}\{\prod_{Z,U}\{(Z \subset X \wedge Z \subset Y \rightarrow Z \subset U) \wedge (X \subset Z \wedge Y \subset Z \rightarrow U \subset Z)\}$
$\wedge \prod_{V}\{V \not\subset Y \rightarrow \sum_{T}\{T \subset X \wedge T \not\subset Y \wedge T \subset V\}\}$
$\wedge \prod_{V}\{V \not\subset X \rightarrow \sum_{T}\{T \subset Y \wedge T \not\subset X \wedge T \subset V\}\}\}$

I shall write down the last axiom of the system AT with the help of the defined term which is introduced by the following definition:

Df.T1. $\mathrm{At}(X) \equiv \sum_{Y}\{X \not\subset Y\} \wedge \prod_{Y,Z}\{Y \subset X \wedge X \not\subset Y \rightarrow Y \subset Z\}$

AT6. $\sum_{Y}\{X \not\subset Y\} \rightarrow \sum_{Y}\{\mathrm{At}(Y) \wedge Y \subset X\}$

Axioms AT1–AT5 do not need any comment.

Df.T1 is a definition of one-element set which corresponds to the definition of atom (Df.B2) while axiom AT6 whose content is close to AB7 of algebra EB, assumes that every non-empty set includes a one-element set.

Definition T1 appears in Tarski [1933]* as definition 28 and expression AT6 which plays an important role in Tarski's considerations is denoted by the symbol (α).[16]

---

* *Editorial Note*: English translation in Tarski [1956, p. 201].

16 There appear certain terminological differences between the above version of the algebra AT and the original version of this algebra given by Tarski. For instance all axioms and definitions are formulated in the metasystem manner in the paper. Instead of the symbol "At (X)" which I have used in Df.T1, the symbol 'ε_X,,' is used in Df.28.

A more considerable difference may be found in the formulation of the axiom AT5 which in the original version does not contain the factor

$$\prod_{V}\{V \not\subset X \rightarrow \sum_{T}\{T \subset Y \wedge T \not\subset X \wedge T \subset V\}\}.$$

After a careful examination it was found that the axiom set given by Tarski does not suffice to axiomatize the algebra of sets. It may, however, be demonstrated that on the grounds of the postulates AT1–AT5 and after adding the definitions of the appropriate concepts of the algebra of sets to the system, all the axioms of the elementary algebra of sets (Boolean algebra) are theses of the algebra AT.

The extensional equality of sets $=$, the sets $\emptyset$ and 1, the operations $+$, $\cdot$, and $'$ as well as relation ε which corresponds to the relation ε of the algebra EB and of the ontology EO, are introduced in the following way in the algebra AT:

Df.T2. $X = Y \equiv X \subset Y \wedge Y \subset X$

Df.T3. $X = \emptyset \equiv \prod_{Y} \{X \subset Y\}$

Df.T4. $X = 1 \equiv \prod_{Y} \{Y \subset X\}$

Df.T5. $Z = X + Y \equiv X \subset Z \wedge Y \subset Z \wedge \prod_{U} \{X \subset U \wedge Y \subset U \rightarrow Z \subset U\}$

Df.T6. $Z = X \cdot Y \equiv Z \subset X \wedge Z \subset Y \wedge \prod_{U} \{U \subset X \wedge U \subset Y \rightarrow U \subset Z\}$

Df.T7. $Y = X' \equiv \prod_{Z,U} \{(Z \subset X \wedge Z \subset Y \rightarrow Z \subset U) \wedge (X \subset Z \wedge Y \subset Z \rightarrow U \subset Z\}$
$\wedge \prod_{V} \{V \not\subset Y \rightarrow \sum_{T} \{T \subset X \wedge T \not\subset Y \wedge T \subset V\}\}$
$\wedge \prod_{V} \{V \subset X \rightarrow \sum_{T} \{T \subset Y \wedge T \not\subset X \wedge T \subset V\}\}$

Df.T8. $X \varepsilon Y \equiv \mathrm{At}(X) \wedge X \subset Y$

The extensional equality of sets defined by definition T2 is an equivalence relation and satisfies the extensionality condition while definitions T3–T7 satisfy the existence and uniqueness condition which may be easily checked. Some of these definitions may be obviously simplified.

Theorem 2.II. *The following systems are equivalent: algebra of sets* AT, *atomic Boolean algebra* EB *and elementary ontology* EO.

Proof. The set of postulates AT1–AT5 and definitions T2–T7 of the system AT is equivalent on the ground of the predicate calculus, to the axiom set AB1a–AB6b and definition B1 of the system EB. The simple proof of this equivalence is here omitted (cf. Tarski [1933, p. 30]*). Definitions T8 and B3 are equiform. Definition T1 as well as axiom AT6 are equivalent to definition B2 and axiom AB7 respectively, on the grounds of the equivalence: $X = \emptyset \equiv \prod_{Y} \{X \subset Y\}$ which is a thesis of both algebras. Thus the algebras AT and EB are equivalent.

Hence and from theorem 2.I it follows that algebra AT is also equivalent to system EO.

It is worth while observing that to prove the equivalence of the system of elementary ontology and the algebra of sets AT, it is enough to make use of definition 1.1 in the first system and to utilize the definition:

$$X \varepsilon Y \equiv X \subset Y \wedge \sum_{Z} \{X \not\subset Z\} \wedge \prod_{Z,U} \{Z \subset X \wedge X \not\subset Z \rightarrow Z \subset U\}$$

* *Editorial Note*: English translation in Tarski [1956].

instead of definitions T1 and T8 of the algebra AT. Axiom AT6 of the algebra AT may be then written down in the following way:

$$\sum_{Y}\{X \not\subset Y\} \rightarrow \sum_{Y}\{Y \varepsilon X\}.$$

III. Decidability and Semantical Completeness of the System of Elementary Ontology

As I have observed in the introduction, it is easier to carry on methodological investigations into ontology when we compare elementary ontology with other deductive systems which are in the majority of cases more familiar. In the previous chapter I have proved that ontology is equivalent to the algebra of sets AT. Utilizing certain properties of this algebra as well as the results contained in the first part of the present paper, I shall present a simple proof of decidability of elementary ontology.[17] In this proof I make use of the so called axiomatic method of rejected formulae and it is a familiar fact that this was introduced to methodological investigations by Łukasiewicz [1961, 1957].[18]

The rejected expressions are preceded with the asterisk ★.

I assume that the following sentences of the calculus EO are rejected by virtue of axioms:

A★4 $\quad \star\sum_{X}\{X \varepsilon X\}$

A★5 $\quad \star\Big(\sum_{X_1 \neq X_n}\ldots\sum\{X_1 \varepsilon X_1 \wedge \ldots \wedge X_n \varepsilon X_n\} \rightarrow \sum_{X_1 \neq X_{n+1}}\ldots\sum\{X_1 \varepsilon X_1 \wedge \ldots \wedge X_{n+1} \varepsilon X_{n+1}\}\Big)$

Axiom A★4, just like axiom A★5, conforms to the intuition underlying the fundaments of ontology that nothing is decided in the system as far as the existence of any particular object is concerned.

According to A★5 every sentence is rejected when it states that if there exist at least n objects then there exist $n+1$ objects. A★5 is not a single axiom but a scheme under which an infinite number of sentences

[17] The algebra of sets AT which I have presented in this paper is one of the two extensions of the algebra of sets which Tarski discussed in his paper [1933, pp. 51-59]. (*Editorial Note*: English translation in Tarski [1956, pp. 199–209].) The decidability proof of this algebra was not given by Tarski. By analysing certain theorems given in Tarski [1933] one can, however, obtain this proof in a much simple way. At the same time the results of the quoted paper of Skolem should be utilized; this paper plays an important role in Tarski's considerations.

[18] This method is being used in the decidability proofs for some deductive systems, e.g., in the decidability proof for the axiomatic system of Aristotle's syllogistic — see Słupecki [1948]. Also the decidability proof for Aristotle's syllogistic with the negation of names is based on the rules of rejection; this proof was provided in Iwanuś [1973a] and is an extension of the result of Słupecki.

falls.[19] It may be proved that A★5 is equivalent to the following expression:

$$\sum_X \{X \succ n\} \rightarrow \sum_X \{X \succ n+1\},$$

where n is an arbitrary natural number distinct from zero.

I furthermore accept two rules of rejection.

($R_0^\star$) *Rule of rejection by detachment*:

If the implication $\alpha \rightarrow \beta$ is a thesis of the system EO and the expression β is rejected, then the expression α is also rejected.

(OΠ★) *Rule of rejection by omitting the universal quantifier*:

If the expression $\prod_X \{\alpha(X)\}$ is rejected, then the expression $\alpha(X)$ is also rejected. $\alpha(X)$ is here a sentential function and apart from the variable X, the other free variables may occur in it too.

$R_0^\star$ is one of the two rules of rejection formulated by Łukasiewicz [1961]. Introduction of the other rule of Łukasiewicz which is called rule of rejection by substitution is unnecessary for the decidability proof presented here.

The intuitions attached to the rule OΠ★ are the following. If the expression $\prod_X \{\alpha(X)\}$ is not a thesis of the system EO, then obviously nor is the expression $\alpha(X)$ a thesis of the system. Hence, if the set of axiomatically rejected expressions and the set of theses of the system are disjoint and moreover the rule of rejection by detachment does not lead from rejected expressions to theses of the system, then the expression $\alpha(X)$ cannot be a thesis of the system if the expression $\prod_X \{\alpha(X)\}$ is rejected.

By the set of rejected expressions of the system EO I understand the least set which includes the set of axiomatically rejected expressions and which is closed under the operations of rejection $R_0^\star$ and OΠ★. A more precise explication of this concept is given in the following two definitions:

Definition 3.I. α is an expression *rejected in the degree* k if and only if α is a meaningful expression of the calculus EO and,

---

[19] Sentences A★4 and A★5 are independent of the axioms of the system EO as I shall try to prove in the sequel. It is worth observing here that if we augment the system EO with all sentences of the form A★4 and A★5 we should then obtain a complete system, i.e., for every sentence of the calculus EO either the sentence itself or its negation would be a thesis of the system (cf. Th.2.II and also in Tarski [1933, pp. 51–58]). (*Editorial Note*: English translation in Tarski [1956, pp. 199–208].)

1° $k = 0$ and α is axiomatically rejected[20] or
2° $k > 0$ and one of the following conditions is satisfied:
(a) α is rejected in the degree $k-1$,
(b) there exists an expression β rejected in the degree $k-1$, and such that the implication $\alpha \to \beta$ is a thesis of the system,
(c) there exists an expression β rejected in the degree $k-1$, and such that the expression $\prod_{X}\{\alpha\}$ is equiform with β, X being at the same time a free variable that occurs in α.

Definition 3.II. α is a *rejected expression* if and only if for some natural number k, α is an expression rejected in the degree k.

According to theorem 3.II below (p. 213), the set of theses of elementary ontology coincides with the set of sentences true in every (empty or non-empty) domain. If by true sentences of the system EO we understand the sentences true in every domain which is a model of this system, then the set of rejected expressions of this system will coincide with the set of false sentences. This will be demonstrated. It should however, be remembered that the set of false sentences in this sense includes sentences true in certain domains.

The system EO enriched by adding to it axioms A★4 and A★5 as well as rules $R_0^{\star}$ and OΠ★ will be symbolized by EO★. It is apparent that the sets of meaningful expressions of the calculi EO and EO★ are identical. I shall demonstrate that the system EO★ presented above is decidable in the sense that every sentential expression of this system is either a thesis or it is rejected.

For the sake of the simpler formulation of some lemmas I make use of the following definitions as a metasystem abbreviation:

Definition 3.III. $$\beta(n) = \begin{cases} \ulcorner \sum_{X}\{X \varepsilon X\}\urcorner, \text{ for } n = 1, \\ \ulcorner \sum \ldots \sum_{X_1 \neq X_n} \{X_1 \varepsilon X_1 \wedge \ldots \wedge X_n \varepsilon X_n\}\urcorner, \\ \text{for } n > 1. \end{cases}$$

The variable n occuring in definition 3.III represents an arbitrary natural number other than zero. Expressions of the form $\beta(n)$ and $\sim \beta(n)$ are called quantitative sentences. It is easy to observe that the former declares the existence of at least n, while the latter of the two declares the existence of at most $n-1$ objects.

The following formulae composed of quantitative sentences and connectives of the sentential calculus are, respectively, the abbreviations of:

---

[20] The expression a is axiomatically rejected if a is equiform with A★4 or with one of the sentences of the system EO★ which fall under the scheme A★5.

$$\sim \beta(n) - \ulcorner \sim \sum_{X_1 \neq X_n} \ldots \sum \{X_1 \varepsilon X_1 \wedge \ldots \wedge X_n \varepsilon X_n\}\urcorner,$$

$$\beta(n) \wedge \beta(m) - \ulcorner \sum_{X_1 \neq X_n} \ldots \sum \{X_1 \varepsilon X_1 \wedge \ldots \wedge X_n \varepsilon X_n\} \wedge \sum_{X_1 \neq X_m} \ldots \sum \{X_1 \varepsilon X_1 \wedge \ldots$$
$$\ldots \wedge X_m \varepsilon X_m\}\urcorner,$$

$\beta(n) \vee \beta(m) -$ „ $\vee$ „

$\beta(n) \rightarrow \beta(m) -$ „ $\rightarrow$ „

$\beta(n) \equiv \beta(m) -$ „ $\equiv$ „

In the formulation of certain lemmas I utilize, as in Part I of the present paper, the symbol '$\Rightarrow$' to denote the metasystem implication. In the decidability proof of the calculus EO★ I make use of the lemmas given in Part I of this paper as well as of the following lemmas. I assume that number variables that occur in lemmas: 3.1a–3.6 and 3.I, 3.II run over the set of natural numbers other than zero.

L3.1a.	$n \geqslant m \Rightarrow (\beta(n) \rightarrow \beta(m))$	{Df.3.III}
b.	$n < m \Rightarrow (\beta(m) \rightarrow \beta(n))$	
L3.2a.	$n \geqslant m \Rightarrow (\beta(n) \equiv \beta(n) \wedge \beta(m))$	{L3.1a–b}
b.	$n < m \Rightarrow (\beta(m) \equiv \beta(n) \wedge \beta(m))$	
c.	$n \geqslant m \Rightarrow (\sim\beta(m) \equiv \sim\beta(n) \wedge \sim\beta(m))$	
d.	$n < m \Rightarrow (\sim\beta(n) \equiv \sim\beta(n) \wedge \sim\beta(m))$	
L3.3a.	$n \geqslant m \Rightarrow (\beta(m) \equiv \beta(n) \vee \beta(m))$	{L3.1a–b}
b.	$n < m \Rightarrow (\beta(n) \equiv \beta(n) \vee \beta(m))$	
c.	$n \geqslant m \Rightarrow (\sim\beta(n) \equiv \sim\beta(n) \vee \sim\beta(m))$	{L3.1a–b}
d.	$n < m \Rightarrow (\sim\beta(m) \equiv \sim\beta(n) \vee \sim\beta(m))$	
L3.4.	$\emptyset' \succ n \equiv \beta(n)$	
Proof. a) (1)	$\emptyset' \succ n$	{a.}
(2)	$a_1 \neq \ldots \neq a_n$	{Df.1.II, 1}
(3)	$a_1, \ldots, a_n \varepsilon \emptyset'$	
(4)	$a_1 \varepsilon a_1 \wedge \ldots \wedge a_n \varepsilon a_n$	{L1.3a, 3}
(5)	$\sum_{X_1 \neq X_n} \ldots \sum \{X_1 \varepsilon X_1 \wedge \ldots \wedge X_n \varepsilon X_n\}$	{2, 4}
(6)	$\beta(n)$	{Df.3.III, 5}
b) (1)	$\beta(n)$	{a.}
(2)	$a_1 \neq \ldots \neq a_n$	{Df.3.III, 1}
(3)	$a_1 \varepsilon a_1 \wedge \ldots \wedge a_n \varepsilon a_n$	
(4)	$a_1, \ldots, a_n \varepsilon \emptyset'$	{L1.3e, 3}
(5)	$\sum_{X_1 \neq X_n} \ldots \sum \{X_1, \ldots, X_n \varepsilon \emptyset'\}$	{2, 4}
(6)	$\emptyset' \succ n$	{Df.1.II, 5}

There holds the following lemma which is of considerable importance for what follows.

Lemma 3.I. *Every sentence a of the calculus* EO (*and thus, of the calculus* EO* *too*) *which contains no defined terms, may be transformed into a sentence equivalent to it that will have one of the following forms:*[21]

(a) $\beta(n)$,
(b) $\sim\beta(n)$,
(c) $\beta(n)\wedge\sim\beta(m)$,
(d) $\beta(n)\wedge\sim\beta(m)\wedge(\beta(i_1)\vee\sim\beta(j_1))\wedge\ldots\wedge(\beta(i\ )\vee\sim\beta(j_k))$,
(e) $\beta(n)\wedge(\beta(i_1)\vee\sim\beta(j_1))\wedge\ldots\wedge(\beta(i_k)\vee\sim\beta(j_k))$,
(f) $\sim\beta(n)\wedge(\beta(i_1)\vee\sim\beta(j_1))\wedge\ldots\wedge(\beta(i_k)\vee\sim\beta(j_k))$,
(g) $(\beta(i_1)\vee\sim\beta(j_1))\wedge\ldots\wedge(\beta(i_k)\vee\sim\beta(j_k))$.

$\beta(n)$, $\beta(m)$, $\beta(i_1), \ldots, \beta(i_k)$, $\beta(j_1), \ldots, \beta(j_k)$ are here quantitative sentences while $m, n, i_1, \ldots, i_k, j_1, \ldots, j_k$ are natural numbers other than zero.

Proof. Let α be an arbitrary sentence of the calculus EO in which no defined terms occur. Assume that α begins with the existential quantifier and thus:

(1) $\alpha = \ulcorner\sum_{X}\{\beta(X)\}\urcorner$,

where $\beta(X)$ is a sentential function in which X occurs as the only free variable.

According to lemma 1.V there holds the equivalence:

(2) $\beta(X)\equiv\beta_1(X)\vee\beta_2(X)\vee\ldots\vee\beta_k(X)$,

where for any $1\leqslant i\leqslant k$: $\beta_i(X)$ is an elementary conjunction of quantitative formulae composed of the constituents in the variable X (cf. p. 186). By virtue of lemma 1.III and of the thesis of ontology: $X = X$ (L1.9a), there exist natural numbers n_1, n_2, n_3 and n_4 different from zero and such that each such conjunction is equivalent to one of the following fifteen expressions:

(1′) $\emptyset'\cdot X \succ n_1\wedge\emptyset'\cdot X'\succ n_2\wedge\sim\emptyset'\cdot X\succ n_3\wedge\sim\emptyset'\cdot X'\succ n_4$
(2′) $\emptyset'\cdot X \succ n_1\wedge\emptyset'\cdot X'\succ n_2\wedge\sim\emptyset'\cdot X\succ n_3$
(3′) $\emptyset'\cdot X \succ n_1\wedge\emptyset'\cdot X'\succ n_2\wedge\sim\emptyset'\cdot X'\succ n_3$
(4′) $\emptyset'\cdot X \succ n_1\wedge\sim\emptyset'\cdot X\succ n_2\wedge\sim\emptyset'\cdot X'\succ n_3$
(5′) $\emptyset'\cdot X' \succ n_1\wedge\sim\emptyset'\cdot X\succ n_2\wedge\sim\emptyset'\cdot X'\succ n_3$
(6′) $\emptyset'\cdot X \succ n_1\wedge\emptyset'\cdot X'\succ n_2$
(7′) $\emptyset'\cdot X \succ n_1\wedge\sim\emptyset'\cdot X\succ n_2$
(8′) $\emptyset'\cdot X \succ n_1\wedge\sim\emptyset'\cdot X'\succ n_2$
(9′) $\emptyset'\cdot X' \succ n_1\wedge\sim\emptyset'\cdot X\succ n_2$
(10′) $\emptyset'\cdot X' \succ n_1\wedge\sim\emptyset'\cdot X'\succ n_2$
(11′) $\sim\emptyset'\cdot X \succ n_1\wedge\sim\emptyset'\cdot X'\succ n_2$

[21] The restriction that the sentence α cannot contain the defined terms is not essential. Lemma 3.I is valid also in the case when α represents an arbitrary sentence of the system EO.

(12') $\emptyset' \cdot X \succ n_1$
(13') $\emptyset' \cdot X' \succ n_1$
(14') $\sim \emptyset' \cdot X \succ n_1$
(15') $\sim \emptyset' \cdot X' \succ n_1$

Expressions (1')–(15') have been obtained from expressions (1)–(15) that were quoted in lemma 1.III by means of puting the constant $\emptyset'$ in place of δ (cf. p. 185). Assume that each of the components of the alternation $\beta_1(X) \vee \beta_2(X) \vee \ldots \vee \beta_k(X)$ has one of the forms (1')–(15').

From (2) and from the law of distribution for the existential quantifier with respect to alternation we have

(3) $$\sum_X \{\beta(X)\} \equiv \sum_X \{\beta_1(X)\} \vee \sum_X \{\beta_2(X)\} \vee \ldots \vee \sum_X \{\beta_k(X)\}.$$

According to lemmas 1.45a–1.59 each of the sentences $\sum_X \{\beta_i(X)\}$, for $1 \leqslant i \leqslant k$, is equivalent to a conjunction or an alternation of formulae of the form: $\emptyset' \succ n$ and $\sim \emptyset' \succ n$. Hence and from lemma 3.4 it follows that the following equivalence is true:

(4) $$\sum_X \{\beta(X)\} \equiv \gamma_1 \vee \gamma_2 \vee \ldots \vee \gamma_l,$$

where each of the expressions $\gamma_1, \gamma_2, \ldots, \gamma_l$ is a conjunction of quantitative sentences.

According to the theorem stating that for every meaningful expression α of the sentential calculus in the normal alternative form there exists an expression β in the normal conjunctive form which is equivalent to it, we may replace the right side of equivalence (4) by the conjunction:

(5) $$\mathring{\gamma}_1 \wedge \mathring{\gamma}_2 \wedge \ldots \wedge \mathring{\gamma}_m.$$

At the same time every component of this conjunction is an alternation of quantitative sentences, i.e. sentences of the form: $\beta(n)$ or $\sim \beta(n)$. Hence and from lemmas 3.2a–3.3d it easily follows that conjunction (5) is equivalent to one of the expressions of the form (a)–(g) quoted in lemma 3.I. This completes the proof of lemma 3.I for these sentences α of the calculus EO which contain no defined terms and begin with the existential quantifier. It may be seen now that in view of de Morgan's Laws for quantifiers and in view of the form of the expressions (a)–(g) quoted in lemma 3.I, lemma 3.I holds also for the sentences beginning with the universal quantifier.

I also make use of the following lemmas characterizing the rejected expressions.

L3.5a. $\star \beta(n)$

Proof. (1) $\sum_{X_1 \neq X_n} \ldots \sum \{X_1 \varepsilon X_n \wedge \ldots \wedge X_n \varepsilon X_n\} \rightarrow \sum_X \{X \varepsilon X\}$ {Thesis of EO}

(2) $\star \sum_X \{X \varepsilon X\}$ {A$\star$4}

(3) $\star \sum_{X_1 \neq X_n} \ldots \sum \{X_1 \varepsilon X_1 \wedge \ldots \wedge X_n \varepsilon X_n\}$ $\{R_0^\star : 1, 2\}$

$\star \beta(n)$ {Df.3.III, 3}

L3.5.b. $\star \sim \beta(n)$

Proof. (1) $\sim \sum_{X_1 \neq X_n} \ldots \sum \{X_1 \varepsilon X_1 \wedge \ldots \wedge X_n \varepsilon X_n\} \to \Big(\sum_{X_1 \neq X_n} \ldots \sum \{X_1 \varepsilon X_1 \wedge \ldots \wedge X_n \varepsilon X_n\} \to \sum_{X_1 \neq X_{n+1}} \ldots \sum \{X_1 \varepsilon X_1 \wedge \ldots \wedge X_{n+1} \varepsilon X_{n+1}\}\Big)$ $\{\sim p \to (p \to q)\}$

(2) $\star \Big(\sum_{X_1 \neq X_n} \ldots \sum \{X_1 \varepsilon X_1 \wedge \ldots \wedge X_n \varepsilon X_n\} \to \sum_{X_1 \neq X_{n+1}} \ldots \sum \{X_1 \varepsilon X_1 \wedge \ldots \wedge X_{n+1} \varepsilon X_{n+1}\}\Big)$ {A★5}

(3) $\star \sim \sum_{X_1 \neq X_n} \ldots \sum \{X_1 \varepsilon X_1 \wedge \ldots \wedge X_n \varepsilon X_n\}$ $\{R_0^\star : 1, 2\}$

$\star \sim \beta(n)$ {Df.3.III. 3}

L3.6. $m < n \Rightarrow \star(\beta(m) \to \beta(n))$

Proof. (1) $m < n$ {a.}

(2) $m+1 \leqslant n$ {1}

(3) $(\beta(m) \to \beta(n)) \to (\beta(m) \to \beta(m+1))$ {L3.1a, 2}

(4) $\star(\beta(m) \to \beta(m+1))$ {A★5, Df.3.III}

$\star(\beta(m) \to \beta(n))$ $\{R_0^\star : 3, 4\}$

L3.7. $\star\alpha \Rightarrow \star(\alpha \wedge \beta)$ $\{R_0^\star, p \cdot q \to p\}$

Let $\alpha(X_1, \ldots, X_n)$ be an arbitrary sentential function of the calculus EO★, $X_1, \ldots, X_n$ being its free variables and let the sentence $\prod_{X_1} \ldots \prod_{X_n} \{\alpha(X_1, \ldots, X_n)\}$ be the generalization of this function.

L3.8a. *The expression* $a(X_1, \ldots, X_n)$ *is a thesis of the system* EO★ *if and only if the sentence* $\prod_{X_1} \ldots \prod_{X_n} \{a(X_1, \ldots, X_n)\}$ *is a thesis of this system.*

b. *The expression* $a(X_1, \ldots, X_n)$ *is rejected in the system* EO★ *if and only if the sentence* $\prod_{X_1} \ldots \prod_{X_n} \{a(X_1, \ldots, X_n)\}$ *is rejected.*

L3.8a is of course, true. The fact that lemma (b) holds may be seen from the fact that except for the primitive rule OΠ★ also the rule of rejection by adjoining the universal quantifier is valid in the system EO★ as a derived rule: if the expression α is rejected then also the expression $\prod_X \{\alpha\}$ is rejected.

Lemma 3.II. *Every sentence of the calculus* EO★ *containing no defined terms is either a thesis or it is rejected.*

Proof. According to lemma 3.I every sentence α of the calculus EO★ containing no defined terms may be transformed into a sentence which is equivalent to α and which has one of the following forms:

(a) $\beta(n)$,
(b) $\sim\beta(n)$,
(c) $\beta(n)\wedge\sim\beta(m)$,
(d) $\beta(n)\wedge\sim\beta(m)\wedge(\beta(i_1)\vee\sim\beta(j_1))\wedge\ldots\wedge(\beta(i_k)\vee\sim\beta(j_k))$,
(e) $\beta(n)\wedge(\beta(i_1)\vee\sim\beta(j_1))\wedge\ldots\wedge(\beta(i_k)\vee\sim\beta(j_k))$,
(f) $\sim\beta(n)\wedge(\beta(i_1)\vee\sim\beta(j_1))\wedge\ldots\wedge(\beta(i_k)\vee\sim\beta(j_k))$,
(g) $(\beta(i_1)\vee\sim\beta(j_1))\wedge\ldots\wedge(\beta(i_k)\vee\sim\beta(j_k))$.

According to lemmas 3.5a–b and 3.7 every sentence of the form (a)–(f) is rejected. To prove lemma 3.II it is enough to demonstrate that every sentence of the form (g) is either a thesis or it is rejected. Every sentence of the form (g) is equivalent to the following conjunction:

(g′) $(\beta(j_1)\to\beta(i_1))\wedge\ldots\wedge(\beta(j_k)\to\beta(i_k))$.

It can easily be seen that:
1) if for every $1\leqslant i\leqslant k: j_l\geqslant i_l$, then by virtue of L3.1a the sentence (g′), and thus the sentence (g) as well, is a thesis of the system EO★;
2) if for some $1\leqslant l\leqslant k: j_l< i_l$, then by virtue of L3.6 the sentence $(\beta(j_l)\to\beta(i_l))$ is rejected. Hence and from L3.7 it follows that also the conjunction (g′), and thus the conjunction (g) as well, is rejected.

Lemma 3.II is then true in the case of (g) too.

Theorem 3.I. *Every sentential expression a of the system* EO★ *is either a thesis or it is rejected.*

Proof. It has already been pointed out in Part I of the present paper that the definitions of elementary ontology satisfy the condition of translatability, i.e., every meaningful expression of this system which is written down with the help of defined terms may be transformed into an expression containing no such terms and equivalent to the former. According to lemmas 3.8a–b for each sentential function $\alpha(X_1,\ldots,X_n)$ there exists a sentence β which is a generalization of this function ($\beta=\ulcorner\prod_{X_1}\ldots\prod_{X_n}\{\alpha(X,\ldots,X_n)\}\urcorner$) and such that if one of these expressions is a thesis then so is the other and when one of them is rejected the other is also rejected. Hence and from lemma 3.II there immediately follows theorem 3.I stating that system EO★ is decidable.

The system EO★ is consistent in the sense that no sentential expression of this system is a thesis and a rejected expression simultaneously.[22]

[22] The proofs of consistency for rejection systems, i.e., systems where there are introduced rejected expressions apart from the accepted ones, consist in most cases in giving an interpretation such that all the theses of the system are true while the rejected expressions are false in it (cf. Słupecki [1948, p. 9]; Łukasiewicz [1957, pp. 126–129]). The consistency of system EO★ cannot be demonstrated in that way since there occur pairs of contradictory sentences among the rejected expressions. It is easy to see that e.g., both the sentence $\sum\{X\varepsilon X\}$ is rejected and so is the sentence $\sim\sum_{X}\{X\varepsilon X\}$.

The proof of consistency for EO★ is here preceded by a proof of semantical completeness of the elementary ontology EO.

Let A be an arbitrary (empty or non-empty) set. By the symbols '$\subseteq$', '$=$' I denote the relation of inclusion and equality between the subsets of the set A, by "$\in$" I symbolize the relation of membership of an element in a set while by '$\{Z\}$' I symbolize the one-element set composed of the element $Z \in A$. I assume that the variables X, Y run over the subsets of the set A.

I introduce the following definition:

(a) $$X \varepsilon^{\star} Y \equiv \sum_{Z \in A}\{X = \{Z\} \wedge X \subseteq Y\}.$$

Conforming to the above definition the relation $\varepsilon^{\star}$ holds between the sets X and Y if and only if X is a one-element set and it is included in Y.

Let $\mathfrak{W} = \langle A, \varepsilon^{\star}\rangle$ be an arbitrary domain where A is a set and $\varepsilon^{\star}$ is the relation holding between the subsets of the set A and defined with the help of definition (a). The domain $\mathfrak{W}$ is obviously a model for the elementary ontology EO (without definitions) with the only primitive term 'ε' of the system being understood as a name of the relation $\varepsilon^{\star}$ (cf. Słupecki [1955, p. 19])*. Symbolize by $\mathrm{E}(\mathfrak{W})$ the set of sentential expressions of the calculus EO (except for defined terms)[23] which are true in the domain $\mathfrak{W}$. Then the following completeness theorem holds for elementary ontology EO:

… α *is a thesis of the system* EO *if and only if for any domain* … $\in \mathrm{E}(\mathfrak{W})$.

…s an arbitrary set and $\subseteq$ as well as $=$ are the relations … the subsets of the set A, then

… $Y = \prod_{Z \subseteq A}\{Z \varepsilon^{\star} X \to Z \varepsilon^{\star} Y\}\}$,

… $Y \equiv X \subseteq Y \wedge \sum_{Z \subseteq A}\{X \not\subseteq Z\} \wedge \prod_{Z,U}\{Z \subseteq Y \wedge X \not\subseteq Z \to Z \subseteq U\}\}$.

…l (2) are the consequences of definitions (a) on the grounds …reover, the following completeness theorem for the system … true; its proof has been given by Tarski ([1933, Th. 24, p. [illegible]])**.

If α is a sentential expression of the algebra AT (containing no defined terms) and the sole primitive term of this algebra, i.e. '$\subset$', is understood as a name of the relation $\subseteq$, then (3) α is a thesis of the algebra AT if and only if for any domain $\mathfrak{W} = \langle A, \subseteq\rangle : \alpha \in \mathrm{E}(\mathfrak{W})$.

* *Editorial Note*: Also in the present volume, p. 71.

** *Editorial Note*: English translation in Tarski [1956, p. 206].

[23] In the above considerations I omit the definitions of ontology EO. There will be no essential change in these considerations if the model $\mathfrak{M}$ is extended to comprise the relations: $\subseteq$, $=$, the empty set and the set theoretic operations of complement, product and union. They will be respectively assigned to the constants '$\subset$', '$=$', '$\emptyset$', '′', '.' and '$+$' introduced with the help of definitions 1.1–1.6 (Section I, p. 169).

From (1), (2) and (3) and by theorem on equivalence of the system of the elementary ontology and the algebra of sets AT (Th.2.II) there easily follows theorem 3.II.

The concept of domain which I made use of after Tarski while formulating theorems 3.II and (3), differs slightly from the familiar concept of domain. In most cases we understand a domain to be an ordered sequence composed of the set B, certain relations $R_1, \ldots, R_n$ holding between the elements but not between the subsets of the set B, and of certain distinguished elements $a_1, \ldots, a_m$ of this set if need be. No major point will change in the formulation of the above theorems if we take instead of the set A the family of all subsets of this set 2^A, and we shall consider $\subseteq$ and $\varepsilon^\star$ as the relations holding between the elements of the family 2^A.

Theorem 3.III. *No sentential expression of the calculus* EO* *is simultaneously a thesis and a rejected expression.*

In the proof of the above theorem I utilize certain properties of the algebra AT as I did in the proof of theorem 3.II. In view of the translatability of definitions of elementary ontology it suffices to limit consideration to expressions in which no defined terms occur.

Proof. I assume that α is a rejected expression. Conforming to definition 3.II, α is a rejected expression provided that for some natural number $k : \alpha$ is rejected in the degree k.

(1) If $k = 0$, then by virtue of definition 3.I, α is an axiomatically rejected expression. No rejected expression is a thesis of the system; that may be seen from the following remark.

The system of ontology and the system AT are equivalent and every sentence axiomatically rejected in EO* is equivalent to the negation of a sentence of the algebra AT which Tarski symbolized by γ_n, where n is a certain natural number. Viz., sentence A*4 is equivalent to the sentence $\sim\gamma_0$, and every sentence of the form A*5, for $n \geqslant 1$, is equivalent to the sentence $\sim\gamma_n$ (cf. Tarski [1933, Df. 30 and 31, p. 53])*. Simple proofs of these equivalences are omitted.

No sentence of the form γ_n and $\sim\gamma_n$ is a thesis of algebra AT. This follows from the completeness theorem for the system AT and from lemma I in Tarski [1933, p. 55]* stating that for any natural number n, the sentence γ_n is true in every n-element domain while the negation of this sentence is true in every domain whose cardinality is different from n. Hence we arrive at the conclusion that if the sentence α is axiomatically rejected in the calculus EO*, then neither α, nor its negation is a thesis of the system EO.

Assume then that:

(2) no sentence β that is rejected in the degree k is a thesis of the system EO, and

(3) sentence α is rejected in the degree $k+1$.

---

* *Editorial Note*: English translation in Tarski [1956].

According to definition 3.I the following cases should be taken into consideration:
(a) α is rejected in the degree k;
(b) there exists an expression β rejected in the degree k and such that the implication $\alpha \to \beta$ is a thesis of the system EO;
(c) there exists an expression β rejected in the degree k and such that $\beta = \ulcorner \prod_X \{\alpha\} \urcorner$.

(a) If α is rejected in the degree k, then according to assumption (2) it is not a thesis of the system EO.
(b) If α has been obtained by rule of rejection $R_0^\star$ from some expression β rejected in the degree k, then by assumption (2), β is not a thesis of the system EO. Hence, from theorem 3.II and from the fact that the implication $\alpha \to \beta$ is a thesis, we obtain that α is not a thesis of the system.
(c) If α has been obtained by rule of rejection $O\Pi^\star$ from some expression β rejected in the degree k, then by assumption (2), β is not a thesis of the system. Then, from the fact that a certain generalization of the expression α is not a thesis of the system EO, it follows that nor is α a thesis of the system.

It is easy to see that line (c) completes the proof of theorem 3.III according to which the elementary ontology EO* is a consistent system.

In the decidability proof for the system EO* which I have presented, I made use of both the rejected formulae and rules of rejection analogous to the investigations of Łukasiewicz. Thus I have been applying two concepts of proof: the proof from assumptions of the accepted formulae and the proof of rejected formulae which I here interpret according to Łukasiewicz's paper [1957]. It is worth observing that system EO* may be simplified without introducing separate rules of rejection apart from the familiar rules of inference. Viz., it is enough to reject axiomatically every expression of the form A*4 and A*5, and to define the set of rejected expressions as the least set:
(1) containing the set of axiomatically rejected expressions and having a property such that:
(2) if a certain axiomatically rejected expression is a consequence of α, then α is included in the set of the rejected expressions.

Only the usual concept of the proof from assumptions may then be used and in the case of the rejected expressions, their generalizations would be treated as the assumptions of the proof. From the above definition it follows that if an axiomatically rejected expression is a consequence of the expression α, then obviously, α is also rejected. Actually, in this way an expression may be rejected not only on the grounds of the axiomatically rejected expressions but also on grounds of expressions earlier rejected in that way.

Bolesław Sobociński

Studies in Leśniewski's Mereology

In this paper I propose to present a few unpublished results concerning Mereology. Although this theory is one of the essential constituents of Leśniewski's system of the foundations of mathematics, it is little known, and little research has been done up to now in this field of study.[1] Consequently Mereology abounds in problems which so far have remained unsolved. Some of these problems will emerge in the last section of the paper.

All the mereological theses in this paper will be written in the Peano-Russellian symbolism with some adaptations to meet the demands of Leśniewski's theories. Thus 'ε' means 'is' in the sense of Leśniewski's Ontology; '$\sim (a)$' means 'not-a' (this is the negation of a noun-expression); '$=$' means 'is identical with', '$A = B$' in Ontology being equivalent to '$A \varepsilon B . B \varepsilon A$'; '$\subset$' is the functor of inclusion; '$\cap$' means 'and' as used between noun-expressions; '$\cup$' means 'or' as used between noun-expressions; '$a - b$' is short for '$a \cap \sim (b)$'; '$\bigvee$' means the same as 'object' (it is the ontological counterpart of 'universal class', '$A \varepsilon \bigvee$' being equivalent to '$A \varepsilon A$'); '$\bigwedge$' means the same as 'non-existent' (it is the ontological counterpart of 'null class', '$A \varepsilon \bigwedge$' being equivalent to '$A \varepsilon A . \sim \sim (A \varepsilon A)$' '$a \circ b$' is equivalent to '$a \subset b . b \subset a$'; '$rf\langle\varphi\rangle$' denotes the domain of φ; '$rl\langle\varphi\rangle$' denotes the converse domain of φ; '$\rightleftarrows\in\varphi\ni$' says that φ is a one-one relation; '$a \infty b$' says that a's are equinumerous with b's (or in other words that there are as many a's as there are b's); '$a \propto b$' says that a's are less numerous than b's; '$a \| b$' says that a's are counternumerous with respect to b's. Since the space at my disposal is limited, the proofs of the main theorems will be indicated in outline only. But I should like to make it quite clear that the missing links in the deductions below are in fact in complete harmony with the laws of Protothetic and Ontology.

It is extremely difficult to give a precise characteristic of Mereology without writing a book instead of a paper. Speaking very generally Mereology can be described as a deductive theory which inquires into the most general relations that may hold among objects (physical objects for in-

---

[1] Mereology was presented by Leśniewski in two Polish papers: Leśniewski [1916] and Leśniewski [1927–1931]. (*Editorial Note*: See also Leśniewski [1983].)

stance). Mereology can be regarded as the theory of collective classes in contradistinction to Ontology which is the theory of distributive classes. One of the principal aims of the theory is to analyse and determine the meaning of the term "class", whose vagueness appears to be the source not only of the Russellian antinomy but also of various other misunderstandings in logic.[2] Behind the somewhat indefinite term 'class' two different concepts can be distinguished: the concept of the distributive class on the one hand and the concept of the collective class on the other. 'Class of' in the distributive sense is a quasi-functor and can be eliminated from a proposition without affecting its logical import. Thus for instance the proposition 'Socrates is an element of the class of mortals' means the same as 'Socrates belongs to the denotation of 'mortal'' or just simply 'Socrates is mortal'. In the case of a distributive class the expression 'is an element of the class of' is reducible to the logical functor 'is' (in the sense of Leśniewski's Ontology). Now 'class of' in the collective sense is a real functor and cannot be reduced to logical terms. Let us consider the expression 'the class of British Dominions'. In the collective sense this expression denotes a physical object consisting of Canada, Australia, New Zealand, etc. This object need not be continuous and gaps may exist between some of its elements. Every British Dominion is an element of this class and so is any part of it, however fancifully cut out. If we interpret 'the class of British Dominions' in the distributive sense then it will be correct for us to say that Canada is an element of the class but we would make a mistake if we supposed that so was Alberta or Saskatchewan. On the contrary, in the collective sense not only Canada is an element of the class of British Dominions but so are Alberta and New Wales, the Rockies and Stewart Island. Moreover the class of all cities in the British Dominions (in the collective sense of the term 'class') is an element of the class of the British Dominions (in the collective sense of the term 'element').

As I have already pointed out, the concept of the collective class cannot be reduced to logical concepts. This was the reason why Leśniewski examined its properties within the framework of a special deductive theory which he constructed in 1916 and which later on was called Mereology. Certain geometrical systems can be based on Mereology and a theory which is akin to it, and which has been constructed independently by Professor Woodger, has been applied in biology.[3] It may be interesting to add that a number of mereological theorems closely correspond to certain theses propounded in the scholastic ontology. The axioms of Mereology contain no presuppositions as regards what kind of objects and how many of them there are in the world. Consequently the theory is free from any metaphysical tendencies.

---

[2] Cf. Leśniewski [1927-1931, Ch. I-III]; B. Sobociński [1949-1950]. Cf. also Kotarbiński [1929]. (*Editorial Note*: English translation – Kotarbiński [1966].)

[3] Cf. Woodger [1937].

The functor 'class of' in the collective sense need not be the primitive term of Mereology. Practically every mereological functor of any significance can be employed in this capacity. Thus for instance Leśniewski's original axiom system for Mereology was based on the functor 'part of' and not on the functor 'class of'. In Mereology the expression 'A is part of B' is not to be understood in its logical or set-theoretical meaning but rather in the classical and scholastic sense whereby a part cannot be equal to the whole. The original axiom system, which dates from 1916, consists of four axioms:[4]

A1 $[A\ B\ C]: A \varepsilon \mathrm{pt}(B) . B \varepsilon \mathrm{pt}(C) . \supset . A \varepsilon \mathrm{pt}(C)$
A2 $[A\ B]: A \varepsilon \mathrm{pt}(B) . \supset . B \varepsilon \sim(\mathrm{pt}(A))$

The first of these axioms establishes the transitivity of 'pt'; the second says that if an object A is part of an object B then the object B is not part of the object A. In the remaining two axioms the following defined terms occur:

D1 $[A\ B] \therefore A \varepsilon A : A = B . \vee . A \varepsilon \mathrm{pt}(B) : \equiv . A \varepsilon \mathrm{el}(B)$
D2 $[A\ a] :: A \varepsilon A : [\exists B] . B \varepsilon a \therefore [B] : B \varepsilon a . \supset . B \varepsilon \mathrm{el}(A) \therefore [B]:$
$B \varepsilon \mathrm{el}(A) . \supset . [\exists C\ D] . C \varepsilon a . D \varepsilon \mathrm{el}(C) . D \varepsilon \mathrm{el}(B) \therefore \equiv . A \varepsilon \mathrm{Kl}(a)$

D1 determines the meaning of the functor 'element of' while D2 is the definition of the functor 'class of' in the collective sense. The adequacy of this definition cannot be discussed here as this would take too much space. With the aid of D1 and D2 the remaining two axioms are given the form of the following expressions:

A3 $[A\ B\ a]: A \varepsilon \mathrm{Kl}(a) . B \varepsilon \mathrm{Kl}(a) . \supset . A = B$
A4 $[A\ a]: A \varepsilon a . \supset . [\exists B] . B \varepsilon \mathrm{Kl}(a)$

A3 says that the class of a's is unique; from A4 we learn that there exists the class of a's provided there exists at least one a.

The consistency of the above axiom system has been proved by Leśniewski by means of an interpretation within the theory of real numbers but the proof has never been published. Even without taking into consideration the latest results obtained in metalogic it is clear that the axiom system is not complete.

The axiom system which I have just presented, is not elegant as it contains defined functors. An equivalent axiom system for 'pt' without defined functors was given by Leśniewski in 1918.[5] It consisted of four axioms.

By deriving the following theses:

D3 $[A\ B]: \sim(A = B) . A \varepsilon \mathrm{el}(B) . \equiv . A \varepsilon \mathrm{pt}(B)$
D4 $[A\ B] \therefore [\exists a] . B \varepsilon \mathrm{Kl}(a) . A \varepsilon a : \equiv . A \varepsilon \mathrm{el}(B)$

---

[4] Cf. Leśniewski [1927-1931, Ch. IV].
[5] Cf. Leśniewski [1927-1931, Ch. VI].

Leśniewski showed that both 'el' and 'Kl' could be used as the only primitive functor of Mereology. An axiom system with 'el' as the primitive term was constructed by Leśniewski in 1920.[6] Again it consisted of four axioms. About 1930 it was simplified by Dr. J. F. Drewnowski but this result has not yet been published. Until recently no axiom system for 'Kl' was known.

The following expression:

D5 $[A\ B]::A\ \varepsilon\ A:[\exists C].C\ \varepsilon\ \mathrm{el}(B)\therefore[C]:C\ \varepsilon\ \mathrm{el}(B).\supset.C\ \varepsilon\ \sim(\mathrm{el}(A))\therefore \equiv.A\ \varepsilon\ \mathrm{extr}(B)$

defines another very important mereological functor. The expression '$A\ \varepsilon\ \mathrm{extr}(B)$' is to be read '$A$ is outside B' which means that A and B have no elements in common. Leśniewski not only proved that the functor 'extr' can be used as the only primitive functor of Mereology but in 1921 he proposed an appropriate axiom system which consisted of two comparatively short theses.[7]

In addition Leśniewski showed that each of the following functors definable in Mereology can be used as the only primitive term of the theory:[8]

D6 $[A\ a]::A\ \varepsilon\ A\therefore[B]:B\ \varepsilon\ \mathrm{el}(A).\supset.[\exists C\ D].C\ \varepsilon\ a.C\ \varepsilon\ \mathrm{el}(A). D\ \varepsilon\ \mathrm{el}(C).D\ \varepsilon\ \mathrm{el}(B)\therefore\equiv.A\ \varepsilon\ \mathrm{cl}(a)$

The class of a's consists of all a's. Definition D6 introduces the concept of 'collection of a's'. A collection of a's is an object which consists of a number of a's. Unlike the class of a's it need not comprise all a's. The functor 'cl' is one of the fundamental mereological constants. Some of its properties will be discussed in the concluding part of the paper.

D7 $[A\ B\ C]:B\ \varepsilon\ \mathrm{el}(C).A\ \varepsilon\ \mathrm{Kl}(\mathrm{el}(C)\cap\mathrm{extr}(B)).\equiv.A\ \varepsilon\ \mathrm{Cmpl}(B\ C)$

The expression '$A\ \varepsilon\ \mathrm{Cmpl}(B\ C)$', which may be read '$A$ is the complement of B with respect to C', means that A together with B, which is outside it, forms a third object C. In other words if from an object C we separate an object B then the remaining part will be A.

D8 $[A\ B\ C]:A\ \varepsilon\ \mathrm{Kl}(B\cup C).B\ \varepsilon\ \mathrm{extr}(C).\equiv.A\ \varepsilon\ B+C$

This definition says that A consists of two objects, B and C, the one of which is outside the other.

D9 $[A\ a]\because A\ \varepsilon\ \mathrm{Kl}(a)::[B\ C]\therefore B\ \varepsilon\ a.C\ \varepsilon\ a.\supset:B=C. \vee.B\ \varepsilon\ \mathrm{extr}(C)::\equiv.A\ \varepsilon\ \mathrm{Sm}(a)$

---

[6] Cf. Leśniewski [1927-1931, Ch. VI I].(*Editorial Note*: See also Leśniewski [1983].)

[7] Cf. Leśniewski [1927-1931, Ch. X].

[8] The theses establishing that each of these four functors can be used as the only primitive term of Mereology, have been proved in Leśniewski [1927-1931].

The expression 'A is the sum of a's', or '$A \varepsilon \mathrm{Sm}(a)$' in symbols, means that a's are discrete with respect to one another and that A is the class of them.

Until now we have no axiom systems for these four functors. It is not difficult to think of other constant terms which could be used as primitive functors of Mereology. In 1934 I found that the functor whose meaning can be determined with the aid of the following definition:

D10 $[a]\because\because[A\ B]\therefore A \varepsilon a . B \varepsilon a . \supset : A = B . \vee . A \varepsilon \mathrm{extr}(B) :: \equiv . \mathrm{dscr}(a)$

has the required property. This follows from the thesis which says that:

B1 $[A\ B] : A \varepsilon A . B \varepsilon B . \sim(A = B) . \mathrm{dscr}(A \cup B) . \equiv . A \varepsilon \mathrm{extr}(B)$

and which can easily be proved in Mereology. The expression 'dscr(a)' may be read 'a's are discrete', which means that a's are outside one another. The functor 'dscr' is the weakest mereological functor known to me which is strong enough to be a single primitive term of the theory.[9] No axiom system has yet been found for this functor. Out of a number of theses which characterize some of its properties, I should like to mention the following three:

B2 $[a\ b] : a \subset b . \mathrm{dscr}(b) . \supset . \mathrm{dscr}(a)$
B3 $[a] :: \mathrm{dscr}(\mathrm{cl}(a)) . \equiv \therefore [A\ B] : A \varepsilon a . B \varepsilon a . \supset . A = B$
B4 $[a\ b] : \mathrm{dscr}(a) . \mathrm{dscr}(b) . \mathrm{cl}(a) \circ \mathrm{cl}(b) . \supset . a \circ b$

It is very unlikely that there exists a functor which is weaker than 'dscr' and which nevertheless could be used as a single primitive term of Mereology. Thus for instance the property in question does not seem to belong to the following constant term:

D11 $[A] : A \varepsilon \mathrm{Kl}(\bigvee) . \equiv . A \varepsilon U$

The expression '$A \varepsilon U$' may be read 'A is the Universe', and definition D11 shows that in Mereology we define the concept of Universe as the collective class of all existing things. There are many theorems in Mereology concerning the Universe. The following one will suffice as an example:

C1 $[A] :: A \varepsilon U . \equiv \therefore A \varepsilon A \therefore [a] : a \subset \mathrm{extr}(A) . \supset . a \subset \bigwedge$

Thesis C1, which was not known to Leśniewski, says that A is the Universe if and only if everything that is outside it is non-existent.

Until 1946 no thesis was known which could be used as a single axiom of Mereology. In 1946 Dr. A. Grzegorczyk made use of a certain construction which occurs in Leśniewski's axiom system for the functor 'extr',

[9] Concerning the concept of 'the weakest primitive term' cf. Lindenbaum [1936].

and succeeded in producing a single axiom for Mereology with the functor 'el' as the primitive term. Grzegorczyk's axiom, which as far as I know has not yet been published, is considerably shorter than the axiom systems known to Leśniewski as it consists of 14 elementary units. It does not, however, satisfy all the methodological demands which Leśniewski set up for single axioms. In connexion with Grzegorczyk's result I constructed in 1948, the following axiom with the functor 'el' as the primitive term:[10]

$$\text{E1}\quad [A\,B]\because\therefore A\,\varepsilon\,\mathrm{el}(B).\equiv\because\therefore B\,\varepsilon\,B\because\therefore[f\,a]::[C]::C\,\varepsilon\,f(a).\equiv\therefore[D]:D\,\varepsilon\,a.\supset.D\,\varepsilon\,\mathrm{el}(C)\therefore[D]:D\,\varepsilon\,\mathrm{el}(C).\supset.[\exists E\,F].E\,\varepsilon\,a.F\,\varepsilon\,\mathrm{el}(D).F\,\varepsilon\,\mathrm{el}(E)\because B\,\varepsilon\,\mathrm{el}(B).B\,\varepsilon\,a\because\supset.A\,\varepsilon\,\mathrm{el}(f(a))$$

El consists of 12 elementary units and satisfies Leśniewski's methodological demands. The inferential equivalence between E1 and the axiom system A1–A4 can be proved without much difficulty. Although there are some reasons to believe that this is not the shortest axiom of Mereology, attempts to find a shorter one have failed so far.

In 1954 and 1955 Dr. C. Lejewski found the following single axioms:[11]

$$\text{E2}\quad [A\,a]\because\therefore A\,\varepsilon\,\mathrm{Kl}(a).\equiv\because A\,\varepsilon\,A\because[f]\because[B\,C]:B\,\varepsilon\,f(C).\equiv.[\exists d].B\,\varepsilon\,d.C\,\varepsilon\,\mathrm{Kl}(d)\therefore A\,\varepsilon\,\mathrm{Kl}(A)\therefore\supset\therefore[B\,b]:B\,\varepsilon\,a.A\,\varepsilon\,b.\supset.B\,\varepsilon\,f(\mathrm{Kl}(b))\therefore[B]:B\,\varepsilon\,f(A).\supset.[\exists C\,D].C\,\varepsilon\,a.D\,\varepsilon\,f(B).D\,\varepsilon\,f(C)$$

E2 is interesting not only because it is a single axiom of Mereology but also because it is the first axiom for the functor 'Kl' ever established.

$$\text{E3}\quad [A\,B]\because\therefore A\,\varepsilon\,\mathrm{extr}(B).\equiv\because\therefore\sim(A\,\varepsilon\,\mathrm{extr}(A))\because\therefore[f]::[C\,a]\because C\,\varepsilon\,f(a).\equiv::C\,\varepsilon\,C::[D]\therefore C\,\varepsilon\,\mathrm{extr}(D).\equiv:[E]:E\,\varepsilon\,a.\supset.E\,\varepsilon\,\mathrm{extr}(D)::\supset::[F]::[\exists b]\therefore A\,\varepsilon\,b.\vee.B\,\varepsilon\,b\therefore[d]:F\,\varepsilon\,d.\supset.\sim(f(b)\,\varepsilon\,f(d))$$

E3 is a single axiom for the functor 'extr'.

The functor 'ov', whose meaning is determined by the following definition:

$$\text{D12}\quad [A\,B]\therefore A\,\varepsilon\,A:[\exists C].C\,\varepsilon\,\mathrm{el}(a).C\,\varepsilon\,\mathrm{el}(B):\equiv.A\,\varepsilon\,\mathrm{ov}(B)$$

can also be used as the primitive term of Mereology. Here is Lejewski's axiom for this functor:

$$\text{E4}\quad [A\,B]\because\therefore A\,\varepsilon\,\mathrm{ov}(B).\equiv\because\therefore A\,\varepsilon\,A.B\,\varepsilon\,B\because\therefore[f]::[D\,a]::D\,\varepsilon\,f(a).\equiv\therefore D\,\varepsilon\,D\therefore[E]:D\,\varepsilon\,\mathrm{ov}(E).\equiv[\exists F].F\,\varepsilon\,a.F\,\varepsilon\,\mathrm{ov}(E)\because B\,\varepsilon\,\mathrm{ov}(B)\because\supset::[\exists C]::[b]\therefore A\,\varepsilon\,b.\vee.B\,\varepsilon\,b:\supset.[\exists d].C\,\varepsilon\,d.f(b)\,\varepsilon\,f(d)$$

The expression '$A\,\varepsilon\,\mathrm{ov}(B)$' may be read '$A$ is an object which overlaps B'. It means that the two objects A and B have an element in common.

---

[10] Cf. Sobociński [1949-1950, p. 53].

[11] Cf. Lejewski [1954a].

Finally Lejewski has proved that the following thesis with the functor 'el' as the primitive term can also be used as a single axiom of Mereology:

E5 $[A\ B]\because\cdot\, A\ \varepsilon\ \mathrm{el}(B) . \equiv :: B\ \varepsilon\ B :: [C\ a\ f] \because\cdot [D] \therefore D\ \varepsilon\ f(a) . \equiv : [E] : [\exists F] F\ \varepsilon\ \mathrm{el}(D) . F\ \varepsilon\ \mathrm{el}(F) . \equiv . [\exists G\ H] . G\ \varepsilon\ a . H\ \varepsilon\ \mathrm{el}(E) . H\ \varepsilon\ \mathrm{el}(G) :: B\ \varepsilon\ \mathrm{el}(B) . B\ \varepsilon\ \mathrm{el}(C) . C\ \varepsilon\ a :: \supset . A\ \varepsilon\ \mathrm{el}(f(a))$

Interestingly enough the axioms discovered by Lejewski, with the exception of E3 are not shorter than my axiom E1. E3 is shorter but it has a methodological defect: it is not organic. In other words it is a mere combination of two theses. Nevertheless it deserves our attention for two reasons. First it has a quite unexpected form. Secondly the fact that it is not organic although comparatively short suggests that there may exist single axioms of Mereology shorter than the ones which have already been discovered.

Now I proceed to the final section of my paper and to begin with I propose to discuss certain properties of the functor 'cl' whose meaning has been established in D6. Close analogy holds between this functor and the concept of sub-set in logic and in the theory of sets. On the other hand there are fundamental differences. In Mereology the expressions 'a' and 'cl(a)' are of the same semantical category whereas in logic, which presupposes the theory of types, elements and sub-sets of a given set cannot belong to the same type. While in the theory of sets a set of elements is something quite different from a set of sets of the elements in question, in Mereology one of the elementary theses says that:

B5 $[A\ a] : A\ \varepsilon\ \mathrm{cl}(a) . \equiv . A\ \varepsilon\ \mathrm{cl}(\mathrm{cl}(a))$

This means that a collection of a's is the same thing as a collection of collections of a's. Another thesis which is characteristic of this functor is the following:

B6 $[A\ a] : A\ \varepsilon\ a . \supset . A\ \varepsilon\ \mathrm{cl}(a)$

The converse does not hold of course. A collection of a's need not necessarily be identical with an a. An elementary proof by induction shows that if the number of a's is n then the number of the collections of a's is 2^n–1. The analogy between 'cl' and the concept of sub-set becomes more pronounced if we consider collections of objects which are discrete with one another. Thus for instance Leśniewski proved that:

F1 $[A\ B\ a] : \mathrm{dscr}(a) . A\ \varepsilon\ a . B\ \varepsilon\ a . \sim(A = B) . \supset . a \propto \mathrm{cl}(a)$

F1 says that if a's are discrete with one another and if there are more a's than one then a's are less numerous than the collections of a's.[12] This

[12] Cf. Leśniewski [1927–1931]. The functors '∞', '$\propto$', and '$\|$' are defined in Leśniewski's Ontology in the same way as in other systems of logic:

thesis is the mereological correlate of Cantor's famous theorem in accordance with which the number of sub-sets of a set exceeds the number of its elements. As is well known in the logic which disregards the theory of types or similar restrictions, and also in the preaxiomatic form of the theory of sets, this theorem of Cantor's leads to a contradiction, which is sometimes called the contradiction of 'the class of all classes'. Since Mereology is a consistent theory, it is quite clear that F1 cannot lead to contradictions. Moreover the thesis is perfectly intuitive. It is in this connexion that I set myself the problem of finding out what other simple theorems concerning the equinumerousness of collections could be proved in Mereology. In 1952 I proved, among others, the following thesis:

F2 $[a\,b]:\mathrm{dscr}(a).\mathrm{dscr}(b).a\infty b.\supset.\mathrm{cl}(a)\infty\mathrm{cl}(b)$

F2 says that if a's are discrete and b's are discrete, and if a's are equinumerous with b's then the collections of a's are equinumerous with the collections of b's.

The proof of this thesis is rather complicated and has not yet been published. Here is its outline:

We introduce the following definition[13]:

D13 $[A\,B\,a\,b\,\varphi]\therefore B\,\varepsilon\,b:[\exists X].X\,\varepsilon\,a.X\,\varepsilon\,\mathrm{el}(A).\varphi\{XB\}:$
$\equiv.B\,\varepsilon\,\Phi a(Aab\varphi)$

and with its aid we prove:

F3 $[A\,a\,b\varphi]::A\,\varepsilon\,\mathrm{cl}(a)-a.\mathrm{dscr}(b)\therefore[V]:V\,\varepsilon\,a.\supset.V\,\varepsilon\,rf\langle\varphi\rangle\therefore$
$[V]:V\,\varepsilon\,rl\langle\varphi\rangle.\supset.V\,\varepsilon\,b\therefore\rightleftarrows\in\varphi\ni\therefore\supset::[\exists B]\therefore B\,\varepsilon\,\mathrm{cl}(b)-b\therefore$
$[V]:V\,\varepsilon\,a.V\,\varepsilon\,\mathrm{el}(A).\supset.[\exists Y].Y\,\varepsilon\,b.Y\,\varepsilon\,\mathrm{el}(B).\varphi\{VY\}\therefore[V]:$
$V\,\varepsilon\,b.V\,\varepsilon\,\mathrm{el}(B).\supset.[\exists X].X\,\varepsilon\,a.X\,\varepsilon\,\mathrm{el}(A).\varphi\{XV\}$

DI $[ab]::[\exists\varphi]:\Psi\in\varphi\ni.\rightleftarrows\in\varphi\ni.a\circ rf\langle\varphi\rangle.b\circ rl\langle\varphi\rangle\therefore\equiv.a\infty b$

DII $[ab]::[\exists n].n\subset b.n\infty a\because[n]:n\subset a.\supset.\sim(b\infty a)\because\equiv.a\propto b$

DIII $[ab]::[n]:n\subset a.\supset.\sim(b\infty n)\therefore[n]:n\subset b.\supset.\sim(a\infty n)\because\equiv.a\|b$

It can easily be seen that the concept of equinumerousness in Ontology corresponds to that of the similarity of classes in Principia Mathematica (Cf. '73), '$a\infty b$' being the counterpart of '$\alpha\mathrm{sm}\beta$'.

The expression '$\Psi\in\varphi\ni$', which occurs in DI, says that the variable functor φ is a relation. This means that its arguments are always noun-expressions designating one object each. An explicit stipulation to this effect has to be included in the definition as in Ontology, contrary to other systems of logic, expressions of the type '$\varphi\in a\,b\ni$' where φ stands for any functor for two arguments each of which is a noun-expression, do not imply that 'a' or 'b' should stand for individual noun-expressions.

The functor 'Ψ' is defined in Ontology as follows:

DIV $[\varphi]\therefore[A\,B]:\varphi\{A\,B\}.\supset.A\,\varepsilon\,A.B\,\varepsilon\,B:\equiv.\Psi\in\varphi\ni$

[13] D13 and D14 can serve as examples of 'creative definitions' in Mereology as neither F3 nor F4 can be derived without them. Creative definitions constitute one of the characteristic features on the systems constructed by Leśniewski. For further information of the subject see Leśniewski [1931]. Cf. also Myhill [1953].

Similarly with the aid of definition:

D14 $[Ababφ] \therefore A \;\varepsilon\; a : [\exists X] . X \;\varepsilon\; b . X \;\varepsilon\; \mathrm{el}(B) . φ\{AX\} :$
$\equiv . A \;\varepsilon\; Φβ(Babφ)$

we prove:

F4 $[B\,a\,b\,φ] :\cdot: B \;\varepsilon\; \mathrm{cl}(b) - b . \mathrm{dscr}(a) \therefore [V] : V \;\varepsilon\; b . \supset . V \;\varepsilon\; rl\langle φ\rangle \therefore$
$[V] : V \;\varepsilon\; rf\langle φ\rangle . \supset . V \;\varepsilon\; a \therefore \rightleftarrows \in φ \ni \therefore \supset :: [\exists\, A] \therefore A \;\varepsilon\; \mathrm{cl}(a) - a \therefore$
$[V] : V \;\varepsilon\; a . V \;\varepsilon\; \mathrm{el}(A) . \supset . [\exists\, Y] . Y \;\varepsilon\; b . Y \;\varepsilon\; \mathrm{el}\,(B) . φ\,\{VY\} \therefore [V] :$
$V \;\varepsilon\; b . V \;\varepsilon\; \mathrm{el}(B) . \supset . [\exists\, X] . X \;\varepsilon\; a . X \;\varepsilon\; \mathrm{el}(A) . φ\,\{XV\}$

Now in order to establish the required equinumerousness we introduce yet another definition:

D15 $[A\,B\,a\,b\,φ] :: A \;\varepsilon\; \mathrm{cl}(a) - a . B \;\varepsilon\; \mathrm{cl}(b) - b \therefore [V] : V \;\varepsilon\; a . V \;\varepsilon\; \mathrm{el}(A) .$
$\supset . [\exists\; Y] . Y \;\varepsilon\; b . Y \;\varepsilon\; \mathrm{el}(B) . φ\{V\;Y\} \therefore [V] : V \;\varepsilon\; b . V \;\varepsilon\; \mathrm{el}(B) .$
$\supset . [\exists\; X] . X \;\varepsilon\; a . X \;\varepsilon\; \mathrm{el}(A) . φ\{X\;V\} \therefore \equiv . Φγ[abφ]\{AB\}$

With the aid of D15 we prove:

F5 $[a\,b\,φ] : \rightleftarrows \in φ \ni . \supset . \rightleftarrows \in Φγ[abφ] \ni$

F5 says that under the assumption that $φ$ is a one-one relation, the relation defined with the aid of D15 is also one-one. From D15, F3, and F4 we have:

F6 $[B\,a\,b\,φ] :: \mathrm{dscr}(a) . \rightleftarrows \in φ \ni . B \;\varepsilon\; \mathrm{cl}(b) - b \therefore [V] : V \;\varepsilon\; b . \supset . V \;\varepsilon\; rl\langle φ\rangle$
$\therefore [V] : V \;\varepsilon\; rf\langle φ\rangle . \supset . V \;\varepsilon\; a \therefore \supset . [\exists\, A] . Φγ[a\,b\,φ]\{AB\}$

and

F7 $[A\,a\,b\,φ] :: \mathrm{dscr}\,(b) . \rightleftarrows \in φ \ni . A \;\varepsilon\; \mathrm{cl}(a) - a \therefore [V] : V \;\varepsilon\; rl\langle φ\rangle .$
$\supset . V \;\varepsilon\; b \therefore [V] : V \;\varepsilon\; a . \supset . V \;\varepsilon\; rf\langle φ\rangle \therefore \supset . [\exists\, B] . Φγ[a\,b\,φ]\{AB\}$

Finally, from D15, F5, F6, F7, and from the definition of equinumerousness we deduce:

F8 $[a\,b] : \mathrm{dscr}(a) . \mathrm{dscr}(b) . a \infty b . \supset . \mathrm{cl}(a) - a \infty \mathrm{cl}(b) - b$

F8 is a stronger thesis than F2. It says that if the conditions of F2 are fulfilled then the collections of a's none of which is itself an a, are equinumerous with the collection of b's none of which is itself a b. F2 is a simple logical consequence of F8. Once we have F2, we can easily prove:

F9 $[a\,b] : \mathrm{dscr}(a) . \mathrm{dscr}(b) . \mathrm{cl}(\mathrm{a}) \| \mathrm{cl}(b) . \supset . a \| b$

This thesis says that if a's are discrete and b's are discrete, and in addition if the collections of a's are counternumerous with respect to the collections of b's then a's are counternumerous with respect to b's. At this junction I should perhaps make it clear that in Mereology we are allowed to operate with the concept of counternumerousness as we have not included the axiom of choice in Ontology, i.e., in the logic on which Mereology is based.

With the aid of F2 we can further prove the following thesis:

F10 $[a\ b]::\text{dscr}(a).\text{dscr}(b).\therefore[d]:d \subset \text{cl}(a).\supset.\sim(\text{cl}(b) \infty d).\therefore \supset \therefore [d]:d \subset a.\supset.\sim(b \infty d)$

But inspite of several attempts I have not been able to prove any of the following four theses, which are closely related to F2 and which seem to be quite intuitive:

G1 $[a\ b]:\text{dscr}(a).\text{dscr}(b).\text{cl}(a) \infty \text{cl}(b).\supset.a \infty b$

G1 says that if *a*'s are discrete and *b*'s are discrete, and in addition if the collections of *a*'s are equinumerous with the collections of *b*'s then *a*'s are equinumerous with *b*'s.

G2 $[a\ b]:\text{dscr}(a).\text{dscr}(b).a \propto b.\supset.\text{cl}(a) \propto \text{cl}(b)$

G2 says that if *a*'s are discrete and *b*'s are discrete, and in addition if *a*'s are less numerous than *b*'s then the collections of *a*'s are less numerous than the collections of *b*'s.

G3 $[a\ b]:\text{dscr}(a).\text{dscr}(b).\text{cl}(a) \propto \text{cl}(b).\supset.a \propto b$

G3 says that if *a*'s are discrete and *b*'s are discerte, and in addition if the collections of *a*'s are less numerous than the collections of *b*'s then *a*'s are less numerous than *b*'s.

G4 $[a\ b]:\text{dscr}(a).\text{dscr}(b).a \| b.\supset.\text{cl}(a) \| \text{cl}(b)$

G4 says that if *a*'s are discrete and *b*'s are discrete, and in addition if *a*'s are counternumerous with respect to *b*'s then the collections of *a*'s are counternumerous with respect to the collections of *b*'s.

The difficulties with which I was faced when trying to deduce G1–G4, prove to be analogous to those which worry us in the theory of sets in connexion with the following formulae:

I. $2^m = 2^n.\supset.m = n$ II. $m < n.\supset 2^m < 2^n$

III. $2^m < 2^n.\supset.m < n$

where m and n are any cardinal numbers. We know now how to prove formula III provided we are allowed to use the axiom of choice[14], and we can prove formulae I and II if we accept the generalised continuum

[14] Cf. Sierpiński [1928, pp. 97–98].

hypothesis. Similarly in Mereology we can easily obtain G3 if we have F2 and if in Ontology we assume the trichotomy for noun-expressions (i.e. the axiom of choice for the semantical category of noun-expressions). But even with the aid of this axiom I do not know how to deduce G1 or G2.

The four expressions G1–G4 can be proved without much difficulty if we accept the following hypothesis:

H1 $[a\ b]::\mathrm{dscr}(a).\mathrm{dscr}(b).\therefore[d]:d\subset a.\supset.\sim(b\infty d).\therefore\supset:[d]:$
$d\subset\mathrm{cl}(a).\supset.\sim(\mathrm{cl}(b)\infty d)$

It is easy to see that in a sense H1 is the converse of F10. Naturally I cannot prove this hypothesis within the framework of Mereology but I can deduce it on the additional assumption that in Mereology G1 and G3, or G2 and G4 are true.

Since our systems of Mereology are not complete, which means that we can strengthen the axioms in one direction or another, our discussion suggests the following unsolved problems:

1) Is H1 independent of the axioms of Mereology?

2) If so, does Mereology remain consistent if H1 is added to the axioms?

3) Is there a simpler expression equivalent to H1?

4) To what extent does the addition of H1 strengthen Mereology?

5) Does the addition of H1 to the axioms of Mereology enable us to derive the trichotomy for noun-expressions in the framework of Ontology?

Robert E. Clay

On the Definition of Mereological Class

Consider mereology axiomatized as in Clay [1965]*. Sobociński has posed the question, "If the usual definition of class, *DM1*, is replaced by

$[Aa] \therefore A \varepsilon \mathbf{Kl}(a) . \equiv : A \varepsilon A : [B] : a \subset \mathbf{el}(B) . \equiv . A \varepsilon \mathbf{el}(B),$

is the resulting system equivalent to the original?". This note gives a negative answer. Theses *A12* and *A13*, together with the two trivial models which follow them, show where the resulting system is weaker than mereology.

Consider the axiom system A consisting of *A1–A6*; *DA1*.

A1 $[A] : A \varepsilon A . \supset . A \varepsilon \mathbf{el}(A)$

A2 $[AB] : A \varepsilon \mathbf{el}(B) . B \varepsilon \mathbf{el}(A) . \supset . A = B$

A3 $[ABC] : A \varepsilon \mathbf{el}(B) . B \varepsilon \mathbf{el}(C) . \supset . A \varepsilon \mathbf{el}(C)$

A4 $[AB] : A \varepsilon \mathbf{el}(B) . \supset . B \varepsilon B$

DA1 $[Aa] \therefore A \varepsilon \mathbf{Cl}(a) . \equiv : A \varepsilon A : [B] : a \subset \mathbf{el}(B) . \equiv . A \varepsilon \mathbf{el}(B)$

A5 $[Aa] : A \varepsilon a . \supset . [\exists B] . B \varepsilon \mathbf{Cl}(a)$

A6 $[ABa] : A \varepsilon \mathbf{Cl}(a) . B \varepsilon \mathbf{Cl}(a) . \supset . A = B$**

DA2 $[Aa] \therefore A \varepsilon \mathbf{Kl}(a) . \equiv : A \varepsilon A : [D] : D \varepsilon a . \supset . D \varepsilon \mathbf{el}(A) : [D] :$
$D \varepsilon \mathbf{el}(A) . \supset . [\exists EF] . E \varepsilon a . F \varepsilon \mathbf{el}(D) . F \varepsilon \mathbf{el}(E)$

A7 $[Aa] : A \varepsilon \mathbf{Kl}(a) . \supset . [\exists B] . B \varepsilon a$ [*DA2*,*A1*]

A8 $[Aa] : A \varepsilon \mathbf{Kl}(a) . \supset . [\exists B] . B \varepsilon \mathbf{Cl}(a)$ [*A7*, *A5*]

A9 $[ABa] : B \varepsilon \mathbf{Kl}(a) . A \varepsilon \mathbf{Cl}(a) . \supset . A \varepsilon \mathbf{el}(B)$

PF $[ABa] : \mathrm{Hp}(2) . \supset .$

3) $a \subset \mathbf{el}(B) .$ [*DA2*, *1*]

$A \varepsilon \mathbf{el}(B) .$ [*DA1*, *2*, *3*]

A10 $[ABDa] : A \varepsilon \mathbf{Cl}(a) . B \varepsilon \mathbf{Kl}(a) . D \varepsilon \mathbf{el}(A) . \supset . [\exists EF] . E \varepsilon a .$
$F \varepsilon \mathbf{el}(D) . F \varepsilon \mathbf{el}(E)$ [*A9*, *DA2*]

---

* Refer to Clay [1965] for the definitions of terms used in this note. (*Editorial Note*: See also the footnote on p. 230.)

** This system is not independent.

$A11$ $[Aa]:A \varepsilon \mathbf{Cl}(a). \supset .a \subset \mathbf{el}(A)$ [$DA1$]

$A12$ $[Aa]:A \varepsilon \mathbf{Cl}(a).!\{\mathbf{Kl}(a)\}. \supset .A \varepsilon \mathbf{Kl}(a)$ [$DA2$, $A11$, $A10$]*

To show that $!\{\mathbf{Kl}(a)\}$ may fail, consider the model for A consisting of four names A, B, C, D with the relations, $A \neq B$, $B \neq C$, $A \neq C$, dscr $\{A \cup B \cup C\}$, $D \varepsilon \mathbf{Cl}(A \cup B \cup C)$. Then $\mathbf{Kl}(A \cup B) \mathrm{o} \wedge$

$A13$ $[Aa]:A \varepsilon \mathbf{Kl}(a). \rightarrow \{\mathbf{Kl}(a)\}. \supset A \varepsilon \mathbf{Cl}(a)$

PF $[Aa]:\mathrm{Hp}(2). \supset .$

$[\exists B]$.

3) $B \varepsilon \mathbf{Cl}(a)$. [$A8$, 1]

4) $B \varepsilon \mathbf{Kl}(a)$. [$A12$, 3]

5) $A = B$. [2, 1, 4]

$A \varepsilon \mathbf{Cl}(a)$ [3, 5]

To show that $\rightarrow\{\mathbf{Kl}(a)\}$ may fail, consider the model for A consisting of the two names A, B with the relations, $A \neq B$, $A \varepsilon \mathbf{el}(B)$. Then $A \varepsilon \mathbf{Kl}(A)$ and $B \varepsilon \mathbf{Kl}(A)$.

---

* *Editorial Note*: In order to facilitate the understanding of the present paper we recall some definitions from Clay [1965]:

$DO1$ $[a]:!\{a\}. \equiv .[\exists A].A \varepsilon a$

$DO2$ $[a] \therefore \rightarrow \{a\}. \equiv :[AB]:A \varepsilon a.B \varepsilon a. \supset .A \varepsilon B$

$DO6$ $[A]:A \varepsilon \wedge . \equiv .A \varepsilon A. \sim (A \varepsilon A)$

$DO9$ $[ab] \therefore a \mathrm{o} b. \equiv :[A]:A \varepsilon a. \equiv .A \varepsilon b$

$DM4$ $[AB] \therefore A \varepsilon \mathbf{ex}(B). \equiv .A \varepsilon A.B \varepsilon B:[C]:C \varepsilon \mathbf{el}(A). \supset . \sim (C \varepsilon \mathbf{el}(B))$

$DM6$ $[a]:: \mathrm{dscr}\{a\}. \equiv \therefore [AB] \therefore A \varepsilon a.B \varepsilon a. \supset :A = B. \vee .A \varepsilon \mathbf{ex}(B).$

Czesław Lejewski

Consistency of Leśniewski's Mereology

According to Sobociński's recollection, the consistency of Mereology was proved by Leśniewski by means of an appropriate interpretation within the framework of the theory of real numbers. His proof was never published, but in a recent paper R. E. Clay has succeeded in reconstructing a version of it.[1] Clay's result amounts to showing that if Leśniewski's Ontology expanded by the addition of the axioms for the real numbers is consistent then Mereology is consistent. Without casting any doubts on the validity of the proof one can hardly fail to note that here we have a case where the consistency of a conceptually simple theory is made to depend on the consistency of a theory which from the point of view of intuition is far from being obvious. What we would like to be in a position to do is to prove the consistency of Mereology relative to a theory which is more obvious than Mereology, or, preferably, relative to a theory which is, in fact, a much weaker subsystem of Mereology. It is with this methodological principle in mind that I propose to outline, in what follows, a new proof of the consistency of the theory under consideration.

Presuppositions of Mereology. Leśniewski's Mereology, which is a theory of part-whole relations, can be based on the following three presuppositions:

PA1. $[p\,q]::p \equiv q. \equiv \therefore [f] \therefore f(p f(p[u].u)). \equiv :[r]:f(q\,r). \equiv .q \equiv p.$

OA1. $[A\,a]:A \varepsilon a. \equiv .[\exists B].A \varepsilon B.B \varepsilon a.$

MA1. $[A\,B] \vdots : \vdots A \varepsilon \mathrm{el}(B). \equiv \vdots \cdot \vdots B \varepsilon B \vdots \cdot \vdots [C\,a] \vdots\vdots [D]::D \varepsilon C. \equiv \therefore [E]:$
$E \varepsilon a. \supset .E \varepsilon \mathrm{el}(D) \therefore [E]:E \varepsilon \mathrm{el}(D). \supset .[\exists FG].F \varepsilon a.G \varepsilon \mathrm{el}(E).$
$G \varepsilon \mathrm{el}(F):\cdot:B \varepsilon \mathrm{el}(B).B \varepsilon a:\cdot: \supset .A \varepsilon \mathrm{el}(C).$

PA1 and OA1 are sufficiently strong to serve as presuppositions of Leśniewski's Ontology while PA1 alone can serve as a single axiom of his Protothetic.[2] In a less rigorous mood we usually talk about OA1 as

[1] See Clay [1968].

[2] See Sobociński [1949-1950, Vol. 2, p. 257] and Lejewski [1963].

a single axiom of Ontology, Protothetic being taken for granted as a logically prior theory. Similarly, we could refer to MA1 as a single axiom of Mereology, assuming, tacitly, the availability of Ontology and, *a fortiori*, the availability of Protothetic.

In addition to the axiomatic presuppositions just mentioned, three kinds of directives or rules of procedure are operative in Mereology. First we have three rules of inference, namely substitution, quantification, and detachment. Secondly, there are two rules of definition: the rule for introducing propositional definitions and the rule for introducing nominal definitions. The former enables us to define (i) constant propositions and constant functors which, directly or indirectly form propositions, while the latter concerns (ii) constant names, singular or general, and functors which, directly or indirectly, form names. Finally, we have two rules of extensionality: the rule of propositional extensionality involving proposition-forming functors, and the rule of nominal extensionality involving names and name-forming functors. The law of extensionality for propositions is embedded in the axiom of Protothetic.

The rules of inference need not be considered in any detail in the present context, but the rules of definition and the rules of extensionality require some amplification. In the first place we must remember that in Leśniewski's theories definitions are not to be construed as abbreviations or typographical conveniences. They serve the purpose of expanding the vocabulary and the categorial variety of the system under construction. Thus in many respects they are akin to axioms. The rules of definition operative in Mereology can be outlined as follows:

RDI. On the assumption that a thesis T is, at a given stage, the last thesis of a system of Mereology, the rule for introducing propositional definitions allows us to add to the system a new thesis of the form

(1) $[\cdots]:\alpha. \equiv .\beta(\cdots)$

provided the following conditions are fulfilled: the definiens, which in Leśniewski's systems presupposing PA1 is written on the left-hand side of the definitional equivalence, and which in (1) is represented by 'α', is, with respect to T, a meaningful propositional expression, i.e. every constant term occurring in the definiens occurs in T or in a thesis preceding T in the system, and every variable occurring in the definiens belongs to a semantic category already available in the system; the definiendum, i.e. the expression represented in (1) by '$\beta(\cdots)$' is either (i) a constant proposition, which does not occur in T or in a thesis preceding T in the system, or (ii) a simple propositional function or a many-link propositional function; the constant functor of it, represented by 'β', occurs neither in T nor in any thesis preceding T in the system, and the arguments of the function are all variables, none of which occurs in the definiendum more than once;

every variable occurring in the definiendum occurs in the definiens as a free variable, and every free variable in the definiens occurs in the definiendum; a universal quantifier binds all free variables occurring in the definitional equivalence.

RDII. On the assumption that a thesis T is, at a given stage, the last thesis of a system of Mereology, the rule for introducing nominal definitions allows us to add to the system a new thesis of the form

(2) $[A \cdots]: \alpha(A \,\varepsilon\, \beta) . \equiv . A \,\varepsilon\, \gamma(\cdots)$

provided the following conditions are fulfilled: the definiens, i.e. the expression represented in (2) by '$\alpha(A \,\varepsilon\, \beta)$', is, with respect to T, a meaningful propositional expression; it is either (i) an expression of the form '$A \,\varepsilon\, \beta$', 'β' standing for a nominal expression, or (ii) a conjunction one of whose conjuncts is of the form '$A \,\varepsilon\, \beta$', or (iii) a conjunction of the kind just described, preceded by a particular quantifier; every constant term occurring in the definiens occurs in T or in a thesis preceding T in the system, and every variable occurring in the definiens belongs to a semantic category already available in the system; the expression represented in (2) by '$\gamma(\cdots)$' is (i) a constant name, which does not occur in T or in a thesis preceding T in the system, or (ii) a simple nominal function or a many-link nominal function; the constant functor of it, represented by 'γ', does not occur in T or in a thesis preceding T in the system, and the arguments of the function are all variables none of which occurs in the function more than once or is equiform with 'A'; every variable occurring in the definiendum, i.e. in the expression represented by '$A \,\varepsilon\, \gamma(\cdots)$' occurs in the definiens as a free variable, and every free variable in the definiens occurs in the definiendum; a universal quantifier binds all free variables occurring in the definitional equivalence.

The following three theses can serve as examples of possible nominal definitions in a system of Mereology:

MD1. $[A] :: A \,\varepsilon\, A \therefore [B] : B \,\varepsilon\, B . \supset . B \,\varepsilon\, \mathrm{el}(A) \therefore \equiv . A \,\varepsilon\, U.$

MD2. $[A\, a] :: A \,\varepsilon\, A \therefore [B] : B \,\varepsilon\, a . \supset B \,\varepsilon\, \mathrm{el}(A) \therefore [B] : B \,\varepsilon\, \mathrm{el}(A) . \supset . [\exists C\, D] . C \,\varepsilon\, a . D \,\varepsilon\, \mathrm{el}(B) . D \,\varepsilon\, \mathrm{el}(C) \therefore \equiv . A \,\varepsilon\, \mathrm{Kl}(a).$

MD3. $[A\, a] :: [\exists b] \therefore A \,\varepsilon\, \mathrm{Kl}(b) \therefore [C] : C \,\varepsilon\, b . \supset . C \,\varepsilon\, a \therefore \equiv . A \,\varepsilon\, \mathrm{cl}(a).$

The rules of extensionality obtaining in Mereology can now be outlined as follows:

REI. On the assumption that a thesis T is, at a given stage, the last thesis of a system of Mereology, the rule of propositional extensionality allows us to add to the system a new thesis of the form

(3) $[\varphi\, \psi] \therefore [\cdots] : \varphi(\cdots) . \equiv . \psi(\cdots) : \equiv : [\Phi] : \Phi(\varphi) . \equiv . \Phi(\psi)$

provided every variable occurring in this new thesis belongs to a semantic category already available in the system, and provided the following further conditions are fulfilled: the two expressions represented in (3) by '$\varphi(\cdots)$' and '$\psi(\cdots)$' respectively are simple propositional functions or many-link propositional functions; they are equiform with each other except for the variable functors 'φ' and 'ψ'; their arguments are all variables, none of which occurs in either function more than once.

REII. On the assumption that a thesis T is, at a given stage, the last thesis of a system of Mereology, the rule of nominal extensionality allows us to add to the system a new thesis of the form:

$$(4)\quad [\varphi\ \psi]\therefore[A\ \cdots]:A\ \varepsilon\ \varphi(\cdots).\equiv.A\ \varepsilon\ \psi(\cdots):\equiv:[\Phi]:\Phi(\varphi).\equiv.\Phi(\psi)$$

provided every variable occurring in the new thesis belongs to a semantic category already available in the system, and provided the following further conditions are fulfilled: the two expressions represented in (4) by '$\varphi(\cdots)$' and '$\psi(\cdots)$' respectively are either (i) nominal variables different from 'A' and from each other or (ii) simple nominal functions or many-link nominal functions, which are equiform with each other except for the variable functors 'φ' and 'ψ'; the arguments of the two functions are all variables, none of which is equiform with 'A' or occurs in either function more than once.

The following three formulae will suffice as examples of theses which could be added to a system of Mereology in virtue of the rule of nominal extensionality:

ME1. $[b\ c]\therefore[A]:A\ \varepsilon\ b.\equiv.A\ \varepsilon\ c:\equiv:[\Phi]:\Phi(b).\equiv.\Phi(c).$
ME2. $[\varphi\ \psi]\therefore[A\ b]:A\ \varepsilon\ \varphi(b).\equiv.A\ \varepsilon\ \psi(c):\equiv:[\Phi]:\Phi(\varphi).\equiv.\Phi(\psi).$
ME3. $[\theta\ \xi]\therefore[A\ b\ c]:A\ \varepsilon\ \theta\langle b\rangle(c).\equiv.A\ \varepsilon\ \xi\langle b\rangle(c):\equiv:[\Phi]:\Phi(\theta).\equiv.\Phi(\xi).$

In outlining the rules of definition and the rules of extensionality no mention was made of the conventions which enable us to identify and differentiate semantic categories. The availability of such conventions was tacitly assumed.

The directives of Ontology differ from those of Mereology in that the notion of meaningfulness which is implicit in the former is related to the ontological vocabulary only, i.e. to the vocabulary which originates from the ontological primitive constants. In the case of Protothetic the notion of meaningfulness implicit in the directives is related to the protothetical vocabulary only. The rules of nominal definition and nominal extensionality, which presuppose ontological vocabulary, have no application in Protothetic.

Consistency of Ontology. The proof that Ontology is consistent if Protothetic is consistent is due to a member of Leśniewski's seminar before the war.[3] It consists in showing that if we interpret ontological

[3] See Słupecki [1955, p. 66] (*Editorial Note*: Also in the present volume, p. 118).

variables as protothetical variables and if, at the same time, we interpret 'ε' as the functor of conjunction then OA1, which is the ontological axiom, becomes a thesis of Protothetic. Under the same interpretation the ontological rules of inference and the rules of propositional definition and propositional extensionality become directives of protothetic while the ontological RDII and REII turn out to be derivatives of the protothetical RDI and REI respectively. Since the proof has never been published, I propose to reconstruct it in outline in what follows.

To begin with let us note that Protothetic implies the classical logic of propositions (or calculus of propositions) and that the familiar rules of inference operative in that logic, including the rules for the use of quantifiers in connection with propositional variables, can easily be shown to be valid in Protothetic. Thus in Protothetic we have:

PT1. $[p\ q]:p.q. \supset .p.p.p.q,$

from which we derive

PT2. $[p\ q]:p.q. \supset .[\exists r].p.r.r.q.$

We also have:

PT3. $[p\ q\ r]:p.r.r.q. \supset .p.q,$

which implies

PT4. $[p\ q]:[\exists r].p.r.r.q. \supset .p.q.$

Now, PT2 and PT4 yield:

PT5. $[p\ q]:p.q. \equiv .[\exists r].p.r.r.q.$

It is easy to see that PT5 is the protothetical analogue of OA1 under the suggested interpretation of 'ε'. It remains to show that the protothetical analogue of any ontological thesis introduced into a system of Ontology in virtue of RDII or REII is also a thesis of Protothetic.

Consider an expression which could be added to a system of Ontology in virtue of RDII. Its protothetical analogue can be represented by

(5) $[p\ \cdots]:\alpha(p.\beta). \equiv p.\gamma(\cdots),$

and it can be proved to be a thesis of Protothetic by the following procedure.

First, by applying RDI within the framework of Protothetic we get a thesis represented by:

(6) $[\cdots]:\alpha(1.\beta). \equiv .\gamma(\cdots).,$

in which the expression represented by '$\gamma(\cdots)$' is equiform with the expression represented by '$\gamma(\cdots)$' in (5), and in which the expression represented by '$\alpha(1.\beta)$' is the result of substituting '1', i.e. a true proposition, for 'p' in the expression represented by '$\alpha(p.\beta)$' in (5). In view of

T6. $[p]:p \equiv .1.p,$

(6) implies:

(7) $[\cdots]{:}\alpha(1.\beta).\equiv.1.\gamma(\cdots).$

Secondly, we note that a protothetical expression which can be represented by:

(8) $[\cdots]{:}\alpha(0.\beta).\equiv.0.\gamma(\cdots),$

and which is the result of substituting '0', i.e. a false proposition, for 'p' in the expression represented by (5), is a thesis of Protothetic. This is obvious if the expression for which '$\alpha(0.\beta)$' stands is of the form '$0.\beta$'. Equally obvious is the case where the expression represented by '$\alpha(0.\beta)$' is a conjunction, whether preceded by a particular quantifier or not, of which the expression represented by '$0.\beta$' is a conjunct.

Now, by applying RDI again we obtain a protothetical thesis of the form:

(9) $[p\cdots]{\therefore}\alpha(p.\beta).\equiv.p.\gamma(\cdots){:}\equiv.+\langle\cdots\rangle(p),$

in which the left-hand side argument of the main equivalence is equiform with the expression represented by (5).

(9) in connection with (8) and (7) yields:

(10) $[\cdots].+\langle\cdots\rangle(0)$

and

(11) $[\cdots].+\langle\cdots\rangle(1).$

Since in Protothetic we have

PT7. $[f]{:}f(0).f(1).\supset.[p].f(p),$

we use it together with (10) and (11) to obtain

(12) $[p\cdots].+\langle\cdots\rangle(p),$

which, in connection with (9), implies (5). This result shows that any thesis introduced into Ontology by applying RDII has, under the proposed interpretation of 'ε', a protothetical analogue which is derivable within the framework of Protothetic from a thesis introduced into the system by means of RDI.

The protothetical analogue of an expression which could be added to a system of Ontology in virtue of REII, can be represented by:

(13) $[\varphi\,\psi]{\therefore}[p\cdots]{:}p.\varphi(\cdots).\equiv.p.\psi(\cdots){:}\equiv{:}[\Phi]{:}\Phi(\varphi).\equiv.\Phi(\psi)$

with '$\varphi(\cdots)$' and '$\psi(\cdots)$' standing for propositional variables or for propositional functions. Considering that in Protothetic we have:

PT8. $[p\,q]{\therefore}p\equiv q.\equiv{:}[r]{:}r.p.\equiv.r.q,$

it can easily be shown that the expression represented by (13) is a thesis. If '$\varphi(\cdots)$' and '$\psi(\cdots)$' in (13) stand for propositional variables then the expression represented by (13) can be derived from PT8 and the law of extensionality for propositions, i.e. from:

PT9. $[p\ q]\therefore p \equiv q. \equiv :[f]:f(p). \equiv .f(q),$

which is derivable from PA1. If '$\varphi(\cdots)$' and '$\psi(\cdots)$' in (13) stand for propositional functions then by applying REI, as restricted to Protothetic, we can introduce into Protothetic a thesis represented by:

(14) $[f\,g]\therefore[\cdots]:f(\cdots). \equiv .g(\cdots): \equiv :[\Phi]:\Phi(f). \equiv .\Phi(g)$

and fulfilling the following condition: the propositional functions represented by '$f(\cdots)$' and '$g(\cdots)$' in (14) are equiform to the propositional functions represented by '$\varphi(\cdots)$' and '$\psi(\cdots)$' in (13). Now, the thesis represented by (14) and PT8 yield the protothetical analogue represented by (13). This completes the outline of the proof that Ontology has an interpretation within the framework of Protothetic, which means that Ontology is consistent if Protothetic is consistent.

Consistency of Mereology. Now, we go on to show that Mereology has an interpretation in Protothetic. In order to convince ourselves of this we interpret nominal variables of Mereology as propositional variables, 'ε' — as the functor of conjunction, and 'el' — as the functor of assertion, which in Protothetic is defined as follows:

PD1. $[p]:p \equiv .\mathrm{as}(p).$

Since our protothetical interpretation of Mereology presupposes the protothetical interpretation of Ontology as discussed above, all that remains to be done is to prove that the protothetical analogue of MA1 is a thesis of Protothetic. For we have shown already that if we interpret 'ε' as the functor of conjunction then OA1 becomes a thesis of Protothetic, and it is obvious from what was said before that under the suggested interpretation of Mereology the mereological rules of inference and the mereological rules of propositional definition and propositional extensionality become directives of Protothetic whereas the mereological RDII and REII become derivatives of the protothetical RDI and REI respectively.

It is not difficult to see that within the framework of Protothetic the following theses hold:

PT10. $[p\ q]:p.\mathrm{as}(q). \supset q.$
PT11. $[u\ s]:u.s. \supset .u.\mathrm{as}(s).$
PT12. $[u\ s]:u.\mathrm{as}(s). \supset .s.s.u.\mathrm{as}(u).u.\mathrm{as}(s).$
PT13. $[u\ s]:u.\mathrm{as}(s). \supset .[\exists v\ w].v.s.w.\mathrm{as}(u).w.\mathrm{as}(v).$ (PT12)

PT14. $[p\,q\,r\,s]:\!:\!:p:\cdot:[t]::t.r. \equiv \therefore [u]:u.s. \supset .u.\mathrm{as}(t) \therefore [u]:u.\mathrm{as}(t).$
$\supset .[\exists v\,w].v.s.w.\mathrm{as}(u).w.\mathrm{as}(v):\cdot: \supset .p.\mathrm{as}(r).$

Proof.

$[p\,q\,r\,s]:\!:\!:$

(1) $p:\cdot:$

(2) $[t]::t.r. \equiv \therefore [u]:u.s. \supset .u.\mathrm{as}(t) \therefore [u]:u.\mathrm{as}(t). \supset$
$[\exists v\,w].v.s.w.\mathrm{as}(u).w.\mathrm{as}(v):\cdot: \supset .$

(3) $r.$ (2, PT11, PT13)

$p.\mathrm{as}(r).$ (1, PD1, 3)

PT15. $[t\,u\,q]:t.u.q. \supset .u.\mathrm{as}(t).$

PT16. $[q\,t] \therefore t \supset :[u]:u.q. \supset .u.\mathrm{as}(t).$ (PT15)

PT17. $[q\,t\,u]:q.u.\mathrm{as}(t). \supset .u.q.u.\mathrm{as}(u).u.\mathrm{as}(u).$

PT18. $[q\,t\,u]:q.u.\mathrm{as}(t). \supset .[\exists v\,w].v.q.w.\mathrm{as}(u).w.\mathrm{as}(v).$ (PT17)

PT19. $[q\,t] \therefore q \supset :[u]:u.\mathrm{as}(t). \supset .[\exists v\,w].v.q.w.\mathrm{as}(u).w.\mathrm{as}(v).$ (PT18)

PT20. $[q\,t]::q \therefore [u]:u.q. \supset .u.\mathrm{as}(t) \therefore \supset .t.q.$

PT21. $[p\,q]:\!:\!:q.q:\cdot\!:[r\,s]:\!:\!:[t]::t.r. \equiv \therefore [u]:u.s. \supset .u.\mathrm{as}(t) \therefore [u]:u.\mathrm{as}(t)$
$. \supset .[\exists v\,w].v.s.w.\mathrm{as}(u).w.\mathrm{as}(v):\cdot:q.\mathrm{as}(q).q.s:\cdot: \supset .p.\mathrm{as}(r):\cdot\!:$
$\supset .p.\mathrm{as}(q).$

Proof.

$[p\,q]:\!:\!:$

(1) $q.q:\cdot\!:$

(2) $[r\,s]:\!:\!:[t]::t.r. \equiv \therefore [u]:u.s. \supset .u.\mathrm{as}(t) \therefore [u]:u.\mathrm{as}(t).$
$\supset .[\exists v\,w].v.s.w.\mathrm{as}(u).w.\mathrm{as}(v):\cdot:q.\mathrm{as}(q).q.s:\cdot:$
$\supset .p.\mathrm{as}(r):\cdot\!: \supset :\cdot:$

(3) $[t]::t.q. \equiv \therefore [u]:u.q. \supset .u.\mathrm{as}(t) \therefore [u]:u.\mathrm{as}(t). \supset .$
$[\exists v\,w].v.q.w.\mathrm{as}(u).w.\mathrm{as}(v):\cdot:$ (PT16, PT19, PT20, 1)

(4) $\mathrm{as}(q).$ (PT11, 1)

$p.\mathrm{as}(q).$ (2, 3, 1, 4, 1)

PT22. $[p\,q]:\!:\!:p.\mathrm{as}(q). \equiv :\cdot\!:q.q:\cdot\!:[r\,s]:\!:\!:[t]::t.r. \equiv \therefore [u]:u.s.$
$\supset .u.\mathrm{as}(t) \therefore [u]:u.\mathrm{as}(t). \supset .[\exists v\,w].v.s.w.\mathrm{as}(u).w.\mathrm{as}(v):\cdot:q.$
$\mathrm{as}(q).q.s:\cdot: \supset .p.\mathrm{as}(r).$ (PT10, PT14, PT21)

PT22 being, under the suggested interpretation, the protothetical analogue of MA1, our proof that Mereology is consistent if Protothetic is consistent has been completed. It complies with the methodological principle mentioned in the beginning of the present paper since Protothetic is, in fact, a weaker subsystem of Mereology.

Robert E. Clay

The Dependence of a Mereological Axiom

In this note we show that in the standard axiom system for mereology which follows, the reflexive axiom, *M2*, is dependent on *M3*, *DM*, *M4* and *M5*.

M1. $[AB]:A \varepsilon \mathbf{el}(B).\supset.B \varepsilon B.$

M2. $[A]:A \varepsilon A.\supset.A \varepsilon \mathbf{el}(A).$

M3. $[ABC]:A \varepsilon \mathbf{el}(B).B \varepsilon \mathbf{el}(C).\supset.A \varepsilon \mathbf{el}(C).$

DM. $[Aa]\therefore A \varepsilon \mathbf{Kl}(a).\equiv:A \varepsilon A:[D]:D \varepsilon a.\supset.D \varepsilon \mathbf{el}(A):[D]:$
$D \varepsilon \mathbf{el}(A).\supset.[\exists EF].E \varepsilon a.F \varepsilon \mathbf{el}(D).F \varepsilon \mathbf{el}(E).$

M4. $[Aa]:A \varepsilon a.\supset.[\exists B].B \varepsilon \mathbf{Kl}(a).$

M5. $[ABa]:A \varepsilon \mathbf{Kl}(a).B \varepsilon \mathbf{Kl}(a).\supset.A \varepsilon B.$

P1. $[AD]:.D \varepsilon \mathbf{el}(A).\supset.[\exists F].F \varepsilon \mathbf{el}(D)$

PF $[AD]:\mathrm{Hyp}(2).\supset.A \varepsilon A$ [*M1*, 1]
    $[\exists B].$
    3) $B \varepsilon \mathbf{Kl}(A).$ [*M4*, 1]
    4) $A \varepsilon \mathbf{el}(B).$ [*DM*, 3, 1]
    5) $D \varepsilon \mathbf{el}(B).$ [*M3*, 2, 4]
    $[\exists F].F \varepsilon \mathbf{el}(D)$ [*DM*, 3, 5]

P2. $[A]:A \varepsilon A.\supset.A \varepsilon \mathbf{Kl}(\mathbf{el}(A))$ [*DM*, $a/\mathbf{el}(A)$, E/D, *P1*]

P3. $[ABD]:A \varepsilon A.B \varepsilon \mathbf{Kl}(A).D \varepsilon \mathbf{el}(B).\supset.[\exists EF].E \varepsilon \mathbf{el}(A).$
$F \varepsilon \mathbf{el}(D).F \varepsilon \mathbf{el}(E)$

PF $[ABD]:\mathrm{Hyp}(3).\supset.$
    $[\exists EF].$
    5) $E \varepsilon A.$
    6) $F \varepsilon \mathbf{el}(D).$ [*DM*, 2, 3] (5–7)
    7) $F \varepsilon \mathbf{el}(E).$
    8) $E = A.$ [5, 1]
    9) $F \varepsilon \mathbf{el}(A).$ [7, 8]
    10) $F \varepsilon \mathbf{el}(B).$ [*M3*, 6, 3]
    $[\exists G].$
    11) $G \varepsilon \mathbf{el}(F).$ [*DM*, 2, 10]

12) $G \varepsilon \mathbf{el}(D)$. [$M3$, 11, 6]

$[\exists EF].E \varepsilon \mathbf{el}(A).F \varepsilon \mathbf{el}(D).F \varepsilon \mathbf{el}(E)$ [9, 12, 11]

P4. $[A]:A \varepsilon A. \supset .A \varepsilon \mathbf{el}(A)$

PF $[A]\therefore A \varepsilon A. \supset :$

$[\exists B]:$

2) $B \varepsilon \mathbf{Kl}(A)$. [$M4$, 1]

3) $A \varepsilon \mathbf{el}(B)$: [$DM$, 2, 1]

4) $[D]:D \varepsilon \mathbf{el}(A). \supset .D \varepsilon \mathbf{el}(B)$: [$M3$, 3]

5) $[D]:D \varepsilon \mathbf{el}(B). \supset .[\exists EF].E \varepsilon \mathbf{el}(A).F \varepsilon \mathbf{el}(D)$.
$F \varepsilon \mathbf{el}(E)$: [$P3$, 1, 2]

6) $B \varepsilon \mathbf{Kl}(\mathbf{el}(A))$. [$DM$, 2, 4, 5]

7) $A \varepsilon \mathbf{Kl}(\mathbf{el}(A))$. [$P2$, 1]

8) $A = B$: [$M5$, 6, 7]

$A \varepsilon \mathbf{el}(A)$ [3, 8]

Robert E. Clay

Relation of Leśniewski's Mereology to Boolean Algebra

It has been stated in Tarski [1956] and 'proved' in Grzegorczyk [1955] that:

(A) The models of mereology and the models of complete Boolean algebra with zero deleted[1] are identical.

Proved has been put in quotes, not because Grzegorczyk's proof is faulty but because the system he describes as mereology is in fact not Leśniewski's mereology.

Leśniewski's first attempt at describing the collective class, i.e. mereology, was done in ordinary language with no rigorous logical foundation. In describing the collective class, he needed to use the notion of distributive class. So as to clearly distinguish and expose the interplay between the two notions of class, he introduced his calculus of names (name being the distributive notion), which is also called ontology, since he used the primitive term 'is'. At this stage, mereology included ontology. Then, in order to have a logically rigorous system, he developed as a basis, a propositional calculus with quantifiers and semantical categories (types), called protothetic. At this final stage, what is properly called mereology includes both protothetic and ontology.

What Grzegorczyk describes as mereology is even weaker than Leśniewski's initial version. To quote from Grzegorczyk [1955]:

> In order to emphasize these formal relations let us consider the systems of axioms of mereology for another of its primitive terms, namely for the term '*ingr*' defined as follows:
>
> A *ingr* $B. \equiv .A$ is a part $B \vee A$ is identical to B.
>
> The proposition 'A *ingr* B' can be read 'A is contained in B' or after Leśniewski, 'A is ingredient of B'.

[1] Deleting zero from a Boolean algebra results in a system without a zero except in the case when the Boolean algebra has exactly two elements.

Notice that by replacing Leśniewski's '*A* is an ingredient of *B*' by '*A ingr B*', Grzegorczyk has eliminated the 'is'. In doing so, he has eliminated the notion of distributive class and consequently the interplay between the two notions of class — aspects of mereology which Leśniewski thought were so important that he constructed his ontology in order to describe them clearly. This elimination of 'is' is the most important deficiency in Grzegorczyk's system. Next he gives the system the weakest possible logical base by making it an elementary theory. It is this weak system that Grzegorczyk proves shares the same models with complete Boolean algebra with zero deleted.

Słupecki [1958] also suffers under the misapprehension that mereology is formally nothing more than a particular elementary theory of partial ordering. He remedies the weak logic by replacing the first-order logic by Chwistek's simple theory of types, but he fails to include ontology or any other theory of the distributive class to use as counterpoint for the notion of collective class. Słupecki unwittingly demonstrates the need for this counterpoint by saying:

> ... a consequence which disagrees with intuition is the nonexistence of finite sets whose number of elements differs from each of the numbers 2^2-1, 2^3-1, 2^4-1, ...

Since cardinality is primarily a distributive notion, one's intuition should not be violated if the collective class cannot describe it.

Since Leśniewski's mereology includes protothetic and ontology and Boolean algebra is usually given some other logical base, statement (A) needs to be put into a precise context. There is also need for a formal definition of complete Boolean algebra with zero deleted. Since statement (A) is in some sense not completely true, we break it up into the following two statements:

(1) Mereology is a complete Boolean algebra with zero deleted.

(2) Complete Boolean algebra with zero deleted is a mereology.

We shall present two alternative ways of introducing partial ordering into Leśniewski's logic and show that in order for (2) to hold we must be unreasonably restrictive in our definition of complete Boolean algebra with zero deleted.

It is commonly required in a definition of Boolean algebra that there exist two distinct elements, namely 0 and 1. Such a definition, together with any reasonable definition of complete Boolean algebra with zero deleted, would require the existence of at least one element, namely 1. Since Leśniewski, due to strong philosophical beliefs, carefully avoids postulating the existence of anything, (1) cannot hold unless we use a definition of Boolean algebra that does not postulate the existence of any nonzero element, but is otherwise equivalent to the usual definition. We use such a definition in what follows.

There seem to be two roads to making (1) precise:

(a) Construct an axiom system, $\mathfrak{A}$ with the following properties:

{$\mathfrak{A}$, zero exists} is an axiom system for complete Boolean algebra and 'zero exists' is independent of $\mathfrak{A}$. Thus $\mathfrak{A}$ makes no claim about the existence or nonexistence of zero. We then call the system {$\mathfrak{A}$, if two elements exist then zero does not exist} a complete Boolean algebra with zero deleted. Then show that mereology satisfies this axiom system.

Using a set-theoretical base such an axiom system is constructed in Clay [1969]. The following single axiom for a complete Boolean algebra with zero deleted is derived:

$$\mathbf{BD'}.\ [AB]\because\therefore A \leqslant B .\equiv \because\because A \in \boldsymbol{U}.B \in \boldsymbol{U}\because\because B \leqslant B .\supset \because\because [ab]\because\because a \subset \boldsymbol{U}.b \subset \boldsymbol{U}.B \in a\therefore [C]\therefore C \in b .\equiv : [D]: D \in a .\supset .D \leqslant C : [D]: D \leqslant C .\supset .[\exists EF].E \in a.F \leqslant D.F \leqslant E :: \supset .[\exists L].b = \{L\}. A \leqslant L,$$ [2]

where $\leqslant$ is a relation on $\boldsymbol{U}$.

(b) Extend the field of mereology to include exactly one more name and extend the ordering so as to yield a complete Boolean algebra.

Before either road can be taken we must decide how to introduce partial ordering into Leśniewski's logic.

(i) The first method is to introduce $\leqslant$ as a name-forming functor of one name argument and to formalize the statement 'A is less than or equal to B' by

$$A\ \varepsilon \leqslant (B)$$

Since every element in the field of the partial ordering must be able to occur to the left of ε, the field must be contained in **V**. Therefore the field is a general name which we shall designate by f. The translation of **BD′** into Leśniewski's logic then is:

$$\mathbf{BDL}.\ [AB]\because\therefore A\ \varepsilon \leqslant (B) .\equiv \because\because A\ \varepsilon\ f.B\ \varepsilon\ f \because\because B\ \varepsilon \leqslant (B) .\supset \because\because [ab] \because\because a \subset f.b \subset f.B\ \varepsilon\ a \therefore [C]\therefore C\ \varepsilon\ b .\equiv : [D]: D\ \varepsilon\ a .\supset .D\ \varepsilon \leqslant (C) : [D]: D\ \varepsilon \leqslant (C) .\supset .[\exists EF].E\ \varepsilon\ a.F\ \varepsilon \leqslant (D).F\ \varepsilon \leqslant (E) :: \supset .A\ \varepsilon \leqslant (b).$$

A single axiom for mereology is:

$$\mathbf{M}.\ [AB]\because\therefore A\ \varepsilon\ \mathbf{el}(B) .\equiv \because\because A\ \varepsilon\ A.B\ \varepsilon\ B \because\because B\ \varepsilon\ \mathbf{el}(B) \supset \because\because [ab] :: B\ \varepsilon\ a \therefore [C]\therefore C\ \varepsilon\ b .\equiv : [D]: D\ \varepsilon\ a .\supset .D\ \varepsilon\ \mathbf{el}(C) : [D]: D\ \varepsilon\ \mathbf{el}(C) .\supset .[\exists EF].E\ \varepsilon\ a.F\ \varepsilon\ \mathbf{el}(D).F\ \varepsilon\ \mathbf{el}(E) :: \supset .A\ \varepsilon\ \mathbf{el}(b).$$

Remembering that $[A]: A\ \varepsilon\ A .\equiv .A\ \varepsilon\ \mathbf{V}$, if one replaces f by **V** and $\leqslant$ by **el** in **BDL** one arrives at **M**. If one considers $\leqslant$ as **el** restricted to f. one gets **BDL** from **M**. Therefore using road (a) we have shown that complete Boolean algebra with zero deleted and mereology have the same models.

Road (b) fails in this method since any extension of **el** to $\leqslant$ would require that we add another individual name to the field of **el**, i.e. to **V**. But **V** is the totality of individual names, so there is no other individual name to add.

---

[2] This axiom is patterned after a single axiom of mereology due to Sobociński and Lejewski.

Remark. Słupecki, in his above-mentioned system, does add a **0** or empty object. This empty object's existence is guaranteed by the addition of the axiom $[\exists X](X = \mathbf{0})$. However, in Leśniewski's mereology, the particular quantifier has no existential import. Here the existence of the name a requires a thesis of the form $[\exists A].A \varepsilon a$. Using Leśniewski's notation of $\mathbf{\Lambda}$ for the empty name, the ontological thesis $[A].\sim(A \varepsilon \mathbf{\Lambda})$ precludes the existence of $\mathbf{\Lambda}$, so it cannot be added to the individual (i.e. unique and existing) names to yield a complete Boolean algebra.

However this method is too restrictive. It is so restrictive that it excludes names under $\subset$ (the containment relation) from being a partially ordered system, much less a complete Boolean algebra. Even if one excludes the difficulty of the empty name, we still have that the nonempty names under $\sqsubset$ (containment for nonempty names) fail to be a partially ordered system. To show this let us suppose we are able to define ε^* and $\leqslant^*$, analogs for ε and $\leqslant$ so that the following statement holds:

(*) $$[ab]:a \sqsubset b. \equiv .a \varepsilon^* \leqslant^*(b).$$

For ε^*, the whole category of names (except for the empty name) must play the role of individual names; so whatever are to play the roles of nonindividual general names must be of a different semantical category. Since all ε^*-names, both individual and general, must be able to fill the second argument place of ε^*, we would have to have two semantical categories being able to fill the same argument place. But this is prohibited in Leśniewski's logic. Therefore, (*) cannot hold. Therefore $\sqsubset$ cannot be considered a partial ordering.

Therefore method (i) must be discarded and with it our proof that mereology and complete Boolean algebra with zero deleted are equivalent.

(ii) The second method is to introduce $\leqslant$ as a predicate forming functor of two arguments, both from the same semantical category, $\leqslant(AB)$.[3] We need to be able to specify the field of the partial ordering. But Leśniewski's logic is not a set theory. It does not have a general collection-forming mechanism. It only has a collection-forming mechanism if the elements we are collecting can be considered as individuals relative to some analog of ε. So the simplest way it has of referring to a collection of elements from a given category is to use the predicate forming functor of one argument which is true precisely of these elements, i.e. the characteristic functor. Thus, the field will be specified by a predicate $\boldsymbol{U}$. In this situation the translation of **BD′** becomes

BD. $[AB]\because\!\cdot\, A \leqslant B. \equiv \mathbin{::}\boldsymbol{U}(A).\boldsymbol{U}(B)\mathbin{::}B \leqslant B. \supset \mathbin{\because\!\cdot}[\alpha, \beta]\mathbin{::}$
$\alpha \subset \boldsymbol{U}.\beta \subset \boldsymbol{U}.\alpha(B)\therefore[C]\therefore\beta(C). \equiv [D]:\alpha(D). \supset .D \leqslant C:[D]:$
$D \leqslant C. \supset .[\exists EF].\alpha(E).F \leqslant D.F \leqslant E\mathbin{::} \supset [\exists L].\beta = \circ((L))^4.A \leqslant L.$[5]

[3] For ease of reading we shall write $A < B$ for $< (AB)$.

[4] $[ab]:\circ((a))\{b\}. \equiv .a \circ b.$

[5] The parentheses in this axiom are meant to be ambiguous so that in reality we have an axiom for each semantical category.

Now using either (a) or (b) we can prove mereology is a complete Boolean algebra with zero deleted.

For (a) let $A\ \varepsilon\ \mathbf{el}(B)$ be $A \leqslant B$. We need the following definition:

$$[Aa]:\varepsilon\,((a))(A).\equiv.A\ \varepsilon\ a.$$

Let $\varepsilon((\mathbf{V}))$ be $\boldsymbol{U}$. Then $\alpha \subset \boldsymbol{U}$ requires that α equal $\varepsilon((a))$ for some name a. Thus $A\ \varepsilon\ a$ transforms to $\alpha(A)$. Also $A\ \varepsilon\ A$ transforms to $\boldsymbol{U}(A)$. So axiom **M** is easily seen to be an instance of **BD**.

For (b) we shall only give the necessary definitions and interpretations. The proof though straightforward is somewhat tedious.

Let $\boldsymbol{U}$ be $\rightharpoonup$ and $\leqslant$ be φ as they are defined below.

$$[a]\therefore\rightharpoonup\{a\}.\equiv:[BC]:B\ \varepsilon\ a.C\ \varepsilon\ a.\supset.B=C.$$
$$[AB]:\varphi\{AB\}.\equiv.\rightharpoonup\{A\}.\rightharpoonup\{B\}.A\subset\mathbf{el}(B).$$

Returning to (a), one can easily see that nonempty names form a complete Boolean algebra with zero deleted. Let $\boldsymbol{U}$ be ! and $\leqslant$ be $\sqsubset$ as they are defined below.

$$[a]:!\{a\}.\equiv[\exists A].A\ \varepsilon\ a,$$
$$[ab]:a\sqsubset b.\equiv.!\{a\}.a\subset b.$$

However, the nonempty names are not a model for mereology since this would require definitions of ε^* and $\mathbf{el}^*$ such that

(**) $$[ab]:a\sqsubset b.\equiv.a\ \varepsilon^*\ \mathbf{el}^*(b)$$

holds.

The same argument that showed (*) is impossible, also applies here. Therefore (2) is false.

However, there is a sense in which (2) is true; we can construct a model of mereology which reflects the structure of any given complete Boolean algebra with zero deleted. In this model the role of a given A such that $\boldsymbol{U}(A)$ is played by the characteristic functor of A, so that the universe of the model is not in the semantical category of A, but in the category of predicate forming functors with arguments in the category of A. We now give this construction.

For the analogs in the category of names for the definitions involved see Clay [1965]. We shall use ( ) to enclose arguments to the same semantical category as A, and $\langle\ \rangle$ to enclose arguments belonging to the same category as $\boldsymbol{U}$. For ease of reading we shall employ the following conventions: $\Sigma\subset\sigma$ for $\subset\langle\Sigma\sigma\rangle$, $B\circ C$ for $\circ(BC)$, $\Phi\circ\Sigma$ for $\circ\langle\Phi\Sigma\rangle$ ($\circ$ stands for equality, i.e. mutual containment.), $\Sigma\eta\sigma$ for $\eta\langle\Sigma\sigma\rangle$, η will be the analog of ε. The following three definitions give the analogs of general names, individuals, and ε respectively.

DN1. $[\sigma]:\mathscr{N}\langle\sigma\rangle.\equiv.\sigma\subset\boldsymbol{U}.$

DN2. $[\Sigma]:.\mathscr{P}\langle\Sigma\rangle.\equiv:[\exists A].\Sigma(A).\mathscr{N}\langle\Sigma\rangle:[AB]:\Sigma(B).\supset.A\circ B.$

DN3. $[\Sigma\sigma]:\Sigma\eta\sigma.\equiv.\mathscr{P}\langle\Sigma\rangle.\mathscr{N}\langle\sigma\rangle.\Sigma\subset\sigma.$

N1. $[\Sigma]:\Sigma\eta\Sigma.\equiv.\mathscr{P}\langle\Sigma\rangle.$ [DN3, DN2]

N2. $[\Sigma\sigma]:\Sigma\eta\sigma.\supset.\Sigma\eta\Sigma.$ [DN3, DN2]

N3. $[A\Sigma\sigma]:\mathscr{P}\langle\Sigma\rangle.\Sigma(A).\sigma(A).\supset.\Sigma\subset\sigma$

Hyp (3). $\supset$:

(4) $[BC]:\Sigma(B).\Sigma(C).\supset.B\circ C:$ [DN2, 1]

(5) $[B]:\Sigma(B).\supset.\sigma(B):$ [4, 2, 3]

$\Sigma\subset\sigma.$ [5]

N4. $[\Sigma\Phi\sigma]:\Sigma\eta\sigma.\Phi\eta\Sigma.\supset.\Phi\circ\Sigma.$

Hyp (2). $\supset$.

(3) $\mathscr{P}\langle\Sigma\rangle.$

(4) $\Phi\subset\Sigma.$ [DN3, 1]

(5) $\mathscr{P}\langle\Phi\rangle.$ [DN3, 2]

$[\exists A].$

(6) $\Phi(A).$ [DN2, 5]

(7) $\Sigma(A).$ [4, 6]

(8) $\Sigma\subset\Phi.$ [N3, 3, 7, 6]

$\Phi\circ\Sigma.$ [4, 8]

N5. $[\Sigma\Phi\sigma]:\Sigma\eta\sigma.\Phi\eta\Sigma.\supset.\Phi\eta\sigma.$ [N4]

N6. $[\Sigma\Phi\Psi\sigma]:\Sigma\eta\sigma.\Phi\eta\Sigma.\Psi\eta\Sigma.\supset.\Phi\eta\Psi.$ [N4, Φ/Ψ]

DN4. $[AB]:\chi((A))(B).\equiv.\boldsymbol{U}(A).A\circ B.$

N7. $[A:\chi((A))(A).\equiv.\boldsymbol{U}(A).$ [DN4].

Thus $\chi((A))$ is the characteristic function for an A to which $\boldsymbol{U}$ applies.

N8. $[A].\mathscr{N}\langle\chi((A))\rangle.$ [DN1, DN4]

N9. $[A]:\boldsymbol{U}(A).\equiv.\mathscr{P}\langle\chi((A))\rangle$ [DN2, N7, N8, DN4, DN1]

N10. $[\sigma A]:\mathscr{N}\langle\sigma\rangle.\sigma(A).\supset.\chi((A))\eta\sigma.$

Hyp (2). $\supset$.

(3) $\boldsymbol{U}(A).$ [DN1, 1, 2]

(4) $\mathscr{P}\langle\chi((A))\rangle.$ [N9, 3]

(5) $\chi((A))\subset\sigma.$ [DN4, 2]

$\chi((A))\eta\sigma.$ [DN3, 4, 1, 5]

N11. $[\sigma A]:\chi((A))\eta\sigma.\supset.\sigma(A).$

Hyp (1). $\supset$.

(2) $\mathscr{P}\langle\chi((A))\rangle.$

(3) $\chi((A))\subset\sigma.$ [DN3, 1]

$\chi((A))(A).$ [N9, 2, N7]

N12. $[\sigma A]:\mathscr{N}\langle\sigma\rangle.\sigma(A).\equiv.\chi((A))\eta\sigma.$ [N10, N11, DN3]

N13. $[A]:\chi((A))(A).\equiv.\chi((A))\eta\chi((A)).$ [N12, N8]

N14. $[\Sigma AB]::\mathscr{N}\langle\Sigma\rangle:[\Phi\Psi]:\Phi\eta\Sigma.\Psi\eta\Sigma.\supset.\Phi\eta\Psi:\Sigma(A).\Sigma(B):\supset.A\circ B.$

Hyp (4). $\supset$.

(5) $\chi((A))\eta\Sigma.$ [N12, 1, 3]

(6) $\chi((B))\eta\Sigma.$ [N12, 1, 4]

(7) $\chi((A))\eta\chi((B))$ [2, 5, 6]

(8) $\chi((B))(A)$. [N12, 7]

$A \circ B$. [DN4, 8]

N15. $[\theta\Sigma\sigma] \therefore \theta\eta\Sigma : [\Phi] : \Phi\eta\Sigma . \supset . \Phi\eta\sigma : [\Phi\Psi] : \Phi\eta\Sigma . \Psi\eta\Sigma . \supset .$
$\Phi\eta\Psi : \supset . \Sigma\eta\sigma$.

Hyp (3) . ⊃ :

(4) $\mathscr{P}\langle\theta\rangle$.
(5) $\mathscr{N}\langle\Sigma\rangle$ [DN3, 1]
(6) $\theta \subset \Sigma$:

(7) $[AB] : \Sigma(A) . \Sigma(B) . \supset . A \circ B$: [N14, 5, 3]

$[\exists C]$.

(8) $\theta(C)$. [DN2, 4]

(9) $\Sigma(C)$. [8, 6]

(10) $\mathscr{P}\langle\Sigma\rangle$. [DN2, 9, 5, 7]

(11) $\Sigma\eta\Sigma$. [N1, 6]

$\Sigma\eta\sigma$ [2, 11]

N16. $[\Sigma\sigma] :. \Sigma\eta\sigma . \equiv : [\exists\theta] . \theta\eta\Sigma : [\Phi] : \Phi\eta\Sigma . \supset .$
$\Phi\eta\sigma : [\Phi\Psi] : \Phi\eta\Sigma . \Psi\eta\Sigma . \supset . \Phi\eta\Psi$. [N2, N5, N6: N15]

This is an analog for a single axiom of ontology. In order to complete the ontological portion of the model we need to prove a thesis corresponding to the law of extensionality and a metatheorem corresponding to the rule of definition.

N17. $[\sigma\tau\Sigma] : \sigma \subset \tau . \mathscr{N}\langle\tau\rangle . \Sigma\eta\sigma . \supset . \Sigma\eta\tau$.

Hyp (3) . ⊃ .

(4) $\mathscr{P}\langle\Sigma\rangle$.
(5) $\Sigma \subset \sigma$. [DN3, 3]

(6) $\Sigma \subset \tau$. [1, 5]

$\Sigma\eta\tau$. [DN3, 4, 2, 6]

N18. $[\sigma\tau] :: \mathscr{N}\langle\tau\rangle . \supset \therefore \sigma \subset \tau . \supset : [\Sigma] : \Sigma\eta\sigma . \supset . \Sigma\eta\tau$. [N17]

N19. $[A\sigma\tau] \therefore \mathscr{N}\langle\sigma\rangle : [\Sigma] : \Sigma\eta\sigma . \supset . \Sigma\eta\tau : \sigma(A) : \supset . \tau(A)$.

Hyp (3) . ⊃ .

(4) $\chi((A))\eta\sigma$. [N12, 1, 3]

(5) $\chi((A))\eta\tau$. [2, 4]

$\tau(A)$. [N12, 5]

N20. $[\sigma\tau] \therefore \mathscr{N}\langle\sigma\rangle : [\Sigma] : \Sigma\eta\sigma . \supset . \Sigma\eta\tau : \supset . \sigma \subset \tau$. [N19]

N21. $[\sigma\tau] :: \mathscr{N}\langle\sigma\rangle . \mathscr{N}\langle\tau\rangle . \supset \therefore \sigma \subset \tau . \equiv : [\Sigma] : \Sigma\eta\sigma . \supset . \Sigma\eta\tau$. [N18, N20]

N22. $[\sigma\tau] :: \mathscr{N}\langle\sigma\rangle . \mathscr{N}\langle\tau\rangle . \supset \therefore \sigma\circ\tau \equiv : [\Sigma] : \Sigma\eta\sigma . \equiv . \Sigma\eta\tau$ [N21]

In what follows, $a_1 \cdots a_n$ represent variables for previously introduced semantical categories.

DN5. $[a_1 \cdots a_n A\sigma] : \mathscr{K}\langle\langle\sigma\rangle\rangle \lessdot a_1 \cdots a_n A \gtrdot . = . \sigma\{a_1 \cdots a_n\}(A)$.

DN6. $[\alpha\lambda] : \mathscr{L}\{\{\alpha\}\} \lessdot \lambda \gtrdot . = . [\exists\sigma] . \lambda\circ\mathscr{K}\langle\langle\sigma\rangle\rangle . \alpha\langle\langle\sigma\rangle\rangle$.

N23. $[\alpha\sigma] : \mathscr{L}\{\{\alpha\}\} \lessdot \mathscr{K}\langle\langle\sigma\rangle\rangle \gtrdot . \equiv . \alpha\langle\langle\sigma\rangle\rangle$. [DN6]

N24. $[\sigma\tau a_1 \cdots a_n] \therefore \mathscr{N}\langle\sigma\{a_1 \cdots a_n\}\rangle . \mathscr{N}\langle\tau\{a_1 \cdots a_n\}\rangle : [\Sigma]$:

$\Sigma\eta\sigma\{a_1 \cdots a_n\}. \equiv .\Sigma\eta\tau\{a_1 \cdots a_n\}: \supset :[\alpha]:\alpha\langle\langle\sigma\rangle\rangle. \equiv .\alpha\langle\langle\tau\rangle\rangle.$

Hyp (3). $\supset$ ∴

(4) $\sigma\{a_1 \cdots a_n\}\circ\tau\{a_1 \cdots a_n\}.$ [N22, 1, 2, 3]

(5) $\mathscr{K}\langle\langle\sigma\rangle\rangle\circ\mathscr{K}\langle\langle\tau\rangle\rangle:$ [DN5, 4]

(6) $[\alpha]:\mathscr{L}\{\{\alpha\}\}\not<\mathscr{K}\langle\langle\sigma\rangle\rangle\not>. \equiv .\mathscr{L}\{\{\alpha\}\}\not<\mathscr{K}\langle\langle\tau\rangle\rangle\not>:$ [5]

$[\alpha]:\alpha\langle\langle\sigma\rangle\rangle. \equiv \alpha\langle\langle\tau\rangle\rangle.$ [6, N23]

DN7. $[a_1 \cdots a_n\Sigma\sigma]:\mathscr{M}(|a_1 \cdots a_n\Sigma|)\langle\sigma\rangle. \equiv .\Sigma\eta\sigma\{a_1 \cdots a_n\}.$

N25. $[a_1 \cdots a_n\sigma\tau\Sigma]$∴ $[\alpha]:\alpha\langle\sigma\rangle. \equiv .\alpha\langle\tau\rangle: \supset :\Sigma\eta\sigma\{a_1 \cdots a_n\}.$
$\equiv .\Sigma\eta\tau\{a_1 \cdots a_n\}.$ [DN7]

N26. $[a_1 \cdots a_n\sigma\tau]::\mathscr{N}\langle\sigma\{a_1 \cdots a_n\}\rangle.\mathscr{N}\langle\tau\{a_1 \cdots a_n\}\rangle. \supset$ ∴ $[\Sigma]:$
$\Sigma\eta\sigma\{a_1 \cdots a_n\}. \equiv .\Sigma\eta\tau\{a_1 \cdots a_n\}: \equiv :[\alpha]:\alpha\langle\langle\sigma\rangle\rangle. \equiv .\alpha\langle\langle\tau\rangle\rangle$
[N24, N25]

N26 is the law of ontological extensionality for η. Next we prove a metatheorem which corresponds to the ontological rule of definition for η.

N27. $[\Sigma]:\mathscr{P}\langle\Sigma\rangle. \supset .[\exists A].\Sigma(A).\Sigma\circ\chi((A)).\boldsymbol{U}(A).$

Hyp (1). $\supset$.

(2) $\mathscr{N}\langle\Sigma\rangle.$ [DN2, 1]

$[\exists A].$

(3) $\Sigma(A).$
(4) $\Sigma \subset \boldsymbol{U}.$ [DN2, 1]

(5) $\boldsymbol{U}(A).$ [3, 4]

(6) $\chi((A))(A).$ [N7, 5]

(7) $\mathscr{P}\langle\chi((A))\rangle.$ [N9, 4]

(8) $\Sigma \subset \chi((A)).$ [N3, 1, 3, 6]

(9) $\chi((A)) \subset \Sigma.$ [N3, 7, 6, 3]

(10) $\Sigma\circ\chi((A)).$ [8, 9]

$[\exists A].\Sigma(A).\Sigma\circ\chi((A)).\boldsymbol{U}(A).$ [3, 10, 5]

Metatheorem. *An expression of the form*

$[a_1 \cdots a_n]$∴ $\mathscr{N}\langle\sigma\{a_1 \cdots a_n\}\rangle. \supset :[\Sigma]:\Sigma\eta\sigma\{\{a_1 \cdots a_n\}\}. \equiv .\Sigma\eta\Sigma.\sigma^*\{\{a_1 \cdots a_n\}\}\langle\Sigma\rangle$

can be introduced as a thesis of the system, provided that the functor $\sigma\{a_1 \cdots a_n\}$ *has been defined by*

(*) $[a_1 \cdots a_n A]:\sigma\{\{a_1 \cdots a_n\}\}(A). \equiv .\sigma^*\{\{a_1 \cdots a_n\}\}\langle\chi((A))\rangle.$

Proof.

*1. $[a_1 \cdots a_n\Sigma]:\Sigma\eta\sigma\{a_1 \cdots a_n\}. \supset .\sigma^*\{a_1 \cdots a_n\}\langle\Sigma\rangle$

Hyp $\supset$.

(2) $\mathscr{P}\langle\Sigma\rangle.$
(3) $\Sigma \subset \sigma\{a_1 \cdots a_n\}.$ [DN3, 1]

$[\exists A].$

(4) $\Sigma(A)$.
(5) $\Sigma\circ\chi((A))$. [N27, 2]
(6) $\sigma\{a_1 \cdots a_n\}(A)$. [3, 4]
(7) $\sigma^*\{\{a_1 \cdots a_n\}\}\langle\chi((A))\rangle$ [*, 6]
$\sigma^*\{\{a_1 \cdots a_n\}\}\langle\Sigma\rangle$. [5, 7]

*2. $[a_1 \cdots a_n \Sigma]:\mathscr{N}\langle\sigma\{a_1 \cdots a_n\}\rangle.\Sigma\eta\Sigma.\sigma^*\{\{a_1 \cdots a_n\}\}\langle\Sigma\rangle. \supset$
$\supset .\Sigma\eta\sigma\{a_1 \cdots a_n\}$.
Hyp (3). $\supset$.
(4) $\mathscr{P}\langle\Sigma\rangle$. [N1, 2]
$[\exists A]$.
(5) $\Sigma(A)$.
(6) $\Sigma\circ\chi((A))$. [N27, 4]
(7) $\sigma^*\{\{a_1 \cdots a_n\}\}\langle\chi((A))\rangle$. [3, 6]
(8) $\sigma\{a_1 \cdots a_n\}(A)$. [*, 7]
(9) $\Sigma \subset \sigma\{a_1 \cdots a_n\}$. [N3, 4, 5, 8]
$\Sigma\eta\sigma\{a_1 \cdots a_n\}$. [DN3, 4, 1, 9]

*3. $[a_1 \cdots a_n]\therefore\mathscr{N}\langle\sigma\{a_1 \cdots a_n\}\rangle. \supset :[\Sigma]:\Sigma\eta\sigma\{a_1 \cdots a_n\}. \equiv \Sigma\eta\Sigma.$
$\sigma^*\{\{a_1 \cdots a_n\}\}\langle\Sigma\rangle$. [N2, *1: *2]

This completes the proof of the ontological portion of the model. Within this model, we have the expected characterization of the η-individuals as given by the following theorem.

N24. $[\Sigma]:\mathscr{P}\langle\Sigma\rangle. = [\exists A].\boldsymbol{U}(A).\Sigma\circ\chi((A))$. [N27, N9]

We now construct the portion of the model involving "element". We wish a definition of element so that one η-individual is an element of another if the argument of the first is less than or equal to the argument of the second, i.e.

$$[\Sigma\Phi]:\Sigma\eta\mathbf{el}\langle\langle\Phi\rangle\rangle. \equiv .\Sigma\eta\Sigma.\Phi\eta\Phi.[\exists BC].\Sigma(B).\Phi(C).B \leqslant C.$$

According to the metatheorem we must introduce

DN8. $[\Phi A]:\mathbf{el}\langle\langle\Phi\rangle\rangle(A). \equiv .\Phi\eta\Phi.[\exists BC].\chi((A))(B).\Phi(C).B \leqslant C$.

So that the following may be introduced as a thesis

N29. $[\Phi]\therefore\mathscr{N}\langle\mathbf{el}\langle\langle\Phi\rangle\rangle\rangle. \supset :[\Sigma]:\Sigma\eta\mathbf{el}\langle\langle\Phi\rangle\rangle. \equiv .\Sigma\eta\Sigma.\Phi\eta\Phi.$
$[\exists BC].\Sigma(B).\Phi(C).B \leqslant C$. [Metatheorem, DN8]

N30. $[\Phi A]:\mathbf{el}\langle\langle\Phi\rangle\rangle(A). \supset .\boldsymbol{U}(A)$.
Hyp (1). $\supset$.
$[\exists B]$.
(2) $\chi((A))(B)$. [DN8, 1]
(3) $\chi((A))(A)$. [DN4, 2]
$\boldsymbol{U}(A)$. [N7, 3]

N31. $[\Phi].\mathscr{N}\langle\mathbf{el}\langle\langle\Phi\rangle\rangle\rangle$. [N30, DN1]

N32. $[\Sigma\Phi]:\Sigma\eta\mathbf{el}\langle\langle\Phi\rangle\rangle.\equiv.\Sigma\eta\Sigma.\Phi\eta\Phi.[\exists BC].\Sigma(B).\Phi(C).$
$B\leqslant C.$ [N29, N31]

N33. $[AB]:\chi((A))\eta\mathbf{el}\langle\langle\chi((B))\rangle\rangle.\supset.A\leqslant B.$
Hyp (1) $.\supset.$
$[\exists CD].$
(2) $\chi((A))(C).$
(3) $\chi((B))(D).$ [N32, 1]
(4) $C\leqslant D.$
(5) $A\circ C.$ [DN4, 2]
(6) $B\circ D.$ [DN4, 3]
$A\leqslant B.$ [4, 5, 6]

N34. $[AB]:A\leqslant B.\supset.\chi((A))\eta\mathbf{el}\langle\langle\chi((B))\rangle\rangle$
Hyp (1) $.\supset.$
(2) $\boldsymbol{U}(A).$
(3) $\boldsymbol{U}(B).$ [**BD**, 1]
(4) $\chi((A))\eta\chi((A)).$ [N9, 2, N1]
(5) $\chi((B))\eta\chi((B)).$ [N9, 3, N1]
(6) $\chi((A))(A).$ [N7, 2]
(7) $\chi((B))(B).$ [N7, 3]
$\chi((A))\eta\mathbf{el}\langle\langle\chi((B))\rangle\rangle.$ [N32, 4, 5, 6, 7, 1]

N35. $[AB]:A\leqslant B.\equiv.\chi((A))\eta\mathbf{el}\langle\langle\chi((B))\rangle\rangle.$ [N33, N34]

N36. $[\sigma BC]\therefore\sigma(B):[D]:\sigma(D).\supset.D\leqslant C:\supset.\boldsymbol{U}(C).$ [**BD**]

Starting with **BD** and using the following theses, DN1, N36, N12, N35, in order, we arrive at

N37. $[AB]::\chi((A))\eta\mathbf{el}\langle\langle\chi((B))\rangle\rangle.\equiv::\boldsymbol{U}(A).\boldsymbol{U}(B)::$
$\chi((B))\eta\mathbf{el}\langle\langle\chi((B))\rangle\rangle.\supset::[\sigma\tau]::\mathcal{N}\langle\tau\rangle.\chi((B))\eta\sigma\therefore[C]\therefore$
$\chi((C))\eta\tau.\equiv:\boldsymbol{U}(C):[D]:\chi((D))\eta\sigma.\supset.\chi((D))\eta\mathbf{el}\langle\chi((C))\rangle:[D]:$
$\chi((D))\eta\mathbf{el}\langle\langle\chi((C))\rangle\rangle.\supset.[\exists EF].\chi((E))\eta\sigma.\chi((F))\eta\mathbf{el}\langle\langle\chi((D))\rangle\rangle.$
$\chi((F))\eta\mathbf{el}\langle\langle\chi((E))\rangle\rangle::\supset.[\exists L].\tau\circ\chi((L)).\chi((A))\eta\mathbf{el}\langle\langle\chi((L))\rangle\rangle.$

N38. $[\Delta]\therefore[\exists\Psi\theta].\Psi\eta\sigma.\theta\eta\mathbf{el}\langle\langle\Delta\rangle\rangle.\theta\eta\mathbf{el}\langle\langle\Psi\rangle\rangle:\supset:$
$[\exists EF].\chi((E))\eta\sigma.\chi((F))\eta\mathbf{el}\langle\langle\Delta\rangle\rangle.\chi((F))\eta\mathbf{el}\langle\langle\chi((E))\rangle\rangle$
Hyp (1) $.\supset.$
(2) $\mathscr{P}\langle\Psi\rangle.$ [DN3, 1]
(3) $\mathscr{P}\langle\theta\rangle.$ [DN3, 2]
$[\exists EF].$
(4) $\Psi\circ\chi((E)).$ [N27, 2]
(5) $\theta\circ\chi((F)).$ [N27, 3]
$[\exists EF].\chi((E))\eta\sigma.\chi((F))\eta\mathbf{el}\langle\langle\Delta\rangle\rangle.\chi((F))\eta\mathbf{el}\langle\langle\chi((E))\rangle\rangle$
[1, 4, 5]

N39. $[\Delta]\therefore[\exists EF].\chi((E))\eta\sigma.\chi((F))\eta\mathbf{el}\langle\langle\Delta\rangle\rangle.\chi((F))\eta\mathbf{el}\langle\langle\chi((E))\rangle\rangle:$
$\equiv:[\exists\Psi\theta].\Psi\eta\sigma.\theta\eta\mathbf{el}\langle\langle\Delta\rangle\rangle.\theta\eta\mathbf{el}\langle\langle\Psi\rangle\rangle.$ [N38]

In what follows $\mathscr{E}\langle\langle\chi((A))\rangle\rangle$ will represent any expression involving $\chi((A))$, but not otherwise involving A. $\mathscr{E}\langle\Sigma\rangle$ represents the expression obtained by replacing $\chi((A))$ by Σ. Thus A does not occur in $\mathscr{E}\langle\Sigma\rangle$.

N40. $[\Delta\sigma]\therefore[D]:\chi((D))\eta\sigma.\mathscr{E}\langle\chi((D))\rangle:\Delta\eta\sigma:\supset.\mathscr{E}\langle\Delta\rangle$

Hyp (2). $\supset$.

(3) $\mathscr{P}\langle\Delta\rangle$. [DN3, 2]

$[\exists D]$.

(4) $\Delta\circ\chi((D))$. [N27, 3]

(5) $\chi((D))\eta\sigma$. [2, 4]

(6) $\mathscr{E}\langle\chi((D))\rangle$. [1, 5]

$\mathscr{E}\langle\Delta\rangle$. [6, 4]

N41. $[\sigma]\therefore[D]:\chi((D))\eta\sigma.\supset.\mathscr{E}\langle\chi((D))\rangle:\equiv:[\Delta]:\Delta\eta\sigma.\supset.\mathscr{E}\langle\Delta\rangle$. [N40]

N42. $[\sigma\Gamma]\therefore[C]:\chi((C))\eta\sigma.\equiv.\boldsymbol{U}(C).\mathscr{E}\langle\chi((C))\rangle:\Gamma\eta\sigma:\supset.\Gamma\eta\Gamma.\mathscr{E}\langle\Gamma\rangle$. [N2, N40]

N43. $[\sigma\Gamma]\therefore[C]:\chi((C))\eta\sigma.\equiv.\boldsymbol{U}(C).\mathscr{E}\langle\chi((C))\rangle:\Gamma\eta\Gamma.\mathscr{E}\langle\Gamma\rangle:\supset.\Gamma\eta\sigma.$

Hyp (3). $\supset$.

(4) $\mathscr{P}\langle\Gamma\rangle$. [N1, 2]

$[\exists C]$.

(5) $\Gamma(C)$.

(6) $\Gamma\circ\chi((C))$.

(7) $\boldsymbol{U}(C)$. [N27, 4] (for 5–7)

(8) $\mathscr{E}\langle\chi((C))\rangle$. [3, 6]

(9) $\chi((C))\eta\sigma$. [1, 7, 8]

$\Gamma\eta\sigma$. [6, 9]

N44. $[\sigma C]\therefore[\Gamma]:\Gamma\eta\sigma\equiv.\Gamma\eta\Gamma.\mathscr{E}\langle\Gamma\rangle:\chi((C))\eta\sigma:\supset.\boldsymbol{U}(C).\mathscr{E}\langle\chi((C))\rangle$

Hyp (2) $\supset$

(3) $\mathscr{P}\langle\chi((C))\rangle$. [DN3, 2]

(4) $\boldsymbol{U}(C)$. [N9, 3]

(5) $\mathscr{E}\langle\chi((C))\rangle$. [1, 2]

$\boldsymbol{U}(C).\mathscr{E}\langle\chi((C))\rangle$. [4, 5]

N45. $[\sigma C]\therefore[\Gamma]:\Gamma\eta\sigma.\equiv.\Gamma\eta\Gamma.\mathscr{E}\langle\Gamma\rangle:\boldsymbol{U}(C).\mathscr{E}\langle\chi((C)):\supset.\chi((C))\eta\sigma$

Hyp (3). $\supset$

(4) $\mathscr{P}\langle\chi((C))\rangle$. [N9, 2]

(5) $\chi((C))\eta\chi((C))$. [N1, 4]

$\chi((C))\eta\sigma$. [1, 5, 3]

N46. $[\sigma]\therefore[C]:\chi((C))\eta\sigma.\equiv.\boldsymbol{U}(C).\mathscr{E}\langle\chi((C))\rangle:\equiv:[\Gamma]:\Gamma\eta\sigma.\equiv.\Gamma\eta\Gamma.\mathscr{E}\langle\Gamma\rangle$. [N42, N43, N44, N45]

N47. $[\Sigma]\therefore[\Phi]:\Sigma\eta\mathbf{el}\langle\langle\Phi\rangle\rangle.\equiv.\Phi\eta\Phi.\mathscr{E}\langle\Phi\rangle:\supset:[B]:\Sigma\eta\mathbf{el}\langle\langle\chi((B))\rangle\rangle.\equiv.\boldsymbol{U}(B).\mathscr{E}\langle\chi((B))\rangle$. [N1, N9]

N48. $[\Sigma]\therefore[B]:\Sigma\eta\mathbf{el}\langle\langle\chi((B))\rangle\rangle.\equiv.\boldsymbol{U}(B).\mathscr{E}\langle\chi((B))\rangle:\supset:[\Phi]:\Sigma\eta\mathbf{el}\langle\langle\Phi\rangle\rangle.\equiv.\Phi\eta\Phi.\mathscr{E}\langle\Phi\rangle$. [N32, N1, N27]

N49. $[\Sigma] \therefore [B] : \Sigma \eta \mathbf{el} \langle\langle \chi((B)) \rangle\rangle . \equiv . \boldsymbol{U}(B) . \mathscr{E} \langle \chi((B)) \rangle : \equiv : [\Phi] :$
$\Sigma \eta \mathbf{el} \langle\langle \Phi \rangle\rangle . \equiv . \Phi \eta \Phi . \mathscr{E} \langle \Phi \rangle .$ [N47, N48]

N50. $[\Gamma \Phi \sigma] \therefore \Phi \eta \sigma : [\Delta] : \Delta \eta \sigma . \supset . \Delta \eta \mathbf{el} \langle\langle \Gamma \rangle\rangle : \supset . \Gamma \eta \Gamma .$ [N32]

Starting with N37, use N39 to eliminate E and F, then N41 twice to eliminate D, then N46 first to eliminate C and then to eliminate A, then N49 to eliminate B and finally N50 to eliminate $\Gamma \eta \Gamma$, we arrive at

N51. $[\Sigma \Phi] \because\therefore \Sigma \eta \mathbf{el} \langle\langle \Phi \rangle\rangle . \equiv :: \Sigma \eta \Sigma : \Phi \eta \Phi :: \Phi \eta \mathbf{el} \langle\langle \Phi \rangle\rangle . \supset \therefore\because [\sigma \tau] \therefore\because$
$\mathscr{N} \langle \tau \rangle . \Phi \eta \sigma \therefore [\Gamma] \therefore \Gamma \eta \tau . \equiv : [\Delta] : \Delta \eta \sigma . \supset . \Delta \eta \mathbf{el} \langle\langle \Gamma \rangle\rangle : [\Delta] :$
$\Delta \eta \mathbf{el} \langle\langle \Gamma \rangle\rangle . \supset . [\exists \Psi \theta] . \Psi \eta \sigma . \mathbf{a} \eta \mathbf{el} \langle\langle \Delta \rangle\rangle . \theta \eta \mathbf{el} \langle\langle \Psi \rangle\rangle :: \supset .$
$\Sigma \eta \mathbf{el} \langle\langle \tau \rangle\rangle .$

In N51 note that the only relativisation of quantifiers to $\mathscr{N}$ needed is for τ. Comparing N51 to **M** we see that the construction of the model is finished.

Bibliography*

AJDUKIEWICZ, Kazimierz

[1928] *Główne zasady metodologii nauk i logiki formalnej* (Main Principles of the Methodology of Science and of Formal Logic), Authorized Script. Edited by M. Presburger, Warszawa (Polish).

[1934] *Logiczne podstawy nauczania* (The Logical Foundations of Teaching), Encyklopedia Wychowania, Vol. II, 1–75, Warszawa-Wilno (Polish). (See also AJDUKIEWICZ [1974].)

[1934a] *Logistyczny antyirracjonalim w Polsce* (The Logistics Antiirrationalism in Poland), Przegląd Filozoficzny, 37, 399–408 (Polish).

[1935] *Die syntaktische Konnexität*, Studia Philosophica, Vol. I, 1–27, Lwów. (English translation by H. Weber in McCALL [1967], 207-231.)

[1935a] *Die Logistische Antiirrationalismus in Polen*, Erkenntnis, Vol. V, 151–161. (This is the German translation of AJDUKIEWICZ [1934a].)

[1951] *On the Notion of Existence*, Studia Philosophica, Poznań, Vol. IV, for 1949-50, pub. 1951, 7–22.

[1974] *Pragmatic Logic*, PWN/Reidel Publishing Company, Warszawa/Dordrecht/ /Boston.

BAR-HILLEL, Yehoshua

[1950] *On Syntactical Categories*, The Journal of Symbolic Logic, Vol. XV, 1–16.

[1953] *A Quasi-Arithmetical Notation for Syntactic Description*, Language, Vol. XXIX, 47–58.

[1957] *Husserl's Conception of a Purely Logical Grammar*, Philosophy and Phenomenological Research, Vol. XVII, 362–369.

BORKOWSKI, Ludwik

[1970] *Logika formalna* (Formal Logic), PWN, Warszawa (Polish).

CANTY, John T.

[1967] *Leśniewski's Ontology and Gödel's Incompleteness Theorem*, unpublished Ph. D. Thesis, University of Notre Dame, Bloomington, Indiana.

[1969] *Ontology: Leśniewski's Logical Language*, in: Foundations of Language, International Journal of Language and Philosophy, Vol. V, 455–469. (Also in the present volume.)

[1969a] *Leśniewski's Terminological Explanation as Recursive Concepts*, Notre Dame Journal of Formal Logic, Vol. X, No. 4, 337–369.

CLAY, Robert E.

[1965] *The Relation of Weakly Discrete to Set and Equinumerosity in Mereology*, Notre Dame Journal of Formal Logic, Vol. VI, No. 4, 325–340.

[1966] *On the Definition of Mereological Class*, Notre Dame Journal of Formal Logic, Vol. VII, No. 4, 359–360. (Also in the present volume.)

---

* *Editorial Note*: This is the working bibliography relating to the present volume. Fuller bibliography is to be found at the end of the volume "Leśniewski's System-Protothetic" (in preparation).

[1968] *The Consistency of Leśniewski's Mereology Relative to Real Number System*, The Journal of Symbolic Logic, Vol. XXXIII, 251–257.

[1969] *Sole Axioms for Partially Ordered Sets*. Logique et Analyse, Vol. XLVIII, 361–375.

[1970] *The Dependence of a Mereological Axiom*, Notre Dame Journal of Formal Logic, Vol. XI, No. 4, 471–472. (Also in the present volume.)

[1974] *Relation of Leśniewski's Mereology to Boolean Algebra*, The Journal of Symbolic Logic, Vol. XXXIX, No. 4, 638–648. (Also in the present volume.)

CURRY, Haskell B.

[1963] *Foundations of Mathematical Logic*, McGraw-Hill Book Co., New York.

FREGE, Gottlob

[1903] *Grundgesetze der Arithmetik*, Jena, 2 volumes (the first volume appeared in 1893), reprinted 1962 by Georg Olms Verlagsbuchhandlung Hildesheim.

GÖDEL, Kurt

[1944] *Russell's Mathematical Logic*, in: The Philosophy of Bertrand Russell (Edited by Paul A. Schilpp), North-Western University Press, Chicago, 125–153.

GRZEGORCZYK, Andrzej

[1955] *The system of Leśniewski in Relation to Contemporary Logical Research*, Studia Logica, Vol. III, 77–95.

[1961] *Zarys logiki matematycznej* (The Outline of Mathematical Logic), first edition PWN, Warszawa (Polish). (English translation of the third Polish edition — GRZEGORCZYK [1974].)

[1974] *The Outline of Mathematical Logic*, North-Holland Publishing Co., Amsterdam-London. (This is the English translation of GRZEGORCZYK [1961].)

HILBERT, David and BERNAYS, Paul

[1934] *Grundlagen der Mathematik*, Vol. I, Springer, Leipzig-Berlin, Vol. II, 1939. Reprinted Ann Arbor, Mich. (J. W. Edwards) 1944.

HOBBES, Thomas

[1839-1845a] *De Corpore*, in: Opera Philosophica, Vol. I (edited by W. Molesworth), London.

[1839-1845b] *Leviathan*, in: Opera Philosophica, Vol. III (edited by W. Molesworth), London.

IWANUŚ, Bogusław

[1973] *On Leśniewski's Elementary Ontology*, Studia Logica, Vol. XXXI, 73–119. (Also in the present volume.)

[1973a] *Proof of Decidability of The Traditional Calculus of Names*, Studia Logica, Vol. XXXII, 131–147.

JORDAN, Zbigniew

[1945] *The Development of Mathematical Logic and of Logical Positivism in Poland Between the Two Wars*, Polish Science and Learning, No. 6, Oxford University Press, London. (The first six (of ten) sections and the relevant parts of the bibliography are reprinted in MCCALL [1967], 346–406.)

KOTARBIŃSKI, Tadeusz

[1929] *Elementy teorii poznania, logiki formalnej i metodologii nauk* (Elements of

the theory of Knowledge, of Formal Logic and the Methodology of Sciences), Lwów (Polish). (English translation — KOTARBIŃSKI [1966].)
[1952] *Humanistyka bez hipostaz* (The Humanities without Hypostases), Myśl Filozoficzna, Vol. I (3) (Polish). (English translation — in Kotarbiński [1966].)
[1966] *Gnosiology. The Scientific Approach to the Theory of Knowledge,* Pergamon Press /Ossolineum, Oxford / Wrocław. (This is the English translation of KOTARBIŃSKI [1929].)

KRUSZEWSKI, Z.
[1925] *Ontologia bez aksjomatów* (Ontology without Axioms), Przegląd Filozoficzny, Vol. XXVIII, 136 (abstract). (Also in the present volume.)

KUBIŃSKI, Tadeusz
[1958] *Nazwy nieostre* (Vague Terms), Studia Logica, Vol. VII, 115–176 (Polish).

KURATOWSKI, Kazimierz and MOSTOWSKI, Andrzej
[1952] *Teoria mnogości* (Set Theory), Monografie Matematyczne, Vol. XXVIII, PWN, Warszawa-Wrocław (Polish). (English translation — KURATOWSKI and MOSTOWSKI [1967].)
[1967] *Set Theory,* North-Holland Publishing Co./PWN, Amsterdam-London/Warszawa. (This is the English translation by Maciej Mączyński of the second Polish edition of KURATOWSKI-MOSTOWSKI [1952].)

LEBIEDIEWA, Swietłana
[1969] *The Systems of Modal Calculus of Names* I, Studia Logica, Vol. XXIV, 83–104.

LEJEWSKI, Czesław
[1954] *Logic and Existence,* The British Journal for the Philosophy of Science, Vol. V, 104–119. (Also in the present volume.)
[1954a] *A Contribution to Leśniewski's Mereology,* Yearbook for 1954–55 of the Polish Society of Arts and Sciences Abroad, Vol. V, London, 1954-55, 43–50.
[1958] *On Leśniewski's Ontology,* Ratio, Vol. I, No. 2, 150–176. (Also in the present volume.)
[1960] *Studies in Axiomatic Foundations of Boolean Algebra,* Notre Dame Journal of Formal Logic, Part, I, Vol. I, 23–47, Part II, *ibid.*, 91–106, Part III, Vol. II, 79–93.
[1963] *A Note on a Problem Concerning Axiomatic Foundations of Mereology,* Notre Dame Journal of Formal Logic, Vol. IV, 135–139.
[1969] *Consistency of Leśniewski's Mereology,* The Journal of Symbolic Logic, Vol. XXXIV, No. 3, 321–328. (Also in the present volume.)

LEŚNIEWSKI, Stanisław
[1916] *Podstawy ogólnej teoryi mnogości,* I (Foundations of General Set Theory I), Prace Polskiego Koła Naukowego w Moskwie, Sekcya matematyczno-przyrodnicza, No.2, Moscow (Polish). (English translation — in LEŚNIEWSKI [1984].)
[1927-1931] *O podstawach matematyki* (On the Foundations of Mathematics,) Przegląd Filozoficzny, Vol. XXX (1927), 164–206; Vol. XXXI (1928), 261–291; Vol. XXXII (1929), 60–101; Vol. XXXIII (1930), 77–105; Vol. XXXIV (1931), 142–170. (Polish). (English translation — LEŚNIEWSKI [1983].)
[1929] *Über Funktionen, deren Felder Gruppen mit Rücksicht auf diese Funktionen sind,* Fundamenta Mathematicae, Vol. XIII, 319–332. (English translation — in LEŚNIEWSKI [1984].)

[1929a] *Grundzüge eines neuen Systems der Grundlagen der Mathematik*, Fundamenta Mathematicae, Vol. XIV, 1–81. (English translation – in LEŚNIEWSKI [1984].)

[1929b] *Über Funktionen, deren Felder Abelsche Gruppen in Bezug auf diese Funktionen sind*, Fundamenta Mathematicae, Vol. XIV, 242–251. (English translation – in LEŚNIEWSKI [1984].)

[1930] *Über die Grundlagen der Ontologie*, Comptes rendus des séances de la Société des Sciences et des Lettres de Varsovie, Classe III, 23 Année, 111–132, Warszawa. (English translation – in LEŚNIEWSKI [1984].)

[1931] *Über Definitionen in der sogenannten Theorie der Deduktion*, Comptes rendus des séances de la Societé des Sciences et des Lettres de Varsovie, Classe III, 24 Année, 289–309, Warszawa. (English translation – in LEŚNIEWSKI [1984] and in MCCALL [1967], 170–187.)

[1938] *Einleitende Bemerkungen zur Forsetzung meiner Mitteilung u.d.t "Grundzüge eines neuen Systems der Grundlagen der Mathematik"*, Collectanea Logica, Vol. I, Warszawa, 1–60. (English translation in LEŚNIEWSKI [1984].) Reprinted in Jahrbuch über die Fortschritte der Mathematik 65, No. 11, for 1939, pub. 1941. Reprinted and translated into English in MCCALL [1967], 116–169.

[1983] *On the Foundations of Mathematics*, Topoi, Vol. II, No. 1, 7–52. (This is the abridged English translation by Vito F. Sinisi of LEŚNIEWSKI [1927-1931].)

[1984] *Collected Works of Stanisław Leśniewski* (edited by Jan Srzednicki, Stanisław J. Surma, and Dene I. Barnett), Synthese Library, D. Reidel Publishing Co./PWN, Dordrecht-Boston/Warszawa, to appear.

[1984a] *Stanisław Leśniewski's Lecture Notes* (edited by Zbigniew Stachniak and Jan Srzednicki), NIPS, Nijhoff Publishing Co./PWN, The Hague/Warszawa, to appear.

LINDENBAUM, Adolf

[1936] *Sur la simplicité formelle des notions*, Actes du Congrés International de Philosophie Scientifique, VII Logique, Paris.

LUSCHEI, Eugene C.

[1962] *The Logical Systems of Leśniewski*, North-Holland Publishing Co., Amsterdam-London.

ŁUKASIEWICZ, Jan

[1951] *Aristotle's Syllogistic*, Oxford.

[1953] *The Principle of Individuation*, Proceedings of the Aristotelian Society, Supplementary Volume XXVII, London, 69–82.

[1957] *Aristotle's Syllogistic from the Standpoint of Modern Formal Logic*, Oxford, 1958.

[1961] *O sylogistyce Arystotelesa* (On Aristotle's Syllogistic), Pisma wybrane. Z zagadnień logiki i filozofii (edited by Jerzy Słupecki), PWN, Warszawa (Polish), 220–227.

[1970] *Selected Works* (edited by Ludwik Borkowski), North-Holland Publishing Co./PWN, Amsterdam-London/Warszawa.

MACHOVER, Maurice

[1966] *Contextual Determinacy in Leśniewski's Grammar*, Studia Logica, Vol. XIX, 47–55.

McCall, Storrs (ed.)
[1967] *Polish Logic 1920-1939*, Clarendon Press, Oxford.

Mostowski, Andrzej
[1948] *Logika matematyczna* (Mathematical Logic), Monografie Matematyczne, Vol. XVIII, PWN, Warszawa-Wrocław (Polish).

Myhill, John R.
[1953] *Arithmetic with Creative Definitions by Induction*, The Journal of Symbolic Logic, Vol. XVIII, 115–118.

Prior, Arthur N.
[1955] *Formal Logic*. Oxford. (Revised 1962).

Quine, Willard Van Orman
[1939] *Designation and Existence*, The Journal of Philosophy, Vol. XXXVI, New York, 701–709. (Also Quine [1949].)
[1940] *Mathematical Logic*, New York (Norton). Revised ed., Harvard University Press, Cambridge (Mass.) 1951.
[1940a] *A Logical Approach to the Ontological Problem*, The Journal of Unified Science, Vol. IX, Chicago.
[1943] *Notes on Existence and Necessity*, The Journal of Philosophy, Vol. XL, New York.
[1947] *Mathematical Logic*, Cambridge, Massachussets (second printing).
[1948] *On What There Is*, The Review of Metaphysics, Vol. II, New Haven, 21–38. (Also reprinted as Quine [1951a].)
[1949] *Designation and Existence*, in: Readings on Philosophical Analysis (edited by H. Feigel and W. Sellars), Appelton-Century Croffs Inc., New York. (Also Quine [1939].)
[1951] *Whitehead and the Rise of Modern Logic*, in: The Philosophy of Alfred North Whitehead (edited by Paul A. Schilpp), Northwestern University Press, Chicago, 127–167.
[1951a] *On What There Is*, Proceedings of the Aristotelian Society, Supplementary Volume XXV, London. (Also Quine [1948].)
[1952] *Methods of Logic*, London.
[1953] *From a Logical Point of View*, Harvard University Press, Cambridge, Massachussets.

Schröder, Ernst
[1891] *Vorlesungen über die Algebra der Logik*, II Band, 1 Abt, Leipzig.

Sierpiński, Wacław
[1929] *Leçon sur les nombres transfinis*, Paris.

Skolem, Thoralf
[1919] *Untersuchungen über die Axiome des Klassenkalküls*, Skrifter utgit av Videnskapsselskapet i Kristiana, I. Matematisk-naturvidenskabelig klasse.

Słupecki, Jerzy
[1946] *Uwagi o sylogistyce Arystotelesa* (Remarks on Aristotle's Syllogictic), Annales Universitatis Mariae Skłodowska-Curie, Vol. I, No. 3, section F, Lublin (Polish), 187–191.

[1948] *Z badań nad sylogistyką Arystotelesa* (From the Investigations into Aristotle's Syllogistic), Prace Wrocławskiego Towarzystwa Naukowego, Series B, No. 6, PIW, Wrocław (Polish). (English translation — SŁUPECKI [1949].)
[1949] *On Aristotle's Syllogistic*, Poznań. (This is the English translation of SŁUPECKI [1948].)
[1953] *S. Leśniewski's Protothetics*, Studia Logica, Vol. I, 44–111.
[1955] *S. Leśniewski's Calculus of Names*, Studia Logica, Vol. III, 7–72. (Also in the present volume.)
[1958] *Towards a Generalized Mereology of Leśniewski*, Studia Logica, Vol. VIII, 131–154.

SŁUPECKI, Jerzy and BORKOWSKI, Ludwik
[1967] *Elements of Mathematical Logic and Set Theory*, Oxford.

SOBOCIŃSKI, Bolesław
[1934] *O kolejnych uproszczeniach aksjomatyki "ontologii" prof. S. Leśniewskiego*, (On the Successive Simplifications of the Axiom-System of Prof. S. Leśniewski's Ontology), Fragmenty Filozoficzne, 143–160 (Polish). (English translation by Z. Jordan in McCALL [1967], 188–200.)
[1949] *An Investigation of Protothetic*, Cahiers de l'Institut d'Etudes Polonaises en Belgique, No. 5, Bruxelles. (Also in McCALL [1967], 201–206.)
[1949-1950] *L'analyse de l'antinomie Russellienne par Leśniewski*, Methodos, Vol. I, (1949), 94–107, 220–228, 308–316; Vol. II (1950), 237–257. (English translation by Robert Clay in the present volume.)
[1954] *Studies in Leśniewski's Mereology*, Yearbook for 1954-55 of the Polish Society of Art and Sciences Abroad, Vol. V, 34–43. (Also in the present volume.)
[1956] *In Memoriam Jan Łukasiewicz (1878-1956)*, Philosophical Studies, Vol. VI, Maynooth (Ireland), 3–49.

TARSKI, Alfred
[1922] *O wyrazie pierwotnym logistyki* (On the Primitive Term of Logistic), Przegląd Filozoficzny, Vol. XXVI, No. 1/2, 68–89 (Polish). A modified French translation appeared in two parts: *Sur le terme primitif de la logistique*, Fundamenta Mathematicae, Vol. IV (1923), 196–200, and *Sur les truth-functions au sens de M. M. Russell et Whitehead*, Fundamenta Mathematicae, Vol. V (1924), 59–74. (English translation in TARSKI [1956], 1–23.)
[1933] *Pojęcie prawdy w językach nauk dedukcyjnych* (The Concept of Truth in the Language of the Deductive Sciences), Prace Towarzystwa Naukowego Warszawskiego, Wydział III, Warszawa, No. 34 (Polish). (English translation in TARSKI [1956], 152-278.)
[1935] *Zur Grandlegung der Boole'sche Algebra I*, Fundamenta Mathematicae, Vol. XXIV, 177–198.
[1956] *Logic, Semantics, and Metamathematics*; *Papers from 1923-1938 by Alfred Tarski*, Clarendon Press, Oxford. Translations by J. H. Woodger.

WHITEHEAD, Alfred N. and RUSSELL, Bertrand
[1913] *Principia Mathematica*, Cambridge.
[1935] *Principia Mathematica* (Vol. I), Cambridge.
[1963] *Principia Mathematica*, Cambridge University Press, Cambridge.

WOODGER, Joseph H.
[1937] *The Axiomatic Method in Biology* (with appendices by A. Tarski and W. F. Floyd), Cambridge University Press, Cambridge-London.

[1952] *Science without Properties*, The British Journal for the Philosophy of Science, Vol. II, 193–217.

[1952a] *Biology and Language: An Introduction to the Methodology of the Biological Sciences Including Medicine*, Cambridge University Press, Cambridge-London.

Index of Names*

* *Editorial Note*: Index of symbols covering the present volume and "Leśniewski's Systems – Protothetic" will appear in the volume "Protothetic".

Nijhoff International Philosophy Series

Other volumes in this series

1. Rotenstreich N: Philosophy, History and Politics – Studies in Contemporary English Philosophy of History. 1967. ISBN 90–247–1743–4
2. Srzednicki JTJ: Elements of Social and Political Philosophy. 1976. ISBN 90–247–1744–2.
3. Tatarkiewicz W: Analysis of Happiness. 1976. ISBN 90–247–1807–4.
4. Twardowski K: On the Content and Object of Presentations – A Psychological Investigation. Translated and with an Introduction by R. Grossman. 1977. ISBN 90–247–1926–7.
5. Tatarkiewicz W: A History of Six Ideas – An Essay in Aesthetics. 1980. ISBN 90–247–2233–0.
6. Noonan HW: Objects and Identity – An Examination of the Relative Identity Thesis and Its Consequences. 1980. ISBN 90–247–2292–6.
7. Crocker L.: Positive Liberty – An Essay in Normative Political Philosophy. 1980. ISBN 90–247–2291–8.
8. Brentano F: The Theory of Categories. Translated by R. M. Chisholm and N. Guterman. 1981. ISBN 90–247–2302–7.
9. Marciszewski W(ed): Dictionary of Logic as Applied in the Study of Language. Concept (Methods) Theories. 1981. ISBN 90–247–2123–7.
10. Ruzsa I: Modal Logic with Descriptions. 1981. ISBN 90–247–2473–2.
11. Hoffman P: The Anatomy of Idealism. Passivity and Activity in Kant, Hegel and Marx. 1982.
12. Gramm M. S.: Direct Realism. A Study of Perception. 1983. ISBN 90–247–2870–3.
13. Leśniewski's Systems. Ontology and Mereology. 1984. ISBN 90–247–2879–7.

ISBN 90–247–2879–7 Nijhoff
ISBN 83–04–00879–3 Ossolineum

ISBN 83-04-00879-3

Zakład Narodowy im. Ossolińskich — Wydawnictwo.
Wrocław 1985.
Nakład: 500 egz.
Objętość: ark. wyd. 17,50, ark. druk. 16,50, ark. A1-22.
Papier druk. sat. kl. III, 80 g, 70 × 100.
Wrocławska Drukarnia Naukowa. Zam. 1088/82.
Cena zł 220.—